Advanced High School Statistics
First Edition

David M Diez
david@openintro.org

Christopher D Barr
Yale School of Management
chris@openintro.org

Mine Çetinkaya-Rundel
Duke University
mine@openintro.org

Leah Dorazio
San Francisco University High School
leah@openintro.org

Contents

Preface

Advanced High School Statistics is ready for use with the AP® Statistics Course.[1]

This book may be downloaded as a free PDF at **openintro.org**.

We hope readers will take away three ideas from this book in addition to forming a foundation of statistical thinking and methods.

(1) Statistics is an applied field with a wide range of practical applications.

(2) You don't have to be a math guru to learn from real, interesting data.

(3) Data are messy, and statistical tools are imperfect. But, when you understand the strengths and weaknesses of these tools, you can use them to learn about the real world.

Textbook overview

The chapters of this book are as follows:

1. **Data collection.** Data structures, variables, and basic data collection techniques.

2. **Summarizing data.** Data summaries and graphics.

3. **Probability.** The basic principles of probability.

4. **Distributions of random variables.** Introduction to key distributions, and how the normal model applies to the sample mean and sample proportion.

5. **Foundation for inference.** General ideas for statistical inference in the context of estimating the population proportion.

6. **Inference for categorical data.** Inference for proportions using the normal and chi-square distributions.

7. **Inference for numerical data.** Inference for one or two sample means using the t distribution, and comparisons of many means using ANOVA.

8. **Introduction to linear regression.** An introduction to regression with two variables.

Instructions are also provided in several sections for using Casio and TI calculators.

Videos

The ▸ icon indicates that a section or topic has a video overview readily available. The icons are hyperlinked in the textbook PDF, and the videos may also be found at

www.openintro.org/stat/videos.php

---

[1]AP® is a trademark registered and owned by the College Board, which was not involved in the production of, and does not endorse, this product.

Examples, exercises, and appendices

Examples and guided practice exercises throughout the textbook may be identified by their distinctive bullets:

● **Example 0.1** Large filled bullets signal the start of an example.

Full solutions to examples are provided and often include an accompanying table or figure.

⊙ **Guided Practice 0.2** Large empty bullets signal to readers that an exercise has been inserted into the text for additional practice and guidance. Students may find it useful to fill in the bullet after understanding or successfully completing the exercise. Solutions are provided for all within-chapter exercises in footnotes.[2]

There are exercises at the end of each chapter that are useful for practice or homework assignments. Many of these questions have multiple parts, and odd-numbered questions include solutions in Appendix A.

Probability tables for the normal, t, and chi-square distributions are in Appendix B, and PDF copies of these tables are also available from **openintro.org** for anyone to download, print, share, or modify.

OpenIntro, online resources, and getting involved

OpenIntro is an organization focused on developing free and affordable education materials. *OpenIntro Statistics*, our first project, is intended for introductory statistics courses at the high school through university levels.

We encourage anyone learning or teaching statistics to visit **openintro.org** and get involved. We also provide many free online resources, including free course software. Most data sets for this textbook are available on the website and through a companion R package.[3] OpenIntro's resources may be used with or without this textbook as a companion.

We value your feedback. If there is a particular component of the project you especially like or think needs improvement, we want to hear from you. Provide feedback through a link provided on the textbook page:

www.openintro.org/stat/textbook.php

Acknowledgements

This project would not be possible without the dedication and volunteer hours of all those involved. No one has received any monetary compensation from this project, and we hope you will join us in extending a *thank you* to the project's volunteers listed at

www.openintro.org/about

and also to the many students, teachers, and other readers who have provided feedback to the project.

---

[2]Full solutions are located down here in the footnote!

[3]Diez DM, Barr CD, Çetinkaya-Rundel M. 2015. openintro: OpenIntro data sets and supplement functions. github.com/OpenIntroOrg/openintro-r-package.

Chapter 1

Data collection

Scientists seek to answer questions using rigorous methods and careful observations. These observations – collected from the likes of field notes, surveys, and experiments – form the backbone of a statistical investigation and are called **data**. Statistics is the study of how best to collect, analyze, and draw conclusions from data. It is helpful to put statistics in the context of a general process of investigation:

1. Identify a question or problem.

2. Collect relevant data on the topic.

3. Analyze the data.

4. Form a conclusion.

Statistics as a subject focuses on making stages 2-4 objective, rigorous, and efficient. That is, statistics has three primary components: How best can we collect data? How should it be analyzed? And what can we infer from the analysis?

Researchers from a wide array of fields have questions or problems that require the collection and analysis of data. Let's consider three examples.

- Climate scientists: how will the global temperature change over the next 100 years?

- Psychology: can a simple reminder about saving money cause students to spend less?

- Political science: what fraction of Americans approve of the job Congress is doing?

What questions from current events or from your own life can you think of that could be answered by collecting and analyzing data? While the questions that can be posed are incredibly diverse, many of these investigations can be addressed with a small number of data collection techniques, analytic tools, and fundamental concepts in statistical inference.

This chapter focuses on collecting data. We'll discuss basic properties of data, common sources of bias that arise during data collection, and several techniques for collecting data through both sampling and experiments. After finishing this chapter, you will have the tools for identifying weaknesses and strengths in data-based conclusions, tools that are essential to be an informed citizen and a savvy consumer of information.

1.1   Case study: using stents to prevent strokes

Section 1.1 introduces a classic challenge in statistics: evaluating the efficacy of a medical treatment. Terms in this section, and indeed much of this chapter, will all be revisited later in the text. The plan for now is simply to get a sense of the role statistics can play in practice.

In this section we will consider an experiment that studies effectiveness of stents in treating patients at risk of stroke.[1] Stents are devices put inside blood vessels that assist in patient recovery after cardiac events and reduce the risk of an additional heart attack or death. Many doctors have hoped that there would be similar benefits for patients at risk of stroke. We start by writing the principal question the researchers hope to answer:

Does the use of stents reduce the risk of stroke?

The researchers who asked this question collected data on 451 at-risk patients. Each volunteer patient was randomly assigned to one of two groups:

Treatment group. Patients in the treatment group received a stent and medical management. The medical management included medications, management of risk factors, and help in lifestyle modification.

Control group. Patients in the control group received the same medical management as the treatment group, but they did not receive stents.

Researchers randomly assigned 224 patients to the treatment group and 227 to the control group. In this study, the control group provides a reference point against which we can measure the medical impact of stents in the treatment group.

Researchers studied the effect of stents at two time points: 30 days after enrollment and 365 days after enrollment. The results of 5 patients are summarized in Table 1.1. Patient outcomes are recorded as "stroke" or "no event", representing whether or not the patient had a stroke at the end of a time period.

Patient	group	0-30 days	0-365 days
1	treatment	no event	no event
2	treatment	stroke	stroke
3	treatment	no event	no event
⋮	⋮	⋮	
450	control	no event	no event
451	control	no event	no event

Table 1.1: Results for five patients from the stent study.

Considering data from each patient individually would be a long, cumbersome path towards answering the original research question. Instead, performing a statistical data analysis allows us to consider all of the data at once. Table 1.2 summarizes the raw data in a more helpful way. In this table, we can quickly see what happened over the entire study. For instance, to identify the number of patients in the treatment group who had a stroke

---

[1]Chimowitz MI, Lynn MJ, Derdeyn CP, et al. 2011. Stenting versus Aggressive Medical Therapy for Intracranial Arterial Stenosis. New England Journal of Medicine 365:993-1003. www.nejm.org/doi/full/10.1056/NEJMoa1105335. NY Times article reporting on the study: www.nytimes.com/2011/09/08/health/research/08stent.html.

within 30 days, we look on the left-side of the table at the intersection of the treatment and stroke: 33.

	0-30 days		0-365 days	
	stroke	no event	stroke	no event
treatment	33	191	45	179
control	13	214	28	199
Total	46	405	73	378

Table 1.2: Descriptive statistics for the stent study.

⊙ **Guided Practice 1.1**   What proportion of the patients in the treatment group had no stroke within the first 30 days of the study? (Please note: answers to all in-text exercises are provided using footnotes.)[2]

We can compute summary statistics from the table. A **summary statistic** is a single number summarizing a large amount of data.[3] For instance, the primary results of the study after 1 year could be described by two summary statistics: the proportion of people who had a stroke in the treatment and control groups.

Proportion who had a stroke in the treatment (stent) group: $45/224 = 0.20 = 20\%$.

Proportion who had a stroke in the control group: $28/227 = 0.12 = 12\%$.

These two summary statistics are useful in looking for differences in the groups, and we are in for a surprise: an additional 8% of patients in the treatment group had a stroke! This is important for two reasons. First, it is contrary to what doctors expected, which was that stents would *reduce* the rate of strokes. Second, it leads to a statistical question: do the data show a "real" difference between the groups?

This second question is subtle. Suppose you flip a coin 100 times. While the chance a coin lands heads in any given coin flip is 50%, we probably won't observe exactly 50 heads. This type of fluctuation is part of almost any type of data generating process. It is possible that the 8% difference in the stent study is due to this natural variation. However, the larger the difference we observe (for a particular sample size), the less believable it is that the difference is due to chance. So what we are really asking is the following: is the difference so large that we should reject the notion that it was due to chance?

While we don't yet have our statistical tools to fully address this question on our own, we can comprehend the conclusions of the published analysis: there was compelling evidence of harm by stents in this study of stroke patients.

Be careful: do not generalize the results of this study to all patients and all stents. This study looked at patients with very specific characteristics who volunteered to be a part of this study and who may not be representative of all stroke patients. In addition, there are many types of stents and this study only considered the self-expanding Wingspan stent (Boston Scientific). However, this study does leave us with an important lesson: we should keep our eyes open for surprises.

---

[2]There were 191 patients in the treatment group that had no stroke in the first 30 days. There were $33 + 191 = 224$ total patients in the treatment group, so the proportion is $191/224 = 0.85$.

[3]Formally, a summary statistic is a value computed from the data. Some summary statistics are more useful than others.

1.2 Data basics

Effective presentation and description of data is a first step in most analyses. This section introduces one structure for organizing data as well as some terminology that will be used throughout this book.

1.2.1 Observations, variables, and data matrices

Table 1.3 displays rows 1, 2, 3, and 50 of a data set concerning 50 emails received during early 2012. These observations will be referred to as the `email50` data set, and they are a random sample from a larger data set that we will see in Section 2.3.

Each row in the table represents a single email or **case**.[4] The columns represent characteristics, called **variables**, for each of the emails. For example, the first row represents email 1, which is not spam, contains 21,705 characters, 551 line breaks, is written in HTML format, and contains only small numbers.

In practice, it is especially important to ask clarifying questions to ensure important aspects of the data are understood. For instance, it is always important to be sure we know what each variable means and the units of measurement. Descriptions of all five email variables are given in Table 1.4.

	spam	num_char	line_breaks	format	number
1	no	21,705	551	html	small
2	no	7,011	183	html	big
3	yes	631	28	text	none
⋮	⋮	⋮	⋮	⋮	⋮
50	no	15,829	242	html	small

Table 1.3: Four rows from the `email50` data matrix.

variable	description
spam	Specifies whether the message was spam
num_char	The number of characters in the email
line_breaks	The number of line breaks in the email (not including text wrapping)
format	Indicates if the email contained special formatting, such as bolding, tables, or links, which would indicate the message is in HTML format
number	Indicates whether the email contained no number, a small number (under 1 million), or a large number

Table 1.4: Variables and their descriptions for the `email50` data set.

The data in Table 1.3 represent a **data matrix**, which is a common way to organize data. Each row of a data matrix corresponds to a unique case, and each column corresponds to a variable. A data matrix for the stroke study introduced in Section 1.1 is shown in Table 1.1 on page 8, where the cases were patients and there were three variables recorded for each patient.

Data matrices are a convenient way to record and store data. If another individual or case is added to the data set, an additional row can be easily added. Similarly, another column can be added for a new variable.

---

[4]A case is also sometimes called a **unit of observation** or an **observational unit**.

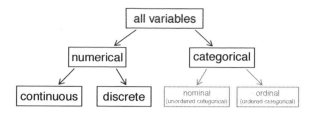

Figure 1.5: Breakdown of variables into their respective types.

⊙ **Guided Practice 1.2** We consider a publicly available data set that summarizes information about the 3,143 counties in the United States, and we call this the `county` data set. This data set includes information about each county: its name, the state where it resides, its population in 2000 and 2010, per capita federal spending, poverty rate, and five additional characteristics. How might these data be organized in a data matrix? Reminder: look in the footnotes for answers to in-text exercises.[5]

Seven rows of the `county` data set are shown in Table 1.6, and the variables are summarized in Table 1.7. These data were collected from the US Census website.[6]

1.2.2 Types of variables

Examine the `fed_spend`, `pop2010`, `state`, and `smoking_ban` variables in the `county` data set. Each of these variables is inherently different from the other three yet many of them share certain characteristics.

First consider `fed_spend`, which is said to be a **numerical** variable since it can take a wide range of numerical values, and it is sensible to add, subtract, or take averages with those values. On the other hand, we would not classify a variable reporting telephone area codes as numerical since their average, sum, and difference have no clear meaning.

The `pop2010` variable is also numerical, although it seems to be a little different than `fed_spend`. This variable of the population count can only take whole non-negative numbers (0, 1, 2, ...). For this reason, the population variable is said to be **discrete** since it can only take numerical values with jumps. On the other hand, the federal spending variable is said to be **continuous**.

The variable `state` can take up to 51 values after accounting for Washington, DC: `AL`, ..., and `WY`. Because the responses themselves are categories, `state` is called a **categorical** variable,[7] and the possible values are called the variable's **levels**.

Finally, consider the `smoking_ban` variable, which describes the type of county-wide smoking ban and takes values `none`, `partial`, or `comprehensive` in each county. This variable seems to be a hybrid: it is a categorical variable but the levels have a natural ordering. A variable with these properties is called an **ordinal** variable. To simplify analyses, any ordinal variables in this book will be treated as categorical variables.

---

[5]Each county may be viewed as a case, and there are eleven pieces of information recorded for each case. A table with 3,143 rows and 11 columns could hold these data, where each row represents a county and each column represents a particular piece of information.

[6]quickfacts.census.gov/qfd/index.html

[7]Sometimes also called a **nominal** variable.

	name	state	pop2000	pop2010	fed.spend	poverty	homeownership	multiunit	income	med.income	smoking_ban
1	Autauga	AL	43671	54571	6.068	10.6	77.5	7.2	24568	53255	none
2	Baldwin	AL	140415	182265	6.140	12.2	76.7	22.6	26469	50147	none
3	Barbour	AL	29038	27457	8.752	25.0	68.0	11.1	15875	33219	none
4	Bibb	AL	20826	22915	7.122	12.6	82.9	6.6	19918	41770	none
5	Blount	AL	51024	57322	5.131	13.4	82.0	3.7	21070	45549	none
⋮	⋮	⋮	⋮	⋮	⋮	⋮	⋮	⋮	⋮	⋮	⋮
3142	Washakie	WY	8289	8533	8.714	5.6	70.9	10.0	28557	48379	none
3143	Weston	WY	6644	7208	6.695	7.9	77.9	6.5	28463	53853	none

Table 1.6: Seven rows from the county data set.

variable	description
name	County name
state	State where the county resides (also including the District of Columbia)
pop2000	Population in 2000
pop2010	Population in 2010
fed_spend	Federal spending per capita
poverty	Percent of the population in poverty
homeownership	Percent of the population that lives in their own home or lives with the owner (e.g. children living with parents who own the home)
multiunit	Percent of living units that are in multi-unit structures (e.g. apartments)
income	Income per capita
med_income	Median household income for the county, where a household's income equals the total income of its occupants who are 15 years or older
smoking_ban	Type of county-wide smoking ban in place at the end of 2011, which takes one of three values: **none**, **partial**, or **comprehensive**, where a **comprehensive** ban means smoking was not permitted in restaurants, bars, or workplaces, and **partial** means smoking was banned in at least one of those three locations

Table 1.7: Variables and their descriptions for the county data set.

● **Example 1.3** Data were collected about students in a statistics course. Three variables were recorded for each student: number of siblings, student height, and whether the student had previously taken a statistics course. Classify each of the variables as continuous numerical, discrete numerical, or categorical.

The number of siblings and student height represent numerical variables. Because the number of siblings is a count, it is discrete. Height varies continuously, so it is a continuous numerical variable. The last variable classifies students into two categories – those who have and those who have not taken a statistics course – which makes this variable categorical.

⊙ **Guided Practice 1.4** Consider the variables group and outcome (at 30 days) from the stent study in Section 1.1. Are these numerical or categorical variables?[8]

1.2.3 Relationships between variables

Many analyses are motivated by a researcher looking for a relationship between two or more variables. A social scientist may like to answer some of the following questions:

(1) Is federal spending, on average, higher or lower in counties with high rates of poverty?

(2) If homeownership is lower than the national average in one county, will the percent of multi-unit structures in that county likely be above or below the national average?

(3) Which counties have a higher average income: those that enact one or more smoking bans or those that do not?

To answer these questions, data must be collected, such as the county data set shown in Table 1.6. Examining summary statistics could provide insights for each of the three questions about counties. Additionally, graphs can be used to visually summarize data and are useful for answering such questions as well.

Scatterplots are one type of graph used to study the relationship between two numerical variables. Figure 1.8 compares the variables fed_spend and poverty. Each point on the plot represents a single county. For instance, the highlighted dot corresponds to County 1088 in the county data set: Owsley County, Kentucky, which had a poverty rate of 41.5% and federal spending of $21.50 per capita. The scatterplot suggests a relationship between the two variables: counties with a high poverty rate also tend to have slightly more federal spending. We might brainstorm as to why this relationship exists and investigate each idea to determine which is the most reasonable explanation.

⊙ **Guided Practice 1.5** Examine the variables in the email50 data set, which are described in Table 1.4 on page 10. Create two questions about the relationships between these variables that are of interest to you.[9]

The fed_spend and poverty variables are said to be associated because the plot shows a discernible pattern. When two variables show some connection with one another, they are called **associated** variables. Associated variables can also be called **dependent** variables and vice-versa.

---

[8]There are only two possible values for each variable, and in both cases they describe categories. Thus, each is a categorical variable.

[9]Two sample questions: (1) Intuition suggests that if there are many line breaks in an email then there would also tend to be many characters: does this hold true? (2) Is there a connection between whether an email format is plain text (versus HTML) and whether it is a spam message?

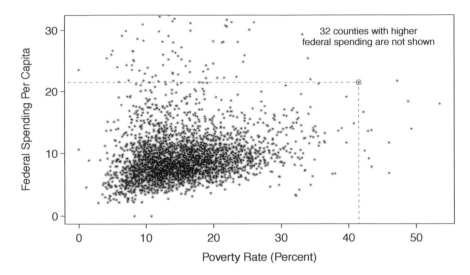

Figure 1.8: A scatterplot showing `fed_spend` against `poverty`. Owsley County of Kentucky, with a poverty rate of 41.5% and federal spending of $21.50 per capita, is highlighted.

● **Example 1.6** The relationship between the homeownership rate and the percent of units in multi-unit structures (e.g. apartments, condos) is visualized using a scatterplot in Figure 1.9. Are these variables associated?

It appears that the larger the fraction of units in multi-unit structures, the lower the homeownership rate. Since there is some relationship between the variables, they are associated.

Because there is a downward trend in Figure 1.9 – counties with more units in multiunit structures are associated with lower homeownership – these variables are said to be **negatively associated**. A **positive association** is shown in the relationship between the `poverty` and `fed_spend` variables represented in Figure 1.8, where counties with higher poverty rates tend to receive more federal spending per capita.

If two variables are not associated, then they are said to be **independent**. That is, two variables are independent if there is no evident relationship between the two.

Associated or independent, not both

A pair of variables are either related in some way (associated) or not (independent). No pair of variables is both associated and independent.

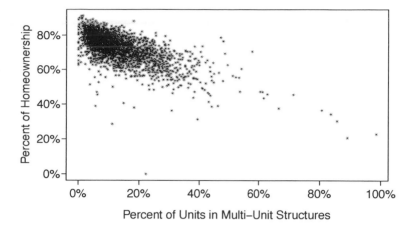

Figure 1.9: A scatterplot of homeownership versus the percent of units that are in multi-unit structures for all 3,143 counties. Interested readers may find an image of this plot with an additional third variable, county population, presented at www.openintro.org/stat/down/MHP.png.

1.3 Overview of data collection principles

The first step in conducting research is to identify topics or questions that are to be investigated. A clearly laid out research question is helpful in identifying what subjects or cases should be studied and what variables are important. It is also important to consider *how* data are collected so that they are reliable and help achieve the research goals.

1.3.1 Populations and samples

Consider the following three research questions:

1. What is the average mercury content in swordfish in the Atlantic Ocean?

2. Over the last 5 years, what is the average time to complete a degree for Duke undergraduate students?

3. Does a new drug reduce the number of deaths in patients with severe heart disease?

Each research question refers to a target **population**. In the first question, the target population is all swordfish in the Atlantic ocean, and each fish represents a case. Often times, it is too expensive to collect data for every case in a population. Instead, a sample is taken. A **sample** represents a subset of the cases and is often a small fraction of the population. For instance, 60 swordfish (or some other number) in the population might be selected, and this sample data may be used to provide an estimate of the population average and answer the research question.

⊙ **Guided Practice 1.7** For the second and third questions above, identify the target population and what represents an individual case.[10]

---

[10](2) Notice that this question is only relevant to students who complete their degree; the average cannot be computed using a student who never finished her degree. Thus, only Duke undergraduate students who have graduated in the last five years are part of the population of interest. Each such student would represent an individual case. (3) A person with severe heart disease represents a case. The population includes all people with severe heart disease.

We collect a sample of data to better understand the characteristics of a population. A **variable** is a characteristic we measure for each individual or case. The overall quantity of interest may be the mean, median, proportion, or some other summary of a population. These population values are called **parameters**. We estimate the value of a parameter by taking a sample and computing a numerical summary called a **statistic** based on that sample. Note that the two p's (population, parameter) go together and the two s's (sample, statistic) go together.

● **Example 1.8**   Earlier we asked the question: what is the average mercury content in swordfish in the Atlantic Ocean? Identify the variable to be measured and the parameter and statistic of interest.

The variable is the level of mercury content in swordfish in the Atlantic Ocean. It will be measured for each individual swordfish. The parameter of interest is the average mercury content in *all* swordfish in the Atlantic Ocean. If we take a sample of 50 swordfish from the Atlantic Ocean, the average mercury content among just those 50 swordfish will be the statistic.

Two statistics we will study are the **mean** (also called the **average**) and **proportion**. When we are discussing a population, we label the mean as μ (the Greek letter, mu), while we label the sample mean as $\bar{x}$ (read as $x\text{-}bar$). When we are discussing a proportion in the context of a population, we use the label p, while the sample proportion has a label of $\hat{p}$ (read as $p\text{-}hat$). Generally, we use $\bar{x}$ to estimate the population mean, μ. Likewise, we use the sample proportion $\hat{p}$ to estimate the population proportion, p.

● **Example 1.9**   Is μ a parameter or statistic? What about $\hat{p}$?

μ is a parameter because it refers to the average of the *entire* population. $\hat{p}$ is a statistic because it is calculated from a sample.

● **Example 1.10**   For the second question regarding time to complete a degree for a Duke undergraduate, is the variable numerical or categorical? What is the parameter of interest?

The characteristic that we record on each individual is the number of years until graduation, which is a numerical variable. The parameter of interest is the average time to degree for all Duke undergraduates, and we use μ to describe this quantity.

⊙ **Guided Practice 1.11**   The third question asked whether a new drug reduces deaths in patients with severe heart disease. Is the variable numerical or categorical? Describe the statistic that should be calculated in this study.[11]

If these topics are still a bit unclear, don't worry. We'll cover them in greater detail in the next chapter.

---

[11]The variable is whether or not a patient with severe heart disease dies within the time frame of the study. This is categorical because it will be a yes or a no. The statistic that should be recorded is the proportion of patients that die within the time frame of the study, and we would use $\hat{p}$ to denote this quantity.

Figure 1.10: In February 2010, some media pundits cited one large snow storm as valid evidence against global warming. As comedian Jon Stewart pointed out, "It's one storm, in one region, of one country."

February 10th, 2010.

1.3.2 Anecdotal evidence

Consider the following possible responses to the three research questions:

1. A man on the news got mercury poisoning from eating swordfish, so the average mercury concentration in swordfish must be dangerously high.

2. I met two students who took more than 7 years to graduate from Duke, so it must take longer to graduate at Duke than at many other colleges.

3. My friend's dad had a heart attack and died after they gave him a new heart disease drug, so the drug must not work.

Each conclusion is based on data. However, there are two problems. First, the data only represent one or two cases. Second, and more importantly, it is unclear whether these cases are actually representative of the population. Data collected in this haphazard fashion are called **anecdotal evidence**.

Anecdotal evidence

Be careful of making inferences based on anecdotal evidence. Such evidence may be true and verifiable, but it may only represent extraordinary cases. The majority of cases and the average case may in fact be very different.

Anecdotal evidence typically is composed of unusual cases that we recall based on their striking characteristics. For instance, we may vividly remember the time when our friend bought a lottery ticket and won $250 but forget most the times she bought one and lost. Instead of focusing on the most unusual cases, we should examine a representative sample of many cases.

1.3.3   Explanatory and response variables

Consider the following question from page 13 for the county data set:

(1) Is federal spending, on average, higher or lower in counties with high rates of poverty?

If we suspect poverty might affect spending in a county, then poverty is the **explanatory** variable and federal spending is the **response** variable in the relationship.[12] If there are many variables, it may be possible to consider a number of them as explanatory variables.

TIP: Explanatory and response variables
To identify the explanatory variable in a pair of variables, identify which of the two is suspected of affecting the other.

$$\text{explanatory variable} \xrightarrow{\text{might affect}} \text{response variable}$$

Caution: Association does not imply causation
Labeling variables as *explanatory* and *response* does not guarantee the relationship between the two is actually causal, even if there is an association identified between the two variables. We use these labels only to keep track of which variable we suspect affects the other.

In many cases, the relationship is complex or unknown. It may be unclear whether variable A explains variable B or whether variable B explains variable A. For example, it is now known that a particular protein called REST is much depleted in people suffering from Alzheimer's disease. While this raises hopes of a possible approach for treating Alzheimer's, it is still unknown whether the lack of the protein causes brain deterioration, whether brain deterioration causes depletion in the REST protein, or whether some third variable causes both brain deterioration and REST depletion. That is, we do not know if the lack of the protein is an explanatory variable or a response variable. Perhaps it is both.[13]

1.3.4   Observational studies versus experiments

There are two primary types of data collection: observational studies and experiments.

Researchers perform an **observational study** when they collect data without interfering with how the data arise. For instance, researchers may collect information via surveys, review medical or company records, or follow a **cohort** of many similar individuals to study why certain diseases might develop. In each of these situations, researchers merely observe or take measurements of things that arise naturally.

When researchers want to investigate the possibility of a causal connection, they conduct an **experiment**. For all experiments, the researchers must impose a treatment. For most studies there will be both an explanatory and a response variable. For instance, we may suspect administering a drug will reduce mortality in heart attack patients over the following year. To check if there really is a causal connection between the explanatory variable and the response, researchers will collect a sample of individuals and split them

---

[12]Sometimes the explanatory variable is called the **independent** variable and the response variable is called the **dependent** variable. However, this becomes confusing since a *pair* of variables might be independent or dependent, so we avoid this language.

[13]nytimes.com/2014/03/20/health/fetal-gene-may-protect-brain-from-alzheimers-study-finds.html

into groups. The individuals in each group are *assigned* a treatment. When individuals are randomly assigned to a group, the experiment is called a **randomized experiment**. For example, each heart attack patient in the drug trial could be randomly assigned into one of two groups: the first group receives a **placebo** (fake treatment) and the second group receives the drug. See the case study in Section 1.1 for another example of an experiment, though that study did not employ a placebo.

● **Example 1.12** Suppose that a researcher is interested in the average tip customers at a particular restaurant give. Should she carry out an observational study or an experiment?

In addressing this question, we ask, "Will the researcher be imposing any treatment?" Because there is no treatment or interference that would be applicable here, it will be an observational study. Additionally, one consideration the researcher should be aware of is that, if customers know their tips are being recorded, it could change their behavior, making the results of the study inaccurate.

TIP: Association ≠ causation
In general, association does not imply causation, and causation can only be inferred from a randomized experiment.

1.4 Observational studies and sampling strategies

1.4.1 Observational studies

Generally, data in observational studies are collected only by monitoring what occurs, while experiments require the primary explanatory variable in a study be assigned for each subject by the researchers.

Making causal conclusions based on experiments is often reasonable. However, making the same causal conclusions based on observational data is treacherous and is not recommended. Observational studies are generally only sufficient to show associations.

◉ **Guided Practice 1.13** Suppose an observational study tracked sunscreen use and skin cancer, and it was found people who use sunscreen are more likely to get skin cancer than people who do not use sunscreen. Does this mean sunscreen *causes* skin cancer?[14]

Some previous research tells us that using sunscreen actually reduces skin cancer risk, so maybe there is another variable that can explain this hypothetical association between sunscreen usage and skin cancer. One important piece of information that is absent is sun exposure. Sun exposure is what is called a **confounding variable** (also called a **lurking variable**, **confounding factor**, or a **confounder**).

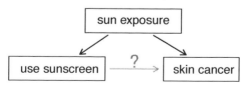

---

[14]No. See the paragraph following the exercise for an explanation.

Confounding variable

A confounding variable is a variable that is associated with both the explanatory *and* response variables. Because of the confounding variable's association with both variables, we do not know if the response is due to the explanatory variable or due to the confounding variable.

Sun exposure is a confounding factor because it is associated with both the use of sunscreen and the development of skin cancer. People who are out in the sun all day are more likely to use sunscreen, and people who are out in the sun all day are more likely to get skin cancer. Research shows us the development of skin cancer is due to the sun exposure. The variables of sunscreen usage and sun exposure are **confounded**, and without this research, we would have no way of knowing which one was the true cause of skin cancer.

● **Example 1.14** In a study that followed 1,169 non-diabetic men and women who had been hospitalized for a first heart attack, the people that reported eating chocolate had increased survival rate over the next 8 years than those that reported eating not eating chocolate.[15] Also, those who ate more chocolate also tended to live longer on average. The researched controlled for several confounding factors, such as age, physical activity, smoking, and many other factors. Can we conclude that the consumption of chocolate caused the people to live longer?

This is an observational study, not a controlled randomized experiment. Even though the researches controlled for many possible variables, there may still be other confounding factors. (Can you think of any that weren't mentioned?) While it is possible that the chocolate had an effect, this study cannot prove that chocolate increased the survival rate of patients.

● **Example 1.15** The authors who conducted the study did warn in the article that additional studies would be necessary to determine whether the correlation between chocolate consumption and survival translates to any causal relationship. That is, they acknowledged that there may be confounding factors. One possible confounding factor not considered was mental health. In context, explain what it would mean for mental health to be a confounding factor in this study.

Mental health would be a confounding factor if, for example, people with better mental health tended to eat more chocolate, and those with better mental health *also* were less likely to die within the 8 year study period. Notice that if better mental health were not associated with eating more chocolate, it would not be considered a confounding factor since it wouldn't explain the observed associated between eating chocolate and having a better survival rate. If better mental health were associated only with eating chocolate and not with a better survival rate, then it would also not be confounding for the same reason. Only if a variable that is associated with both the explanatory variable of interest (chocolate) and the outcome variable in the study (survival during the 8 year study period) can it be considered a confounding factor.

While one method to justify making causal conclusions from observational studies is to exhaust the search for confounding variables, there is no guarantee that all confounding variables can be examined or measured.

---

[15]Janszky et al. 2009. Chocolate consumption and mortality following a first acute myocardial infarction: the Stockholm Heart Epidemiology Program. Journal of Internal Medicine 266:3, p248-257.

In the same way, the `county` data set is an observational study with confounding variables, and its data cannot be used to make causal conclusions.

⊙ **Guided Practice 1.16** Figure 1.9 shows a negative association between the home-ownership rate and the percentage of multi-unit structures in a county. However, it is unreasonable to conclude that there is a causal relationship between the two variables. Suggest one or more other variables that might explain the relationship visible in Figure 1.9.[16]

Observational studies come in two forms: prospective and retrospective studies. A **prospective study** identifies individuals and collects information as events unfold. For instance, medical researchers may identify and follow a group of similar individuals over many years to assess the possible influences of behavior on cancer risk. One example of such a study is The Nurses' Health Study, started in 1976 and expanded in 1989.[17] This prospective study recruits registered nurses and then collects data from them using questionnaires. **Retrospective studies** collect data after events have taken place, e.g. researchers may review past events in medical records. Some data sets, such as `county`, may contain both prospectively- and retrospectively-collected variables. Local governments prospectively collect some variables as events unfolded (e.g. retails sales) while the federal government retrospectively collected others during the 2010 census (e.g. county population counts).

1.4.2 Sampling from a population

We might try to estimate the time to graduation for Duke undergraduates in the last 5 years by collecting a sample of students. All graduates in the last 5 years represent the *population*, and graduates who are selected for review are collectively called the *sample*. In general, we always seek to *randomly* select a sample from a population. The most basic type of random selection is equivalent to how raffles are conducted. For example, in selecting graduates, we could write each graduate's name on a raffle ticket and draw 100 tickets. The selected names would represent a random sample of 100 graduates.

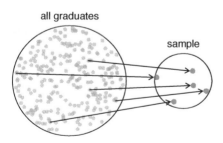

Figure 1.11: In this graphic, five graduates are randomly selected from the population to be included in the sample.

Why pick a sample randomly? Why not just pick a sample by hand? Consider the following scenario.

---

[16]Answers will vary. Population density may be important. If a county is very dense, then this may require a larger fraction of residents to live in multi-unit structures. Additionally, the high density may contribute to increases in property value, making homeownership infeasible for many residents.

[17]www.channing.harvard.edu/nhs

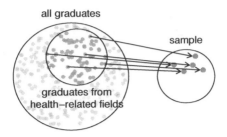

Figure 1.12: Instead of sampling from all graduates equally, a nutrition major might inadvertently pick graduates with health-related majors disproportionately often.

● **Example 1.17**   Suppose we ask a student who happens to be majoring in nutrition to select several graduates for the study. What kind of students do you think she might collect? Do you think her sample would be representative of all graduates?

Perhaps she would pick a disproportionate number of graduates from health-related fields. Or perhaps her selection would be well-representative of the population. When selecting samples by hand, we run the risk of picking a *biased* sample, even if that bias is unintentional or difficult to discern.

If the student majoring in nutrition picked a disproportionate number of graduates from health-related fields, this would introduce selection bias into the sample. **Selection bias** occurs when some individuals of the population are inherently more likely to be included in the sample than others. In the example, this bias creates a problem because a degree in health-related fields might take more or less time to complete than a degree in other fields. Suppose that it takes longer. Since graduates from health-related fields would be more likely to be in the sample, the selection bias would cause her to *overestimate* the parameter.

Sampling randomly resolves the problem of selection bias. The most basic random sample is called a **simple random sample**, which is equivalent to using a raffle to select cases. This means that each case in the population has an equal chance of being included and there is no implied connection between the cases in the sample.

A common downfall is a **convenience sample**, where individuals who are easily accessible are more likely to be included in the sample. For instance, if a political survey is done by stopping people walking in the Bronx, this will not represent all of New York City. It is often difficult to discern what sub-population a convenience sample represents.

Similarly, a **volunteer sample** is one in which people's responses are solicited and those who choose to participate, respond. This is a problem because those who choose to participate may tend to have different opinions than the rest of the population, resulting in a biased sample.

⊙ **Guided Practice 1.18**   We can easily access ratings for products, sellers, and companies through websites. These ratings are based only on those people who go out of their way to provide a rating. If 50% of online reviews for a product are negative, do you think this means that 50% of buyers are dissatisfied with the product?[18]

---

[18]Answers will vary. From our own anecdotal experiences, we believe people tend to rant more about products that fell below expectations than rave about those that perform as expected. For this reason, we suspect there is a negative bias in product ratings on sites like Amazon. However, since our experiences may not be representative, we also keep an open mind.

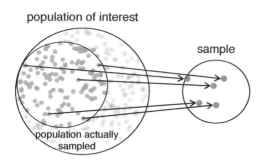

Figure 1.13: Due to the possibility of non-response, surveys studies may only reach a certain group within the population. It is difficult, and often times impossible, to completely fix this problem.

The act of taking a random sample helps minimize bias; however, bias can crop up in other ways. Even when people are picked at random, e.g. for surveys, caution must be exercised if the **non-response** is high. For instance, if only 30% of the people randomly sampled for a survey actually respond, then it is unclear whether the results are **representative** of the entire population. This **non-response bias** can skew results.

Even if a sample has no selection bias and no non-response bias, there is an additional type of bias that often crops up and undermines the validity of results, known as response bias. **Response bias** refers to a broad range of factors that influence how a person responds, such as question wording, question order, and influence of the interviewer. This type of bias can be present even when we collect data from an entire population in what is called a **census**. Because response bias is often subtle, one must pay careful attention to how questions were asked when attempting to draw conclusions from the data.

● **Example 1.19** Suppose a high school student wants to investigate the student body's opinions on the food in the cafeteria. Let's assume that she manages to survey every student in the school. How might response bias arise in this context?

There are many possible correct answers to this question. For example, students might respond differently depending upon who asks the question, such as a school friend or someone who works in the cafeteria. The wording of the question could introduce response bias. Students would likely respond differently if asked "Do you like the food in the cafeteria?" versus "The food in the cafeteria is pretty bad, don't you think?"

TIP: Watch out for bias
Selection bias, non-response bias, and response bias can still exist within a random sample. Always determine how a sample was chosen, ask what proportion of people failed to respond, and critically examine the wording of the questions.

When there is no bias in a sample, increasing the sample size tends to increase the precision and reliability of the estimate. When a sample is biased, it may be impossible to decipher helpful information from the data, even if the sample is very large.

⊙ **Guided Practice 1.20** A researcher sends out questionnaires to 50 randomly selected households in a particular town asking whether or not they support the addition of a traffic light in their neighborhood. Because only 20% of the questionnaires are returned, she decides to mail questionnaires to 50 more randomly selected households in the same neighborhood. Comment on the usefulness of this approach.[19]

1.4.3 Simple, systematic, stratified, cluster, and multistage sampling

Almost all statistical methods for observational data rely on a sample being random and unbiased. When a sample is collected in a biased way, these statistical methods will not generally produce reliable information about the population.

The idea of a simple random sample was introduced in the last section. Here we provide a more technical treatment of this method and introduce four new random sampling methods: systematic, stratified, cluster, and multistage.[20] Figure 1.14 provides a graphical representation of simple versus systematic sampling while Figure 1.15 provides a graphical representation of stratified, cluster, and multistage sampling.

Simple random sampling is probably the most intuitive form of random sampling. Consider the salaries of Major League Baseball (MLB) players, where each player is a member of one of the league's 30 teams. For the 2010 season, N, the population size or total number of players, is 828. To take a simple random sample of $n = 120$ of these baseball players and their salaries, we could number each player from 1 to 828. Then we could randomly select 120 numbers between 1 and 828 (without replacement) using a random number generator or random digit table. The players with the selected numbers would comprise our sample.

Two properties are always true in a simple random sample:

1. Each case in the population has an equal chance of being included in the sample.

2. Each *group* of n cases has an equal chance of making up the sample.

The statistical methods in this book focus on data collected using simple random sampling. Note that Property 2 – that each group of n cases has an equal chance making up the sample – is not true for the remaining four sampling techniques. As you read each one, consider why.

Though less common than simple random sampling, **systematic sampling** is sometimes used when there exists a convenient list of all of the individuals of the population. Suppose we have a roster with the names of all the MLB players from the 2010 season. To take a systematic random sample, number them from 1 to 828. Select one random number between 1 and 828 and let that player be the first individual in the sample. Then, depending on the desired sample size, select every 10th number or 20th number, for example, to arrive at the sample.[21] If there are no patterns in the salaries based on the numbering then this could be a reasonable method.

---

[19]The researcher should be concerned about non-response bias, and sampling more people will not eliminate this issue. The same type of people that did not respond to the first survey are likely not going to respond to the second survey. Instead, she should make an effort to reach out to the households from the original sample that did not respond and solicit their feedback, possibly by going door-to-door.

[20]Systematic and Multistage sampling are not part of the AP syllabus.

[21]If we want a sample of size $n = 138$, it would make sense to select every 6th player since $828/138 = 6$. Suppose we randomly select the number 810. Then player 810, 816, 822, 828, 6, 12, $\cdots$, 798, and 804 would make up the sample.

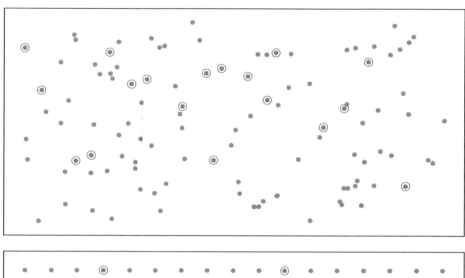

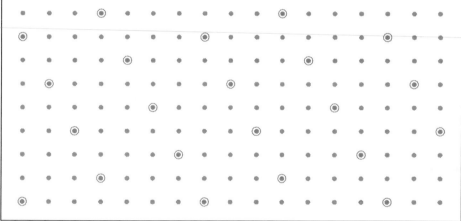

Figure 1.14: Examples of simple random sampling and systematic sampling. In the top panel, simple random sampling was used to randomly select 18 cases. In the lower panel, systematic random sampling was used to select every 7th individual.

● **Example 1.21** A systematic sample is not the same as a simple random sample. Provide an example of a sample that can come from a simple random sample but not from a systematic random sample.

Answers can vary. If we take a sample of size 3, then it is possible that we could sample players numbered 1, 2, and 3 in a simple random sample. Such a sample would be impossible from a systematic sample. Property 2 of simple random samples does not hold for other types of random samples.

Sometimes there is a variable that is known to be associated with the quantity we want to estimate. In this case, a stratified random sample might be selected. **Stratified sampling** is a divide-and-conquer sampling strategy. The population is divided into groups called **strata**. The strata are chosen so that similar cases are grouped together and a sampling method, usually simple random sampling, is employed to select a certain number or a certain proportion of the whole within each stratum. In the baseball salary example, the 30 teams could represent the strata; some teams have a lot more money (we're looking at you, Yankees).

● **Example 1.22** For this baseball example, briefly explain how to select a stratified random sample of size $n = 120$.

Each team can serve as a stratum, and we could take a simple random sample of 4 players from each of the 30 teams, yielding a sample of 120 players.

Stratified sampling is inherently different than simple random sampling. For example, the stratified sampling approach described would make it impossible for the entire Yankees team to be included in the sample.

● **Example 1.23** Stratified sampling is especially useful when the cases in each stratum are very similar *with respect to the outcome of interest*. Why is it good for cases within each stratum to be very similar?

We should get a more stable estimate for the subpopulation in a stratum if the cases are very similar. These improved estimates for each subpopulation will help us build a reliable estimate for the full population. For example, in a simple random sample, it is possible that just by random chance we could end up with proportionally too many Yankees players in our sample, thus overestimating the true average salary of all MLB players. A stratified random sample can assure proportional representation from each team.

Next, let's consider a sampling technique that randomly selects groups of people. **Cluster sampling** is much like simple random sampling, but instead of randomly selecting *individuals*, we randomly select groups or **clusters**. Unlike stratified sampling, cluster sampling is most helpful when there is a lot of case-to-case variability within a cluster but the clusters themselves don't look very different from one another. That is, we expect strata to be self-similar (homogeneous), while we expect clusters to be diverse (heterogeneous).

Sometimes cluster sampling can be a more economical random sampling technique than the alternatives. For example, if neighborhoods represented clusters, this sampling method works best when each neighborhood is very diverse. Because each neighborhood itself encompasses diversity, a cluster sample can reduce the time and cost associated with data collection, because the interviewer would need only go to some of the neighborhoods rather than to all parts of a city, in order to collect a useful sample.

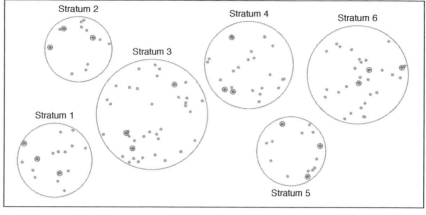

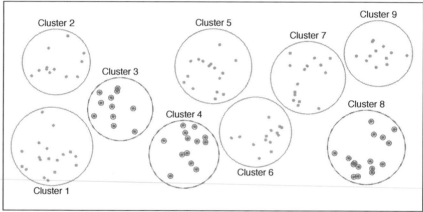

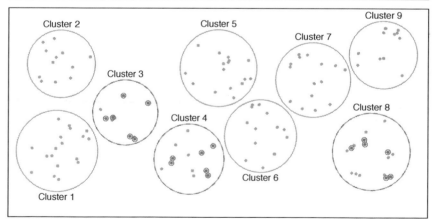

Figure 1.15: Examples of stratified, cluster, and multistage sampling. In the top panel, stratified sampling was used: cases were grouped into strata, and then simple random sampling was employed within each stratum. In the middle panel, cluster sampling was used, where data were binned into nine cluster and three clusters were randomly selected. In the bottom panel, multistage sampling was used. Data were binned into the nine clusters, three of the cluster were randomly selected, and then six cases were randomly sampled in each of the three selected clusters.

Multistage sampling, also called **multistage cluster sampling**, is a two (or more) step strategy. The first step is to take a cluster sample, as described above. Then, instead of including all of the individuals in these clusters in our sample, a second sampling method, usually simple random sampling, is employed within each of the selected clusters. In the neighborhood example, we could first randomly select some number of neighborhoods and then take a simple random sample from just those selected neighborhoods. As seen in Figure 1.15, stratified sampling requires observations to be sampled from *every* stratum. Multistage sampling selects observations *only* from those clusters that were randomly selected in the first step.

It is also possible to have more than two steps in multistage sampling. Each cluster may be naturally divided into subclusters. For example, each neighborhood could be divided into streets. To take a three-stage sample, we could first select some number of clusters (neighborhoods), and then, within the selected clusters, select some number of subclusters (streets). Finally, we could select some number of individuals from each of the selected streets.

● **Example 1.24** Suppose we are interested in estimating the proportion of students at a certain school that have part-time jobs. It is believed that older students are more likely to work than younger students. What sampling method should be employed? Describe how to collect such a sample to get a sample size of 60.

Because grade level affects the likelihood of having a part-time job, we should take a stratified random sample. To do this, we can take a simple random sample of 15 students from each grade. This will give us equal representation from each grade. Note: in a simple random sample, just by random chance we might get too many students who are older or younger, which could make the estimate too high or too low. Also, there are no well-defined clusters in this example. We wouldn't want to use the grades as clusters and sample everyone from a couple of the grades. This would create too large a sample and would not give us the nice representation from each grade afforded by the stratified random sample.

● **Example 1.25** Suppose we are interested in estimating the malaria rate in a densely tropical portion of rural Indonesia. We learn that there are 30 villages in that part of the Indonesian jungle, each more or less similar to the next. Our goal is to test 150 individuals for malaria. What sampling method should be employed?

A simple random sample would likely draw individuals from all 30 villages, which could make data collection extremely expensive. Stratified sampling would be a challenge since it is unclear how we would build strata of similar individuals. However, multistage cluster sampling seems like a very good idea. First, we might randomly select half the villages, then randomly select 10 people from each. This would probably reduce our data collection costs substantially in comparison to a simple random sample and would still give us reliable information.

Caution: Advanced sampling techniques require advanced methods
The methods of inference covered in this book generally only apply to simple random samples. More advanced analysis techniques are required for systematic, stratified, cluster, and multistage random sampling.

1.5 Experiments

In the last section we investigated observational studies and sampling strategies. While these are effective tools for answering certain research questions, often times researchers want to measure the effect of a treatment. In this case, they must carry out an experiment. Just as randomization is essential in sampling in order to avoid selection bias, randomization is essential in the context of experiments to determine which subjects will receive which treatments. If the researcher chooses which patients are in the treatment and control groups, she may unintentionally place healthier or sicker patients in one group or the other, biasing the experiment either for or against the treatment.

1.5.1 Reducing bias in human experiments

Randomized experiments are essential for investigating cause and effect relationships, but they do not ensure an unbiased perspective in all cases. Human studies are perfect examples where bias can unintentionally arise. Here we reconsider a study where a new drug was used to treat heart attack patients.[22] In particular, researchers wanted to know if the drug reduced deaths in patients.

These researchers designed a randomized experiment because they wanted to draw causal conclusions about the drug's effect. Study volunteers[23] were randomly placed into two study groups. One group, the **treatment group**, received the drug. The other group, called the **control group**, did not receive any drug treatment. In an experiment, the explanatory variable is also called a **factor**. Here the factor is receiving the drug treatment. It has two **levels**: yes and no, thus it is categorical. The response variable is whether or not patients died within the time frame of the study. It is also categorical.

Put yourself in the place of a person in the study. If you are in the treatment group, you are given a fancy new drug that you anticipate will help you. On the other hand, a person in the other group doesn't receive the drug and sits idly, hoping her participation doesn't increase her risk of death. These perspectives suggest there are actually two effects: the one of interest is the effectiveness of the drug, and the second is an emotional effect that is difficult to quantify.

Researchers aren't usually interested in the emotional effect, which might bias the study. To circumvent this problem, researchers do not want patients to know which group they are in. When researchers keep the patients uninformed about their treatment, the study is said to be **blind** or **single-blind**. But there is one problem: if a patient doesn't receive a treatment, she will know she is in the control group. The solution to this problem is to give fake treatments to patients in the control group. A fake treatment is called a **placebo**, and an effective placebo is the key to making a study truly blind. A classic example of a placebo is a sugar pill that is made to look like the actual treatment pill. Often times, a placebo results in a slight but real improvement in patients. This effect has been dubbed the **placebo effect**.

The patients are not the only ones who should be blinded: doctors and researchers can accidentally bias a study. When a doctor knows a patient has been given the real treatment, she might inadvertently give that patient more attention or care than a patient that she knows is on the placebo. To guard against this bias, which again has been found to have a measurable effect in some instances, most modern studies employ a **double-blind**

---

[22]Anturane Reinfarction Trial Research Group. 1980. Sulfinpyrazone in the prevention of sudden death after myocardial infarction. New England Journal of Medicine 302(5):250-256.

[23]Human subjects are often called **patients**, **volunteers**, or **study participants**.

setup where researchers who interact with subjects and are responsible for measuring the response variable are, just like the subjects, unaware of who is or is not receiving the treatment.[24]

⊙ **Guided Practice 1.26**   Look back to the study in Section 1.1 where researchers were testing whether stents were effective at reducing strokes in at-risk patients. Is this an experiment? Was the study blinded? Was it double-blinded?[25]

1.5.2   Principles of experimental design

Well-conducted experiments are built on three main principles.

Direct Control. Researchers assign treatments to cases, and they do their best to **control** any other differences in the groups. They want the groups to be as identical as possible *except for the treatment*, so that at the end of the experiment any difference in response between the groups can be attributed to the treatment and not to some other confounding or lurking variable. For example, when patients take a drug in pill form, some patients take the pill with only a sip of water while others may have it with an entire glass of water. To control for the effect of water consumption, a doctor may ask all patients to drink a 12 ounce glass of water with the pill.

Direct control refers to variables that the researcher can control, or make the same. A researcher can directly control the appearance of the treatment, the time of day it is taken, etc. She cannot directly control variables such as gender or age. To control for these other types of variables, she might consider blocking, which is described in Section 1.5.3.

Randomization. Researchers randomize patients into treatment groups to account for variables that cannot be controlled. For example, some patients may be more susceptible to a disease than others due to their dietary habits. Randomizing patients into the treatment or control group helps *even out* the effects of such differences, and it also prevents accidental bias from entering the study.

Replication. The more cases researchers observe, the more accurately they can estimate the effect of the explanatory variable on the response. In an experiment with six subjects, even if there is randomization, it is quite possible for the three healthiest people to be in the same treatment group. In a randomized experiment with 100 people, it is virtually impossible for the healthiest 50 people to end up in the same treatment group. In a single study, we **replicate** by imposing the treatment on a sufficiently large number of subjects or experimental units. A group of scientists may also replicate an entire study to verify an earlier finding. However, each study should ensure a sufficiently large number of subjects because, in many cases, there is no opportunity or funding to carry out the entire experiment again.

It is important to incorporate these design principles into any experiment. If they are lacking, the inference methods presented in the following chapters will not be applicable and their results may not be trustworthy. In the next section we will consider three types of experimental design.

---

[24]There are always some researchers involved in the study who do know which patients are receiving which treatment. However, they do not interact with the study's patients and do not tell the blinded health care professionals who is receiving which treatment.

[25]The researchers assigned the patients into their treatment groups, so this study was an experiment. However, the patients could distinguish what treatment they received, so this study was not blind. The study could not be double-blind since it was not blind.

1.5.3 Completely randomized, blocked, and matched pairs design

A **completely randomized experiment** is one in which the subjects or experimental units are randomly assigned to each group in the experiment. Suppose we have three treatments, one of which may be a placebo, and 300 subjects. To carry out a completely randomized design, we could randomly assign each subject a unique number from 1 to 300, then subjects with numbers 1-100 would get treatment 1, subjects 101-200 would get treatment 2, and subjects 201- 300 would get treatment 3. Note that this method of randomly allocating subjects to treatments in not equivalent to taking a simple random sample. Here we are not sampling a subset of a population; we are randomly *splitting* subjects into groups.

While it might be ideal for the subjects to be a random sample of the population of interest, that is rarely the case. Subjects must volunteer to be part of an experiment. However, because randomization is incorporated in the splitting of the groups, we can still use statistical techniques to check for a causal connection, though the precise population for which the conclusion applies may be unclear. For example, if an experiment to determine the most effective means to encourage individuals to vote is carried out only on college students, we may not be able to generalize the conclusions of the experiment to all adults in the population.

Researchers sometimes know or suspect that another variable, other than the treatment, influences the response. Under these circumstances, they may carry out a **blocked experiment**. In this design, they first group individuals into **blocks** based on the identified variable and then randomize subjects within each block to the treatment groups. This strategy is referred to as **blocking**. For instance, if we are looking at the effect of a drug on heart attacks, we might first split patients in the study into low-risk and high-risk blocks. Then we can randomly assign half the patients from each block to the control group and the other half to the treatment group, as shown in Figure 1.16. At the end of the experiment, we would incorporate this blocking into the analysis. By blocking by risk of patient, we control for this possible confounding factor. Additionally, by randomizing subjects to treatments within each block, we attempt to even out the effect of variables that we cannot block or directly control.

● **Example 1.27** An experiment will be conducted to compare the effectiveness of two methods for quitting smoking. Identify a variable that the researcher might wish to use for blocking and describe how she would carry out a blocked experiment.

The researcher should choose the variable that is most likely to influence the response variable - whether or not a smoker will quit. A reasonable variable, therefore, would be the number of years that the smoker has been smoking. The subjects could be separated into three blocks based on number of years of smoking and each block randomly divided into the two treatment groups.

Even in a blocked experiment with randomization, other variables that affect the response can be distributed unevenly among the treatment groups, thus biasing the experiment in one direction. A third type of design, known as **matched pairs** addresses this problem. In a matched pairs experiment, pairs of people are matched on as many variables as possible, so that the comparison happens between very similar cases. This is actually a special type of blocked experiment, where the blocks are of size two.

An alternate form of matched pairs involves each subject receiving *both* treatments. Randomization can be incorporated by randomly selecting half the subjects to receive

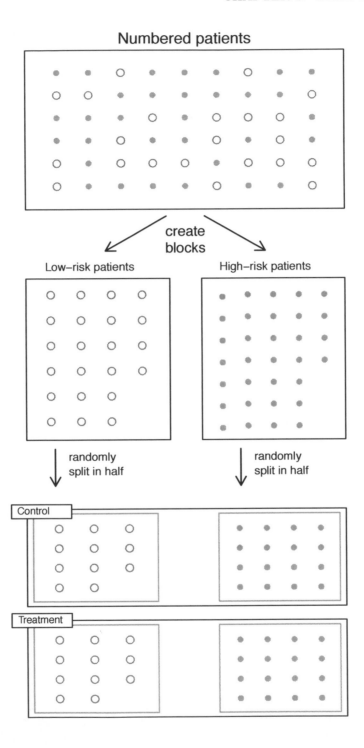

Figure 1.16: Blocking using a variable depicting patient risk. Patients are first divided into low-risk and high-risk blocks, then each block is evenly separated into the treatment groups using randomization. This strategy ensures an equal representation of patients in each treatment group from both the low-risk and high-risk categories.

treatment 1 first, followed by treatment 2, while the other half receives treatment 2 first, followed by treatment.

⊙ **Guided Practice 1.28** How and why should randomization be incorporated into a matched pairs design?[26]

⊙ **Guided Practice 1.29** Matched pairs sometimes involves each subject receiving both treatments at the same time. For example, if a hand lotion was being tested, half of the subjects could be randomly assigned to put Lotion A on the left hand and Lotion B on the right hand, while the other half of the subjects would put Lotion B on the left hand and Lotion A on the right hand. Why would this be a better design than a completely randomized experiment in which half of the subjects put Lotion A on both hands and the other half put Lotion B on both hands?[27]

Because it is essential to identify the type of data collection method used when choosing an appropriate inference procedure, we will revisit sampling techniques and experiment design in the subsequent chapters on inference.

1.5.4 Testing more than one variable at a time

Some experiments study more than one factor (explanatory variable) at a time, and each of these factors may have two or more levels (possible values). For example, suppose a researcher plans to investigate how the type and volume of music affect a person's performance on a particular video game. Because these two factors, `type` and `volume`, could interact in interesting ways, we do not want to do two separate experiments testing one factor at time. Instead, we want to do an experiment in which we test all the *combinations* of the factors. Let's say that `volume` has two levels (soft and loud) and that `type` has three levels (dance, classical, and punk). Then, we would want to carry out the experiment at each of the six (2 x 3 = 6) combinations: soft dance, soft classical, soft punk, loud dance, loud classical, loud punk. Each of the these combinations is a **treatment**. Therefore, this experiment will have 2 factors and 6 treatments. In order to replicate each treatment 10 times, one would need to play the game 60 times.

⊙ **Guided Practice 1.30** A researcher wants to compare the effectiveness of four different drugs. She also wants to test each of the drugs at two doses: low and high. Describe the factors, levels, and treatments of this experiment.[28]

As the number of factors and levels increases, the number of treatments become large and the analysis of the resulting data becomes more complex, requiring the use of advanced statistical methods. We will investigate only one factor at a time in this book.

---

[26]Assume that all subjects received treatment 1 first, followed by treatment 2. If the variable being measured happens to increase naturally over the course of time, it would appear as though treatment 2 had a greater effect than it really did.

[27]The dryness of people's skins varies from person to person, but probably less so from one person's right hand to left hand. With the matched pairs design, we are able control for this variability by comparing each person's right hand to her left hand, rather than comparing some people's hands to other people's hands (as you would in a completely randomized experiment).

[28]There are two factors: type of drug, which has four levels, and dose, which has 2 levels. There will be 4 x 2 = 8 treatments: drug 1 at low dose, drug 1 at high dose, drug 2 at low dose, and so on.

1.6   Exercises

1.6.1   Case study

1.1   Migraine and acupuncture. A migraine is a particularly painful type of headache, which patients sometimes wish to treat with acupuncture. To determine whether acupuncture relieves migraine pain, researchers conducted a completely randomized controlled study where 89 females diagnosed with migraine headaches were randomly assigned to one of two groups: treatment or control. 43 patients in the treatment group received acupuncture that is specifically designed to treat migraines. 46 patients in the control group received placebo acupuncture (needle insertion at nonacupoint locations). 24 hours after patients received acupuncture, they were asked if they were pain free. Results are summarized in the contingency table below.[29]

		Pain free Yes	No	Total
Group	Treatment	10	33	43
	Control	2	44	46
	Total	12	77	89

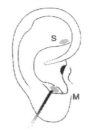

Figure from the original paper displaying the appropriate area (M) versus the inappropriate area (S) used in the treatment of migraine attacks.

(a) What percent of patients in the treatment group were pain free 24 hours after receiving acupuncture? What percent in the control group?

(b) Based on your findings in part (a), which treatment appears to be more effective for sinusitis?

(c) Do the data provide convincing evidence that there is a real pain reduction for those patients in the treatment group? Or do you think that the observed difference might just be due to chance?

1.2   Sinusitis and antibiotics, Part I. Researchers studying the effect of antibiotic treatment for acute sinusitis compared to symptomatic treatments randomly assigned 166 adults diagnosed with acute sinusitis to one of two groups: treatment or control. Study participants received either a 10-day course of amoxicillin (an antibiotic) or a placebo similar in appearance and taste. The placebo consisted of symptomatic treatments such as acetaminophen, nasal decongestants, etc. At the end of the 10-day period patients were asked if they experienced significant improvement in symptoms. The distribution of responses are summarized below.[30]

		Self-reported significant improvement in symptoms Yes	No	Total
Group	Treatment	66	19	85
	Control	65	16	81
	Total	131	35	166

(a) What percent of patients in the treatment group experienced a significant improvement in symptoms? What percent in the control group?

(b) Based on your findings in part (a), which treatment appears to be more effective for sinusitis?

(c) Do the data provide convincing evidence that there is a difference in the improvement rates of sinusitis symptoms? Or do you think that the observed difference might just be due to chance?

---

[29]G. Allais et al. "Ear acupuncture in the treatment of migraine attacks: a randomized trial on the efficacy of appropriate versus inappropriate acupoints". In: *Neurological Sci.* 32.1 (2011), pp. 173–175.

[30]J.M. Garbutt et al. "Amoxicillin for Acute Rhinosinusitis: A Randomized Controlled Trial". In: *JAMA: The Journal of the American Medical Association* 307.7 (2012), pp. 685–692.

1.6.2 Data basics

1.3 Identify study components, Part I. Identify (i) the cases, (ii) the variables and their types, and (iii) the main research question in the studies described below.

(a) Researchers collected data to examine the relationship between pollutants and preterm births in Southern California. During the study air pollution levels were measured by air quality monitoring stations. Specifically, levels of carbon monoxide were recorded in parts per million, nitrogen dioxide and ozone in parts per hundred million, and coarse particulate matter (PM_{10}) in $\mu g/m^3$. Length of gestation data were collected on 143,196 births between the years 1989 and 1993, and air pollution exposure during gestation was calculated for each birth. The analysis suggested that increased ambient PM_{10} and, to a lesser degree, CO concentrations may be associated with the occurrence of preterm births.[31]

(b) The Buteyko method is a shallow breathing technique developed by Konstantin Buteyko, a Russian doctor, in 1952. Anecdotal evidence suggests that the Buteyko method can reduce asthma symptoms and improve quality of life. In a scientific study to determine the effectiveness of this method, researchers recruited 600 asthma patients aged 18-69 who relied on medication for asthma treatment. These patients were split into two research groups: one practiced the Buteyko method and the other did not. Patients were scored on quality of life, activity, asthma symptoms, and medication reduction on a scale from 0 to 10. On average, the participants in the Buteyko group experienced a significant reduction in asthma symptoms and an improvement in quality of life.[32]

1.4 Identify study components, Part II. Identify (i) the cases, (ii) the variables and their types, and (iii) the main research question of the studies described below.

(a) Researchers studying the relationship between honesty, age and self-control conducted an experiment on 160 children between the ages of 5 and 15. Participants reported their age, sex, and whether they were an only child or not. The researchers asked each child to toss a fair coin in private and to record the outcome (white or black) on a paper sheet, and said they they would only reward children who report white. Half the students were explicitly told not to cheat and the others were not given any explicit instructions. In the no instruction group probability of cheating was found to be uniform across groups based on child's characteristics. In the group that was explicitly told to not cheat, girls were less likely to cheat, and while rate of cheating didn't vary by age for boys, it decreased with age for girls.[33]

(b) In a study of the relationship between socio-economic class and unethical behavior, 129 University of California undergraduates at Berkeley were asked to identify themselves as having low or high social-class by comparing themselves to others with the most (least) money, most (least) education, and most (least) respected jobs. They were also presented with a jar of individually wrapped candies and informed that they were for children in a nearby laboratory, but that they could take some if they wanted. Participants completed unrelated tasks and then reported the number of candies they had taken. It was found that those in the upper-class rank condition took more candy than did those in the lower-rank condition.[34]

---

[31]B. Ritz et al. "Effect of air pollution on preterm birth among children born in Southern California between 1989 and 1993". In: *Epidemiology* 11.5 (2000), pp. 502–511.

[32]J. McGowan. "Health Education: Does the Buteyko Institute Method make a difference?" In: *Thorax* 58 (2003).

[33]Alessandro Bucciol and Marco Piovesan. "Luck or cheating? A field experiment on honesty with children". In: *Journal of Economic Psychology* 32.1 (2011), pp. 73–78.

[34]P.K. Piff et al. "Higher social class predicts increased unethical behavior". In: *Proceedings of the National Academy of Sciences* (2012).

1.5 Fisher's irises. Sir Ronald Aylmer Fisher was an English statistician, evolutionary biologist, and geneticist who worked on a data set that contained sepal length and width, and petal length and width from three species of iris flowers (*setosa*, *versicolor* and *virginica*). There were 50 flowers from each species in the data set.[35]

(a) How many cases were included in the data?

(b) How many numerical variables are included in the data? Indicate what they are, and if they are continuous or discrete.

(c) How many categorical variables are included in the data, and what are they? List the corresponding levels (categories).

Photo by Ryan Claussen
(http://flic.kr/p/6QTcuX)
CC BY-SA 2.0 license

1.6 Smoking habits of UK residents. A survey was conducted to study the smoking habits of UK residents. Below is a data matrix displaying a portion of the data collected in this survey. Note that "£" stands for British Pounds Sterling, "cig" stands for cigarettes, and "N/A" refers to a missing component of the data.[36]

	sex	age	marital	grossIncome	smoke	amtWeekends	amtWeekdays
1	Female	42	Single	Under £2,600	Yes	12 cig/day	12 cig/day
2	Male	44	Single	£10,400 to £15,600	No	N/A	N/A
3	Male	53	Married	Above £36,400	Yes	6 cig/day	6 cig/day
⋮	⋮	⋮	⋮	⋮	⋮	⋮	⋮
1691	Male	40	Single	£2,600 to £5,200	Yes	8 cig/day	8 cig/day

(a) What does each row of the data matrix represent?

(b) How many participants were included in the survey?

(c) Indicate whether each variable in the study is numerical or categorical. If numerical, identify as continuous or discrete. If categorical, indicate if the variable is ordinal.

1.6.3   Overview of data collection principles

1.7 Generalizability and causality, Part I. Identify the population of interest and the sample in the studies described in Exercise 1.3. Comment on whether or not the results of the study can be generalized to the population and if the findings of the study can be used to establish causal relationships.

1.8 Generalizability and causality, Part II. Identify the population of interest and the sample in the studies described in Exercise 1.4. Comment on whether or not the results of the study can be generalized to the population and if the findings of the study can be used to establish causal relationships.

1.9 Relaxing after work. The 2010 General Social Survey asked the question, "After an average work day, about how many hours do you have to relax or pursue activities that you enjoy?" to a random sample of 1,155 Americans. The average relaxing time was found to be 1.65 hours. Determine which of the following is an observation, a variable, a sample statistic, or a population parameter.

(a) An American in the sample.

(b) Number of hours spent relaxing after an average work day.

(c) 1.65

(d) Average number of hours all Americans spend relaxing after an average work day.

---

[35]R.A Fisher. "The Use of Multiple Measurements in Taxonomic Problems". In: *Annals of Eugenics* 7 (1936), pp. 179–188.

[36]Stats4Schools, Smoking.

1.10 Cats on YouTube. Suppose you want to estimate the percentage of videos on YouTube that are cat videos. It is impossible for you to watch all videos on YouTube so you use a random video picker to select 1000 videos for you. You find that 2% of these videos are cat videos. Determine which of the following is an observation, a variable, a sample statistic, or a population parameter.

(a) Percentage of all videos on YouTube that are cat videos

(b) 2%

(c) A video in your sample

(d) Whether or not a video is a cat video

1.11 GPA and study time. A survey was conducted on 218 undergraduates from Duke University who took an introductory statistics course in Spring 2012. Among many other questions, this survey asked them about their GPA and the number of hours they spent studying per week. The scatterplot below displays the relationship between these two variables.

(a) What is the explanatory variable and what is the response variable?

(b) Describe the relationship between the two variables. Make sure to discuss unusual observations, if any.

(c) Is this an experiment or an observational study?

(d) Can we conclude that studying longer hours leads to higher GPAs?

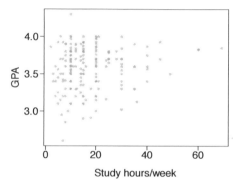

1.12 Income and education. The scatterplot below shows the relationship between per capita income (in thousands of dollars) and percent of population with a bachelor's degree in 3,143 counties in the US in 2010.

(a) What are the explanatory and response variables?

(b) Describe the relationship between the two variables. Make sure to discuss unusual observations, if any.

(c) Can we conclude that having a bachelor's degree increases one's income?

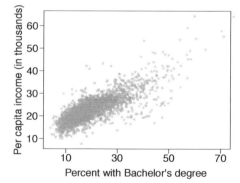

1.6.4 Observational studies and sampling strategies

1.13 Propose a sampling strategy, Part I. A large college class has 160 students. All 160 students attend the lectures together, but the students are divided into 4 groups, each of 40 students, for lab sections administered by different teaching assistants. The professor wants to conduct a survey about how satisfied the students are with the course, and he believes that the lab section a student is in might affect the student's overall satisfaction with the course.

(a) What type of study is this?

(b) Suggest a sampling strategy for carrying out this study.

1.14   Propose a sampling strategy, Part II. On a large college campus first-year students and sophomores live in dorms located on the eastern part of the campus and juniors and seniors live in dorms located on the western part of the campus. Suppose you want to collect student opinions on a new housing structure the college administration is proposing and you want to make sure your survey equally represents opinions from students from all years.

(a) What type of study is this?

(b) Suggest a sampling strategy for carrying out this study.

1.15   Internet use and life expectancy. The scatterplot below shows the relationship between estimated life expectancy at birth as of 2012[37] and percentage of internet users in 2010[38] in 208 countries.

(a) Describe the relationship between life expectancy and percentage of internet users.

(b) What type of study is this?

(c) State a possible confounding variable that might explain this relationship and describe its potential effect.

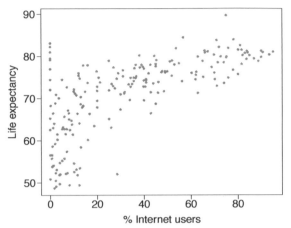

1.16   Stressed out, Part I. A study that surveyed a random sample of otherwise healthy high school students found that they are more likely to get muscle cramps when they are stressed. The study also noted that students drink more coffee and sleep less when they are stressed.

(a) What type of study is this?

(b) Can this study be used to conclude a causal relationship between increased stress and muscle cramps?

(c) State possible confounding variables that might explain the observed relationship between increased stress and muscle cramps.

1.17   Random digit dialing. The Gallup Poll uses a procedure called random digit dialing, which creates phone numbers based on a list of all area codes in America in conjunction with the associated number of residential households in each area code. Give a possible reason the Gallup Poll chooses to use random digit dialing instead of picking phone numbers from the phone book.

1.18   Haters are gonna hate, study confirms. A study published in the *Journal of Personality and Social Psychology* asked a group of 200 randomly sampled men and women to evaluate how they felt about various subjects, such as camping, health care, architecture, taxidermy, crossword puzzles, and Japan in order to measure their dispositional attitude towards mostly independent stimuli. Then, they presented the participants with information about a new product: a microwave oven. This microwave oven does not exist, but the participants didn't know this, and were given three positive and three negative fake reviews. People who reacted positively to the subjects on the dispositional attitude measurement also tended to react positively to the microwave oven, and those who reacted negatively also tended to react negatively to it. Researcher concluded that

---

[37]CIA Factbook, Country Comparison: Life Expectancy at Birth, 2012.

[38]ITU World Telecommunication/ICT Indicators database, World Telecommunication/ICT Indicators Database, 2012.

"some people tend to like things, whereas others tend to dislike things, and a more thorough understanding of this tendency will lead to a more thorough understanding of the psychology of attitudes." [39]

(a) What are the cases?

(b) What is (are) the response variable(s) in this study?

(c) What is (are) the explanatory variable(s) in this study?

(d) Does the study employ random sampling?

(e) Is this an observational study or an experiment? Explain your reasoning.

(f) Can we establish a causal link between the explanatory and response variables?

(g) Can the results of the study be generalized to the population at large?

1.19 Family size. Suppose we want to estimate household size, where a "household" is defined as people living together in the same dwelling, and sharing living accommodations. If we select students at random at an elementary school and ask them what their family size is, will this be a good measure of household size? Or will our average be biased? If so, will it overestimate or underestimate the true value?

1.20 Flawed reasoning. Identify the flaw(s) in reasoning in the following scenarios. Explain what the individuals in the study should have done differently if they wanted to make such strong conclusions.

(a) Students at an elementary school are given a questionnaire that they are asked to return after their parents have completed it. One of the questions asked is, "Do you find that your work schedule makes it difficult for you to spend time with your kids after school?" Of the parents who replied, 85% said "no". Based on these results, the school officials conclude that a great majority of the parents have no difficulty spending time with their kids after school.

(b) A survey is conducted on a simple random sample of 1,000 women who recently gave birth, asking them about whether or not they smoked during pregnancy. A follow-up survey asking if the children have respiratory problems is conducted 3 years later, however, only 567 of these women are reached at the same address. The researcher reports that these 567 women are representative of all mothers.

(c) A orthopedist administers a questionnaire to 30 of his patients who do not have any joint problems and finds that 20 of them regularly go running. He concludes that running decreases the risk of joint problems.

1.21 City council survey. A city council has requested a household survey be conducted in a suburban area of their city. The area is broken into many distinct and unique neighborhoods, some including large homes, some with only apartments, and others a diverse mixture of housing structures. Identify the sampling methods described below, and comment on whether or not you think they would be effective in this setting.

(a) Randomly sample 50 households from the city.

(b) Divide the city into neighborhoods, and sample 20 households from each neighborhood.

(c) Divide the city into neighborhoods, randomly sample 10 neighborhoods, and sample all households from those neighborhoods.

(d) Divide the city into neighborhoods, randomly sample 10 neighborhoods, and then randomly sample 20 households from those neighborhoods.

(e) Sample the 200 households closest to the city council offices.

---

[39]Justin Hepler and Dolores Albarracín. "Attitudes without objects - Evidence for a dispositional attitude, its measurement, and its consequences". In: *Journal of personality and social psychology* 104.6 (2013), p. 1060.

1.22   Sampling strategies. A statistics student who is curious about the relationship between the amount of time students spend on social networking sites and their performance at school decides to conduct a survey. Various research strategies for collecting data are described below. In each, name the sampling method proposed and any bias you might expect.

(a) He randomly samples 40 students from the study's population, gives them the survey, asks them to fill it out and bring it back the next day.

(b) He gives out the survey only to his friends, making sure each one of them fills out the survey.

(c) He posts a link to an online survey on Facebook and asks his friends to fill out the survey.

(d) He randomly samples 5 classes and asks a random sample of students from those classes to fill out the survey.

(e) He stands outside the student center and asks every third person that walks out the door to fill out the survey.

1.23   Reading the paper. Below are excerpts from two articles published in the *NY Times*:

(a) An article titled *Risks: Smokers Found More Prone to Dementia* states the following:[40]

> "Researchers analyzed data from 23,123 health plan members who participated in a voluntary exam and health behavior survey from 1978 to 1985, when they were 50-60 years old. 23 years later, about 25% of the group had dementia, including 1,136 with Alzheimers disease and 416 with vascular dementia. After adjusting for other factors, the researchers concluded that pack-a-day smokers were 37% more likely than nonsmokers to develop dementia, and the risks went up with increased smoking; 44% for one to two packs a day; and twice the risk for more than two packs."

Based on this study, can we conclude that smoking causes dementia later in life? Explain your reasoning.

(b) Another article titled *The School Bully Is Sleepy* states the following:[41]

> "The University of Michigan study, collected survey data from parents on each child's sleep habits and asked both parents and teachers to assess behavioral concerns. About a third of the students studied were identified by parents or teachers as having problems with disruptive behavior or bullying. The researchers found that children who had behavioral issues and those who were identified as bullies were twice as likely to have shown symptoms of sleep disorders."

A friend of yours who read the article says, "The study shows that sleep disorders lead to bullying in school children." Is this statement justified? If not, how best can you describe the conclusion that can be drawn from this study?

1.24   Shyness on Facebook. Given the anonymity afforded to individuals in online interactions, researchers hypothesized that shy individuals might have more favorable attitudes toward Facebook, and that shyness might be positively correlated with time spent on Facebook. They also hypothesized that shy individuals might have fewer Facebook "friends" as they tend to have fewer friends than non-shy individuals have in the offline world. 103 undergraduate students at an Ontario university were surveyed via online questionnaires. The study states "Participants were recruited through the university's psychology participation pool. After indicating an interest in the study, participants were sent an e-mail containing the study's URL." Are the results of this study generalizable to the population of all Facebook users?[42]

---

[40]R.C. Rabin. "Risks: Smokers Found More Prone to Dementia". In: *New York Times* (2010).

[41]T. Parker-Pope. "The School Bully Is Sleepy". In: *New York Times* (2011).

[42]E.S. Orr et al. "The influence of shyness on the use of Facebook in an undergraduate sample". In: *CyberPsychology & Behavior* 12.3 (2009), pp. 337–340.

1.6.5 Experiments

1.25 Stressed out, Part II. In a study evaluating the relationship between stress and muscle cramps half the subjects are randomly assigned to be exposed to increased stressed by being placed into an elevator that falls rapidly and stops abruptly and the other half are left at no or baseline stress.

(a) What type of study is this?

(b) Can this study be used to conclude a causal relationship between increased stress and muscle cramps?

1.26 Light and exam performance. A study is designed to test the effect of light level on exam performance of students. The researcher believes that light levels might have different effects on males and females, so wants to make sure both are equally represented in each treatment. The treatments are fluorescent overhead lighting, yellow overhead lighting, no overhead lighting (only desk lamps).

(a) What is the response variable?

(b) What is the explanatory variable? What are its levels?

(c) What is the blocking variable? What are its levels?

1.27 Vitamin supplements. In order to assess the effectiveness of taking large doses of vitamin C in reducing the duration of the common cold, researchers recruited 400 healthy volunteers from staff and students at a university. A quarter of the patients were assigned a placebo, and the rest were evenly divided between 1g Vitamin C, 3g Vitamin C, or 3g Vitamin C plus additives to be taken at onset of a cold for the following two days. All tablets had identical appearance and packaging. The nurses who handed the prescribed pills to the patients knew which patient received which treatment, but the researchers assessing the patients when they were sick did not. No significant differences were observed in any measure of cold duration or severity between the four medication groups, and the placebo group had the shortest duration of symptoms.[43]

(a) Was this an experiment or an observational study? Why?

(b) What are the explanatory and response variables in this study?

(c) Were the patients blinded to their treatment?

(d) Was this study double-blind?

(e) Participants are ultimately able to choose whether or not to use the pills prescribed to them. We might expect that not all of them will adhere and take their pills. Does this introduce a confounding variable to the study? Explain your reasoning.

1.28 Light, noise, and exam performance. A study is designed to test the effect of light level and noise level on exam performance of students. The researcher believes that light and noise levels might have different effects on males and females, so wants to make sure both are equally represented in each treatment. The light treatments considered are fluorescent overhead lighting, yellow overhead lighting, no overhead lighting (only desk lamps). The noise treatments considered are no noise, construction noise, and human chatter noise.

(a) What type of study is this?

(b) How many factors are considered in this study? Identify them, and describe their levels.

(c) What is the role of the sex variable in this study?

1.29 Music and learning. You would like to conduct an experiment in class to see if students learn better if they study without any music, with music that has no lyrics (instrumental), or with music that has lyrics. Briefly outline a design for this study.

---

[43]C. Audera et al. "Mega-dose vitamin C in treatment of the common cold: a randomised controlled trial". In: *Medical Journal of Australia* 175.7 (2001), pp. 359–362.

1.30 Soda preference. You would like to conduct an experiment in class to see if your class-mates prefer the taste of regular Coke or Diet Coke. Briefly outline a design for this study.

1.31 Exercise and mental health. A researcher is interested in the effects of exercise on mental health and he proposes the following study: Use stratified random sampling to ensure representative proportions of 18-30, 31-40 and 41-55 year olds from the population. Next, randomly assign half the subjects from each age group to exercise twice a week, and instruct the rest not to exercise. Conduct a mental health exam at the beginning and at the end of the study, and compare the results.

(a) What type of study is this?

(b) What are the treatment and control groups in this study?

(c) Does this study make use of blocking? If so, what is the blocking variable?

(d) Does this study make use of blinding?

(e) Comment on whether or not the results of the study can be used to establish a causal relationship between exercise and mental health, and indicate whether or not the conclusions can be generalized to the population at large.

(f) Suppose you are given the task of determining if this proposed study should get funding. Would you have any reservations about the study proposal?

1.32 Chia seeds and weight loss. Chia Pets – those terra-cotta figurines that sprout fuzzy green hair – made the chia plant a household name. But chia has gained an entirely new reputation as a diet supplement. In one 2009 study, a team of researchers recruited 38 men and divided them randomly into two groups: treatment or control. They also recruited 38 women, and they randomly placed half of these participants into the treatment group and the other half into the control group. One group was given 25 grams of chia seeds twice a day, and the other was given a placebo. The subjects volunteered to be a part of the study. After 12 weeks, the scientists found no significant difference between the groups in appetite or weight loss.[44]

(a) What type of study is this?

(b) What are the experimental and control treatments in this study?

(c) Has blocking been used in this study? If so, what is the blocking variable?

(d) Has blinding been used in this study?

(e) Comment on whether or not we can make a causal statement, and indicate whether or not we can generalize the conclusion to the population at large.

1.33 Multitaskers, Part I. Researchers studying the effect of TV watching while studying on school performance conducted the following studies. For each study determine (i) the type of study (observational or experiment), (ii) if there is random sampling, (iii) if there is random assignment, (iv) state the scope of the conclusions of the study, and (v) note if stratifying, blocking, or neither of these techniques were used in the study.

(a) Researchers randomly sampled 100 high school students and asked them whether or not they watched TV while studying. They found that the mean grade point average of students who did not watch TV while studying was significantly higher than the mean grade point average of students who did watch TV while studying.

(b) Researchers randomly sampled 50 female and 50 male high school students and asked them whether or not they watched TV while studying. They found that the mean grade point average of both males and females who did not watch TV while doing homework was significantly higher than the mean grade point average of students who did watch TV while doing homework. **(Note that there is a part (c) on the next page.)**

---

[44]D.C. Nieman et al. "Chia seed does not promote weight loss or alter disease risk factors in overweight adults". In: *Nutrition Research* 29.6 (2009), pp. 414–418.

(c) Researchers randomly sampled 100 high school students and randomly assigned them into two study groups. Throughout the school year, one group was told to study in a room with a TV on while the other was told to study in silence. At the end of the year the researchers compared the grade point averages of the two groups and found that the mean grade point average of students who did not watch TV while studying was significantly higher than the grade point average of students who did watch TV while studying.

1.34 Multitaskers, Part II. Researchers investigating the effect of studying while watching TV on school performance conducted the following studies. For each study determine (i) the type of study (observational or experiment), (ii) if there is random sampling, (iii) if there is random assignment, (iv) state the scope of the conclusions of the study, and (v) note if stratifying, blocking, or neither of these techniques were used in the study.

(a) Researchers randomly sampled 50 female and 50 male high school students. Half of the females and half of the males were randomly assigned to study in a room with a TV and the remainder studied without a TV. They found that the mean grade point average of both males and females who did not watch TV while studying was significantly higher than those who did watch TV while studying.

(b) Researchers surveyed the first 100 students who showed up to prom. They found that the mean grade point average of students who did not watch TV while studying was higher than the mean grade point average for students who did watch TV while studying.

(c) Researchers surveyed the first 50 male and 50 female students who showed up to prom. They found that the mean grade point average of both males and females who did not watch TV while doing homework was higher than the mean grade point average of students who did watch TV while studying.

1.35 Multitaskers, Part III. Suppose a friend of yours wants to design his own study for evaluating the effect of TV watching on school performance. He proposes comparing the grade point averages of everyone in his homeroom class who do and do not watch TV while doing homework, and extending the results of this study to draw conclusions about all high schoolers. Indicate any mistakes with this design, keeping in mind that the goal of the study is to assess the causal relationship between TV watching and school performance for high schoolers.

1.36 Multitaskers, Part IV. Suppose two friends of yours want to design their own study for evaluating the effect of TV watching on school performance. They propose the following designs. Indicate any mistakes with these designs, keeping in mind that the goal of the study is to assess the causal relationship between TV watching and school performance for high schoolers.

(a) Randomly sample 100 students from the entire school. Reserve two classrooms for a study session and the first 50 people that show up get assigned to the room where the TV will be on and the rest to the room where they can study in silence for a final exam. Then compare the average grades of the two groups at the end of the semester.

(b) Age and involvement in extracurricular activities may be factors that affect the academic performance of a student, so these characteristics should be blocked for when studying the effect of watching TV while studying. To achieve this aim, sample 10 firs-years, 10 sophomores, 10 juniors, and 10 seniors, 5 of which are heavily involved with extracurricular activities and 5 of which are not from each class. Then ask them whether they watch TV while they study, and compare the average GPAs of those who do and do not.

1.37 Running on electrolytes. Suppose you would like to design a study evaluating whether consuming a sports drink that replenishes electrolytes can make you run faster. You were able to recruit 50 students to participate in your study. Describe how you can use matched pairs design and blinding for this study.

1.38 Improving life satisfaction. In a study evaluating the effectiveness of a positive psychology group intervention, forty middle schoolers who were identified as being less than delighted with their lives (reported life satisfaction scores between 1 and 6 on a 7-point scale) were randomly assigned to receive the intervention (treatment) or not receive the intervention (control). These students were first matched on attributes such as sex, socioeconomic group, race/ethnicity, and age, such that each student in the treatment group had a matched counterpart in the control group. Researchers found that life satisfaction of students in the intervention group increased significantly, while the control group declined during the same period (although this change was not statistically significant).[45]

(a) What type of study is this?

(b) What type of design is used in this study?

(c) Can the results of this study be used to establish a causal link between the intervention and increased life satisfaction in middle schoolers?

1.39 Alfalfa plants. Researchers were interested in the effect that acid has on the growth rate of alfalfa plants. They created three treatment groups in an experiment: low acid ($+$), high acid ($\triangle$), and control ($\bigcirc$). The alfalfa plants were grown in a Styrofoam cups arranged near a window and the height of the alfalfa plants was measured after five days of growth. The experiment consisted of 6 cups for each of the 3 treatments, for a total of 18 observations. Which of the following designs is preferable? Note that the dotted line indicates the location of the window. Explain your reasoning.

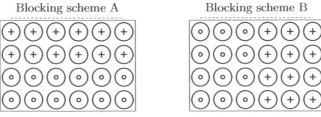

1.40 Chocolate chip cookies. We would like to compare two cookie recipes, one from a popular recipe website and another from the back of a bag of chocolate chips. We will bake 24 cookies, 12 of each type, and then have a friend rate each cookie on a scale of 1 to 10. Both sets of cookies are supposed to be baked for the same amount of time and at the same temperature: 9 minutes at 350F. We will use an old oven available to students in the school, which tends to overheat a little near the oven's back.

(a) Would we bias our results if we cooked one type of cookie first and the other second?

(b) We instead decide to bake each batch simultaneously on the same tray, and decide to block for proximity to the back of the oven. Two blocking schemes shown below are under consideration. For each scheme, cookies made with the recipe from the website are indicated with a $+$ and cookies made with the other recipe are indicated with a $\bigcirc$. The dashed line marks the back of the oven. Which of the blocking schemes, A or B, is better for this experiment? Explain your answer.

(c) How can the blocking scheme you chose in the previous part be improved?

---

[45]Shannon M Suldo et al. "Increasing middle school students' life satisfaction: Efficacy of a positive psychology group intervention". In: *Journal of happiness studies* 15.1 (2014), pp. 19–42.

Chapter 2

Summarizing data

After collecting data, the next stage in the investigative process is to summarize the data. Graphical displays allow us to visualize and better understand the important features of a data set.

2.1 Examining numerical data

In this section we will focus on numerical variables. The `email50` and `county` data sets from Section 1.2 provide rich opportunities for examples. Recall that outcomes of numerical variables are numbers on which it is reasonable to perform basic arithmetic operations. For example, the `pop2010` variable, which represents the populations of counties in 2010, is numerical since we can sensibly discuss the difference or ratio of the populations in two counties. On the other hand, area codes and zip codes are not numerical, but rather they are categorical variables.

2.1.1 Scatterplots for paired data

Sometimes researchers wish to see the relationship between two variables. When we talk of a relationship or an association between variables, we are interested in how one variable behaves as the other variable increases or decreases.

A **scatterplot** provides a case-by-case view of data that illustrates the relationship between two numerical variables. In Figure 1.8 on page 14, a scatterplot was used to examine how federal spending and poverty were related in the `county` data set. Another scatterplot is shown in Figure 2.1, comparing the number of line breaks (`line_breaks`) and number of characters (`num_char`) in emails for the `email50` data set. In any scatterplot, each point represents a single case. Since there are 50 cases in `email50`, there are 50 points in Figure 2.1.

● **Example 2.1** A scatterplot requires paired data. What does **paired data** mean?

We say observations are *paired* when the two observations correspond to each other. In unpaired data, there is no such correspondence. Here the two observations correspond to a particular email.

The variable that is suspected to be the response variable is plotted on the vertical (y) axis and the variable that is suspected to be the explanatory variable is plotted on the

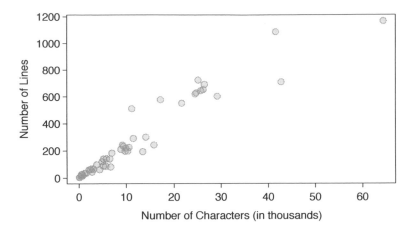

Figure 2.1: A scatterplot of `line_breaks` versus `num_char` for the `email50` data.

horizontal (x) axis. In this example, the variables could be switched since either variable could reasonably serve as the explanatory variable or the response variable.

TIP: Drawing scatterplots
(1) Decide which variable should go on each axis, and draw and label the two axes.
(2) Note the range of each variable, and add tick marks and scales to each axis.
(3) Plot the dots as you would on an x, y-coordinate plane.

The association between two variables can be **positive** or **negative**, or there can be no association. Positive association means that larger values of the first variable are associated with larger values of the second variable. Additionally, the association can follow a linear trend or a curved (nonlinear) trend.

⊙ **Guided Practice 2.2** What would it mean for two variables to have a *negative* association? What about *no* association?[1]

⊙ **Guided Practice 2.3** What does the scatterplot in Figure 2.1 reveal about the email data?[2]

● **Example 2.4** Consider a new data set of 54 cars with two variables: vehicle price and weight.[3] A scatterplot of vehicle price versus weight is shown in Figure 2.2. What can be said about the relationship between these variables?

The relationship is evidently nonlinear, as highlighted by the dashed line. This is different from previous scatterplots we've seen, such as Figure 1.8 on page 14 and Figure 2.1, which show relationships that are very linear.

---

[1] Negative association implies that larger values of the first variable are associated with smaller values of the second variable. No association implies that the values of the second variable tend to be independent of changes in the first variable.

[2] The association between the number of characters in an email and the number of lines in an email is positive (when one is larger, the other tends to be larger as well). As the number of characters increases, number of lines increases is an approximately linear fashion.

[3] Subset of data from www.amstat.org/publications/jse/v1n1/datasets.lock.html

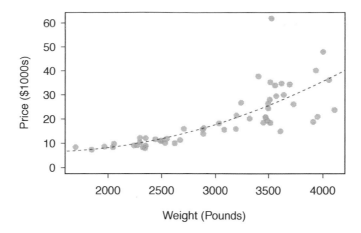

Figure 2.2: A scatterplot of `price` versus `weight` for 54 cars.

⊙ **Guided Practice 2.5** Describe two variables that would have a horseshoe shaped (i.e. "U"-shaped) association in a scatterplot.[4]

2.1.2 Stem-and-leaf plots and dot plots

Sometimes two variables is one too many: only one variable may be of interest. In these cases we want to focus not on the association between two variables, but on the distribution of a single variable. The term **distribution** refers to the values that a variable takes and the frequency of these values. Let's take a closer look at the `email50` data set and focus on the number of characters in each email. To simplify the data, we will round the numbers and record the values in thousands. Thus, 22105 is recorded as 22.

22	0	64	10	6	26	25	11	4	14
7	1	10	2	7	5	7	4	14	3
1	5	43	0	0	3	25	1	9	1
2	9	0	5	3	6	26	11	25	9
42	17	29	12	27	10	0	0	1	16

Table 2.3: The number of characters, in thousands, for the data set of 50 emails.

Rather than look at the data as a list of numbers, which makes the distribution difficult to discern, we will organize it into a table called a **stem-and-leaf plot** shown in Figure 2.4. In a stem-and-leaf plot, each number is broken into two parts. The first part is called the **stem** and consists of the beginning digit(s). The second part is called the **leaf** and consists of the final digits(s). The stems are written in a column in ascending order, and the leaves that match up with those stems are written on the corresponding row. Figure 2.4 shows a stem-and-leaf plot of the number of characters in 50 emails. The stem represents the ten thousands place and the leaf represents the thousands place. For

---

[4]Consider a variable that represents something that is only good in moderation. Water consumption fits this description since water becomes toxic when consumed in excessive quantities. If health was represented on the vertical axis and water consumption on the horizontal axis, then we would create an upside down "U" shape.

example, 1 | 2 corresponds to 12 thousand. When making a stem-and-leaf plot, remember to include a legend that describes what the stem and what the leaf represent. Without this, there is no way of knowing if 1 | 2 represents 1.2, 12, 120, 1200, etc.

```
0 | 0000001111122333445555667777999
1 | 0001124467
2 | 25556679
3 |
4 | 23
5 |
6 | 4
```

Legend: 1 | 2 = 12,000

Figure 2.4: A stem-and-leaf plot of the number of characters in 50 emails.

⊙ **Guided Practice 2.6** There are a lot of numbers on the first row of the stem-and-leaf plot. Why is this the case?[5]

When there are too many numbers on one row or there are only a few stems, we *split* each row into two halves, with the leaves from 0-4 on the first half and the leaves from 5-9 on the second half. The resulting graph is called a **split stem-and-leaf plot**. Figure 2.5 shows the previous stem-and-leaf redone as a split stem-and-leaf.

```
0 | 000000111112233344
0 | 55566777999
1 | 00011244
1 | 67
2 | 2
2 | 5556679
3 |
3 |
4 | 23
4 |
5 |
5 |
6 | 4
```

Legend: 1 | 2 = 12,000

Figure 2.5: A split stem-and-leaf.

⊙ **Guided Practice 2.7** What is the smallest number in this data set? What is the largest?[6]

---

[5]There are a lot of numbers on the first row because there are a lot of values in the data set less than 10 thousand.

[6]The smallest number is less than 1 thousand, and the largest is 64 thousand. That is a big range!

Another simple graph for numerical data is a dot plot. A **dot plot** uses dots to show the **frequency**, or number of occurrences, of the values in a data set. The higher the stack of dots, the greater the number occurrences there are of the corresponding value. An example using the same data set, number of characters from 50 emails, is shown in Figure 2.6.

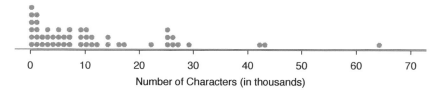

Figure 2.6: A dot plot of `num_char` for the `email50` data set.

⊙ **Guided Practice 2.8**   Imagine rotating the dot plot 90 degrees clockwise. What do you notice?[7]

These graphs make it easy to observe important features of the data, such as the location of clusters and presence of gaps.

● **Example 2.9**   Based on both the stem-and-leaf and dot plot, where are the values clustered and where are the gaps for the `email50` data set?

There is a large cluster in the 0 to less than 20 thousand range, with a peak around 1 thousand. There are gaps between 30 and 40 thousand and between the two values in the 40 thousands and the largest value of approximately 64 thousand.

Additionally, we can easily identify any observations that appear to be unusually distant from the rest of the data. Unusually distant observations are called **outliers**. Later in this chapter we will provide numerical rules of thumb for identifying outliers. For now, it is sufficient to identify them by observing gaps in the graph. In this case, it would be reasonable to classify the emails with character counts of 42 thousand, 43 thousand, and 64 thousand as outliers since they are numerically distant from most of the data.

Outliers are extreme

An **outlier** is an observation that appears extreme relative to the rest of the data.

TIP: Why it is important to look for outliers
Examination of data for possible outliers serves many useful purposes, including

1. Identifying asymmetry in the distribution.

2. Identifying data collection or entry errors. For instance, we re-examined the email purported to have 64 thousand characters to ensure this value was accurate.

3. Providing insight into interesting properties of the data.

---

[7]It has a similar shape as the stem-and-leaf plot! The values on the horizontal axis correspond to the stems and the number of dots in each interval correspond the number of leaves needed for each stem.

⊙ **Guided Practice 2.10**   The observation 64 thousand, a suspected outlier, was found to be an accurate observation. What would such an observation suggest about the nature of character counts in emails?[8]

⊙ **Guided Practice 2.11**   Consider a data set that consists of the following numbers: 12, 12, 12, 12, 12, 13, 13, 14, 14, 15, 19. Which graph would better illustrate the data: a stem-and-leaf plot or a dot plot? Explain.[9]

2.1.3   Histograms

Stem-and-leaf plots and dot plots are ideal for displaying data from small samples because they show the exact values of the observations and how frequently they occur. However, they are impractical for larger samples. For larger samples, rather than showing the frequency of every value, we prefer to think of the value as belonging to a *bin*. For example, in the `email50` data set, we create a table of counts for the number of cases with character counts between 0 and 5,000, then the number of cases between 5,000 and 10,000, and so on. Such a table, shown in Table 2.7, is called a **frequency table**. Observations that fall on the boundary of a bin (e.g. 5,000) are generally allocated to the lower bin.[10] These binned counts are plotted as bars in Figure 2.9 into what is called a **histogram** or **frequency histogram**, which resembles the stacked dot plot shown in Figure 2.6.

Characters (in thousands)	0-5	5-10	10-15	15-20	20-25	25-30	⋯	55-60	60-65
Count	19	12	6	2	3	5	⋯	0	1

Table 2.7: The counts for the binned `num_char` data.

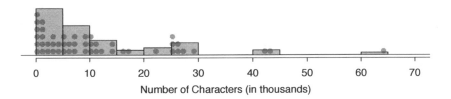

Figure 2.8: A histogram of `num_char`. This histogram is drawn over the corresponding dot plot.

⊙ **Guided Practice 2.12**   What can you see in the dot plot and stem-and-leaf plot that you cannot see in the frequency histogram?[11]

---

[8]That occasionally there may be very long emails.

[9]Because all the values begin with 1, there would be only one stem (or two in a split stem-and-leaf). This would not provide a good sense of the distribution. For example, the gap between 15 and 19 would not be visually apparent. A dot plot would be better here.

[10]This is called *left inclusive*.

[11]Character counts for individual emails.

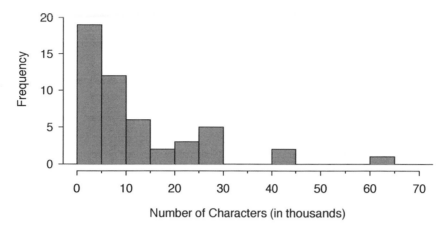

Figure 2.9: A histogram of num_char. This histogram uses bins or class intervals of width 5.

TIP: Drawing histograms

1. The variable is always placed on the horizontal axis. Before drawing the histogram, label both axes and draw a scale for each.

2. Draw bars such that the height of the bar is the frequency of that bin and the width of the bar corresponds to the bin width.

Histograms provide a view of the **data density**. Higher bars represent where the data are relatively more common. For instance, there are many more emails between 0 and 10,000 characters than emails between 10,000 and 20,000 in the data set. The bars make it easy to see how the density of the data changes relative to the number of characters.

● **Example 2.13** How many emails had fewer than 10 thousand characters?

The height of the bars corresponds to frequency. There were 19 cases from 0 to less than 5 thousand and 12 cases from 5 thousand to less than 10 thousand, so there were $19 + 12 = 31$ emails with fewer than 10 thousand characters.

● **Example 2.14** Approximately how many emails had fewer than 1 thousand chacters?

Based just on this histogram, we cannot know the exact answer to this question. We only know that 19 emails had between 0 and 5 thousand characters. If the number of emails is evenly distribution on this interval, then we can estimate that approximately $19/5 \approx 4$ emails fell in the range between 0 and 1 thousand.

● **Example 2.15** What *percent* of the emails had 10 thousand or more characters?

From the first example, we know that 31 emails had fewer than 10 thousand characters. Since there are 50 emails in total, there must be 19 emails that have 10 thousand or more characters. To find the percent, compute $19/50 = 0.38 = 38\%$.

Sometimes questions such as the ones above can be answered more easily with a **cumulative frequency histogram**. This type of histogram shows cumulative, or total, frequency achieved by each bin, rather than the frequency in that particular bin.

Characters (in thousands)	0-5	5-10	10-15	15-20	20-25	25-30	30-35	⋯	55-60	60-65
Cumulative Frequency	19	31	37	39	42	47	47	⋯	49	50

Table 2.10: The cumulative frequencies for the binned `num_char` data.

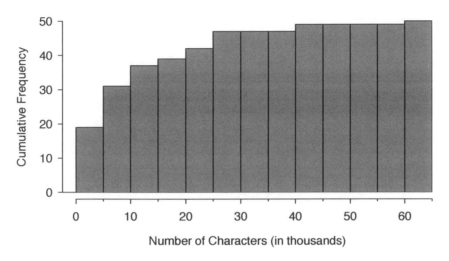

Figure 2.11: A cumulative frequency histogram of `num_char`. This histogram uses bins or class intervals of width 5.

● **Example 2.16**   How many of the emails had fewer than 20 thousand characters?

By tracing the height of the 15-20 thousand bin over to the vertical axis, we can see that it has a height just under 40 on the cumulative frequency scale. Therefore, we estimate that ≈39 of the emails had fewer than 30 thousand characters. Note that, unlike with a regular frequency histogram, we do not add up the height of the bars in a cumulative frequency histogram because each bar already represents a cumulative sum.

● **Example 2.17**   Using the cumulative frequency histogram, how many of the emails had 10-15 thousand characters?

To answer this question, we do a subtraction. ≈39 had fewer than 15-20 thousand emails and ≈37 had fewer than 10-15 thousand emails, so ≈2 must have had between 10-15 thousand emails.

● **Example 2.18**   Approximately 25 of the emails had fewer than how many characters?

This time we are given a cumulative frequency, so we start at 25 on the vertical axis and trace it across to see which bin it hits. It hits the 5-10 thousand bin, so 25 of the emails had fewer than a value somewhere between 5 and 10 thousand characters.

Knowing that 25 of the emails had fewer than a value between 5 and 10 thousand characters is useful information, but it is even more useful if we know what percent of the total 25 represents. Knowing that there were 50 total emails tells us that $25/50 = 0.5 = 50\%$ of the emails had fewer than a value between 5 and 10 thousand characters. When we want to know what fraction or percent of the data meet a certain criteria, we use relative frequency instead of frequency. **Relative frequency** is a fancy term for percent or proportion. It tells us how large a number is relative to the total.

Just as we constructed a frequency table, frequency histogram, and cumulative frequency histogram, we can construct a relative frequency table, relative frequency histogram, and cumulative relative frequency histogram.

⊙ **Guided Practice 2.19** How will the *shape* of the relative frequency histograms differ from the frequency histograms?[12]

Caution: Pay close attention to the vertical axis of a histogram
We can misinterpret a histogram if we forget to check whether the vertical axis represents frequency, relative frequency, cumulative frequency, or cumulative relative frequency.

2.1.4 Describing Shape

Frequency and relative frequency histograms are especially convenient for describing the **shape** of the data distribution. Figure 2.9 shows that most emails have a relatively small number of characters, while fewer emails have a very large number of characters. When data trail off to the right in this way and have a longer right tail, the shape is said to be **right skewed**.[13]

Data sets with the reverse characteristic – a long, thin tail to the left – are said to be **left skewed**. We also say that such a distribution has a long left tail. Data sets that show roughly equal trailing off in both directions are called **symmetric**.

Long tails to identify skew

When data trail off in one direction, the distribution has a **long tail**. If a distribution has a long left tail, it is left skewed. If a distribution has a long right tail, it is right skewed.

⊙ **Guided Practice 2.20** Take a look at the dot plot in Figure 2.6. Can you see the skew in the data? Is it easier to see the skew in the frequency histogram, the dot plot, or the stem-and-leaf plot?[14]

---

[12]The shape will remain exactly the same. Changing from frequency to relative frequency involves dividing all the frequencies by the same number, so only the vertical scale (the numbers on the y-axis) change.

[13]Other ways to describe data that are skewed to the right: **skewed to the right**, **skewed to the high end**, or **skewed to the positive end**.

[14]The skew is visible in all three plots. However, it is not easily visible in the cumulative frequency histogram.

⊙ **Guided Practice 2.21**   Would you expect the distribution of number of pets per household to be right skewed, left skewed, or approximately symmetric? Explain.[15]

In addition to looking at whether a distribution is skewed or symmetric, histograms, stem-and-leaf plots, and dot plots can be used to identify modes. A **mode** is represented by a prominent peak in the distribution.[16] There is only one prominent peak in the histogram of num_char.

Figure 2.12 shows histograms that have one, two, or three prominent peaks. Such distributions are called **unimodal**, **bimodal**, and **multimodal**, respectively. Any distribution with more than 2 prominent peaks is called multimodal. Notice that in Figure 2.9 there was one prominent peak in the unimodal distribution with a second less prominent peak that was not counted since it only differs from its neighboring bins by a few observations.

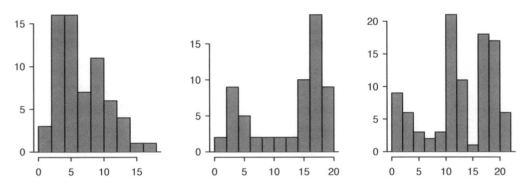

Figure 2.12: Counting only prominent peaks, the distributions are (left to right) unimodal, bimodal, and multimodal.

⊙ **Guided Practice 2.22**   Height measurements of young students and adult teachers at a K-3 elementary school were taken. How many modes would you anticipate in this height data set?[17]

TIP: Looking for modes
Looking for modes isn't about finding a clear and correct answer about the number of modes in a distribution, which is why *prominent* is not rigorously defined in this book. The important part of this examination is to better understand your data and how it might be structured.

---

[15]We suspect most households would have 0, 1, or 2 pets but that a smaller number of households will have 3, 4, 5, or more pets, so there will be greater density over the small numbers, suggesting the distribution will have a long right tail and be right skewed.

[16]Another definition of mode, which is not typically used in statistics, is the value with the most occurrences. It is common to have *no* observations with the same value in a data set, which makes this other definition useless for many real data sets.

[17]There might be two height groups visible in the data set: one of the students and one of the adults. That is, the data are probably bimodal.

2.2 Numerical summaries and box plots ▶️

2.2.1 Measures of center

In the previous section, we saw that modes can occur anywhere in a data set. Therefore, mode is not a measure of **center**. We understand the term *center* intuitively, but quantifying what is the center can be a little more challenging. This is because there are different definitions of center. Here we will focus on the two most common: the mean and median.

The **mean**, sometimes called the average, is a common way to measure the center of a distribution of data. To find the mean number of characters in the 50 emails, we add up all the character counts and divide by the number of emails. For computational convenience, the number of characters is listed in the thousands and rounded to the first decimal.

$$\bar{x} = \frac{21.7 + 7.0 + \cdots + 15.8}{50} = 11.6 \tag{2.23}$$

The sample mean is often labeled $\bar{x}$. The letter x is being used as a generic placeholder for the variable of interest, `num_char`, and the bar on the x communicates that the average number of characters in the 50 emails was 11,600.

$\bar{x}$
sample
mean

Mean

The sample mean of a numerical variable is computed as the sum of all of the observations divided by the number of observations:

$$\bar{x} = \frac{1}{n} \sum x_i = \frac{x_1 + x_2 + \cdots + x_n}{n} \tag{2.24}$$

where $\sum$ is the capital Greek letter sigma and $\sum x_i$ means take the sum of all the individual x values. $x_1, x_2, \ldots, x_n$ represent the n observed values.

⊙ **Guided Practice 2.25** Examine Equations (2.23) and (2.24) above. What does x_1 correspond to? And x_2? What does x_i represent?[18]

⊙ **Guided Practice 2.26** What was n in this sample of emails?[19]

The `email50` data set represents a sample from a larger population of emails that were received in January and March. We could compute a mean for this population in the same way as the sample mean, however, the population mean has a special label: μ. The symbol μ is the Greek letter *mu* and represents the average of all observations in the population. Sometimes a subscript, such as $_x$, is used to represent which variable the population mean refers to, e.g. μ_x.

μ
population
mean

---

[18]x_1 corresponds to the number of characters in the first email in the sample (21.7, in thousands), x_2 to the number of characters in the second email (7.0, in thousands), and x_i corresponds to the number of characters in the ith email in the data set.

[19]The sample size was $n = 50$.

● **Example 2.27**   The average number of characters across all emails can be estimated using the sample data. Based on the sample of 50 emails, what would be a reasonable estimate of μ_x, the mean number of characters in all emails in the email data set? (Recall that email50 is a sample from email.)

The sample mean, 11,600, may provide a reasonable estimate of μ_x. While this number will not be perfect, it provides a *point estimate* of the population mean. In Chapter 5 and beyond, we will develop tools to characterize the reliability of point estimates, and we will find that point estimates based on larger samples tend to be more reliable than those based on smaller samples.

● **Example 2.28**   We might like to compute the average income per person in the US. To do so, we might first think to take the mean of the per capita incomes across the 3,143 counties in the county data set. What would be a better approach?

The county data set is special in that each county actually represents many individual people. If we were to simply average across the income variable, we would be treating counties with 5,000 and 5,000,000 residents equally in the calculations. Instead, we should compute the total income for each county, add up all the counties' totals, and then divide by the number of people in all the counties. If we completed these steps with the county data, we would find that the per capita income for the US is $27,348.43. Had we computed the *simple* mean of per capita income across counties, the result would have been just $22,504.70!

Example 2.28 used what is called a **weighted mean**, which will not be a key topic in this textbook. However, we have provided an online supplement on weighted means for interested readers:

www.openintro.org/stat/down/supp/wtdmean.pdf

The median provides another measure of center. The **median** splits an ordered data set in half. There are 50 character counts in the email50 data set (an even number) so the data are perfectly split into two groups of 25. We take the median in this case to be the average of the two middle observations: $(6{,}768 + 7{,}012)/2 = 6{,}890$. When there are an odd number of observations, there will be exactly one observation that splits the data into two halves, and in this case that observation is the median (no average needed).

Median: the number in the middle

In an ordered data set, the **median** is the observation right in the middle. If there are an even number of observations, the median is the average of the two middle values.

Graphically, we can think of the mean as the balancing point. The median is the value such that 50% of the *area* is to the left of it and 50% of the *area* is to the right of it.

● **Example 2.29**   Based on the data, why is the mean greater than the median in this data set?

Consider the three largest values of 42 thousand, 43 thousand, and 64 thousand. These values drag up the mean because they substantially increase the sum (the total). However, they do not drag up the median because their magnitude does not change the location of the middle value.

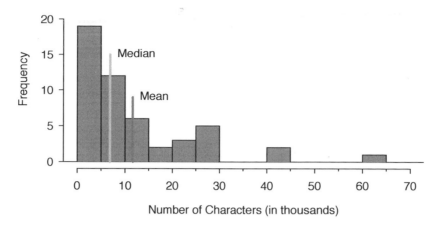

Figure 2.13: A histogram of num_char with its mean and median shown.

The mean follows the tail

In a right skewed distribution, the mean is greater than the median.
In a left skewed distribution, the mean is less than the median.
In a symmetric distribution, the mean and median are approximately equal.

⊙ **Guided Practice 2.30** Consider the distribution of individual income in the United States. Which is greater: the mean or median? Why?[20]

2.2.2 Standard deviation as a measure of spread

The U.S. Census Bureau reported that in 2012, the median family income was \$62,241 and the mean family income was \$82,743.[21]

Is a family income of \$60,000 far from the mean or somewhat close to the mean? In order to answer this question, it is not enough to know the center of the data set and its **range** (maximum value - minimum value). We must know about the variability of the data set within that range. Low variability or small spread means that the values tend to be more clustered together. High variability or large spread means that the values tend to be far apart.

● **Example 2.31** Is it possible for two data sets to have the same range but different spread? If so, give an example. If not, explain why not.

Yes. An example is: 1, 1, 1, 1, 1, 9, 9, 9, 9, 9 and 1, 5, 5, 5, 5, 5, 5, 5, 5, 5, 9.

The first data set has a larger spread because values tend to be farther away from each other while in the second data set values are clustered together at the mean.

---

[20]Because a small percent of individuals earn extremely large amounts of money while the majority earn a modest amount, the distribution is skewed to the right. Therefore, the mean is greater than the median.
[21]www.census.gov/hhes/www/income/

Here, we introduce the standard deviation as a measure of spread. Though its formula is a bit tedious to calculate by hand, the standard deviation is very useful in data analysis and roughly describes how far away, on average, the observations are from the mean.

We call the distance of an observation from its mean its **deviation**. Below are the deviations for the 1st, 2nd, 3rd, and 50th observations in the num_char variable. For computational convenience, the number of characters is listed in the thousands and rounded to the first decimal.

$$x_1 - \bar{x} = 21.7 - 11.6 = 10.1$$
$$x_2 - \bar{x} = 7.0 - 11.6 = -4.6$$
$$x_3 - \bar{x} = 0.6 - 11.6 = -11.0$$
$$\vdots$$
$$x_{50} - \bar{x} = 15.8 - 11.6 = 4.2$$

s^2
sample
variance

If we square these deviations and then take an average, the result is about equal to the sample **variance**, denoted by s^2:

$$
\begin{aligned}
s^2 &= \frac{10.1^2 + (-4.6)^2 + (-11.0)^2 + \cdots + 4.2^2}{50 - 1} \\
&= \frac{102.01 + 21.16 + 121.00 + \cdots + 17.64}{49} \\
&= 172.44
\end{aligned}
$$

We divide by $n - 1$, rather than dividing by n, when computing the variance; you need not worry about this mathematical nuance for the material in this textbook. Notice that squaring the deviations does two things. First, it makes large values much larger, seen by comparing 10.1^2, $(-4.6)^2$, $(-11.0)^2$, and 4.2^2. Second, it gets rid of any negative signs.

The **standard deviation** is defined as the square root of the variance:

s
sample
standard
deviation

$$s = \sqrt{172.44} = 13.13$$

The standard deviation of the number of characters in an email is about 13.13 thousand. A subscript of x may be added to the variance and standard deviation, i.e. s_x^2 and s_x, as a reminder that these are the variance and standard deviation of the observations represented by $x_1, x_2, ..., x_n$. The x subscript is usually omitted when it is clear which data the variance or standard deviation is referencing.

Calculating the standard deviation

The standard deviation is the square root of the variance. It is roughly the average distance of the observations from the mean.

$$s = \sqrt{\frac{1}{n-1} \sum (x_i - \bar{x})^2} \tag{2.32}$$

The variance is useful for mathematical reasons, but the standard deviation is easier to interpret because it has the same units as the data set. The units for variance will be the units squared (e.g. meters^2). Formulas and methods used to compute the variance and

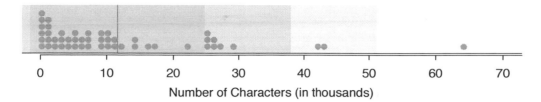

Number of Characters (in thousands)

Figure 2.14: In the `num_char` data, 40 of the 50 emails (80%) are within 1 standard deviation of the mean, and 47 of the 50 emails (94%) are within 2 standard deviations. Usually about 68% (or approximately 2/3) of the data are within 1 standard deviation of the mean and 95% are within 2 standard deviations, though this rule of thumb is less accurate for skewed data, as shown in this example.

standard deviation for a population are similar to those used for a sample.[22] However, like the mean, the population values have special symbols: σ^2 for the variance and σ for the standard deviation. The symbol σ is the Greek letter *sigma*.

σ^2
population
variance

TIP: thinking about the standard deviation
It is useful to think of the standard deviation as the average distance that observations fall from the mean.

σ
population
standard
deviation

The **empirical rule** tells us that usually about 68% of the data will be within one standard deviation of the mean and about 95% will be within two standard deviations of the mean. However, as seen in Figures 2.14 and 2.15, these percentages are not strict rules.[23]

⊙ **Guided Practice 2.33** On page 53, the concept of shape of a distribution was introduced. A good description of the shape of a distribution should include modality and whether the distribution is symmetric or skewed to one side. Using Figure 2.15 as an example, explain why such a description is important.[24]

● **Example 2.34** Earlier we reported that the mean family income in the U.S. in 2012 was $82,743. Estimating the standard deviation of income as approximately $50,000, is a family income of $60,000 unusually far from the mean or relatively close to the mean?

Because $60,000 is less that one standard deviation from the mean, it is relatively close to the mean. If the value were more than 2 standard deviations away from the mean, we would consider it far from the mean.

When describing any distribution, comment on the three important characteristics of center, spread, and shape. Also note any especially unusual cases.

---

[22]The only difference is that the population variance has a division by n instead of $n - 1$.

[23]We will learn where these two numbers come from in Chapter 4 when we study the normal distribution.

[24]Figure 2.15 shows three distributions that look quite different, but all have the same mean, variance, and standard deviation. Using modality, we can distinguish between the first plot (bimodal) and the last two (unimodal). Using skewness, we can distinguish between the last plot (right skewed) and the first two. While a picture, like a histogram, tells a more complete story, we can use modality and shape (symmetry/skew) to characterize basic information about a distribution.

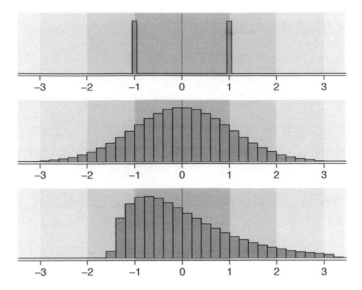

Figure 2.15: Three very different population distributions with the same mean $\mu = 0$ and standard deviation $\sigma = 1$.

● **Example 2.35** In the data's context (the number of characters in emails), describe the distribution of the num_char variable using the histogram in Figure 2.16.

The distribution of email character counts is unimodal and very strongly skewed to the right. Many of the counts fall near the mean at 11,600, and most fall within one standard deviation (13,130) of the mean. There is one exceptionally long email with about 65,000 characters.

In this chapter we use standard deviation as a descriptive statistic to describe the variability in a given data set. In Chapter 5 we will use the standard deviation to assess how close a sample mean is to the population mean.

2.2.3   Box plots and quartiles

A **box plot** summarizes a data set using five summary statistics while also plotting unusual observations, called outliers. Figure 2.17 provides a box plot of the num_char variable from the email50 data set.

The five summary statistics used in a box plot are known as the **five-number summary**, which consists of the minimum, the maximum, and the three quartiles (Q_1, Q_2, Q_3) of the data set being studied.

Q_2 represents the **second quartile**, which is equivalent to the 50th percentile (i.e. the median). Previously, we saw that Q_2 (the median) for the email50 data set was the average of the two middle values: $\frac{6,768+7,012}{2} = 6,890$.

Q_1 represents the **first quartile**, which is the 25th percentile, and is the median of the smaller half of the data set. There are 25 values in the lower half of the data set, so Q_1 is the middle value: 2,454 characters. Q_3 represents the **third quartile**, or 75th percentile, and is the median of the larger half of the data set: 15,829 characters.

We calculate the variability in the data using the range of the middle 50% of the data: $Q_3 - Q_1 = 13,375$. This quantity is called the **interquartile range** (IQR, for short).

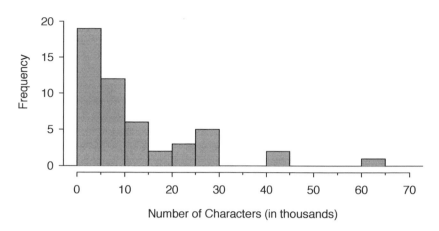

Figure 2.16: A copy of Figure 2.9.

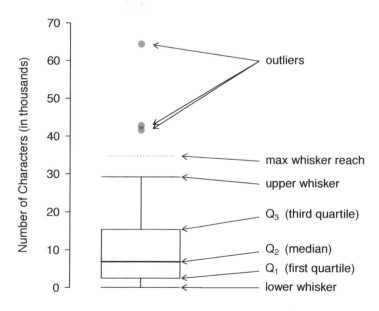

Figure 2.17: A labeled box plot for the nuber of characters in 50 emails. The median (6,890) splits the data into the bottom 50% and the top 50%.

It, like the standard deviation, is a measure of variability or **spread** in data. The more variable the data, the larger the standard deviation and IQR tend to be.

Interquartile range (IQR)

The IQR is the length of the box in a box plot. It is computed as

$$IQR = Q_3 - Q_1$$

where Q_1 and Q_3 are the 25th and 75th percentiles.

Outliers in the context of a box plot

When in the context of a box plot, define an **outlier** as an observation that is more than $1.5 \times IQR$ above Q_3 or $1.5 \times IQR$ below Q_1. Such points are marked using a dot or asterisk in a box plot.

To build a box plot, draw an axis (vertical or horizontal) and draw a scale. Draw a dark line denoting Q_2, the median. Next, draw a line at Q_1 and at Q_3. Connect the Q_1 and Q_3 lines to form a rectangle. The width of the rectangle corresponds to the IQR and the middle 50% of the data is in this interval.

Extending out from the rectangle, the **whiskers** attempt to capture all of the data remaining outside of the box, except outliers. In Figure 2.17, the upper whisker does not extend to the last three points, which are beyond $Q_3 + 1.5 \times IQR$ and are outliers, so it extends only to the last point below this limit.[25] The lower whisker stops at the lowest value, 33, since there are no additional data to reach. Outliers are each marked with a dot or asterisk. In a sense, the box is like the body of the box plot and the whiskers are like its arms trying to reach the rest of the data.

● **Example 2.36** Compare the box plot to the graphs previously discussed: stem-and-leaf plot, dot plot, frequency and relative frequency histogram. What can we learn more easily from a box plot? What can we learn more easily from the other graphs?

It is easier to immediately identify the quartiles from a box plot. The box plot also more prominently highlights outliers. However, a box plot, unlike the other graphs, does not show the *distribution* of the data. For example, we cannot generally identify modes using a box plot.

● **Example 2.37** Is it possible to identify skew from the box plot?

Yes. Looking at the lower and upper whiskers of this box plot, we see that the lower 25% of the data is squished into a shorter distance than the upper 25% of the data, implying that there is greater density in the low values and a tail trailing to the upper values. This box plot is right skewed.

---

[25]You might wonder, isn't the choice of $1.5 \times IQR$ for defining an outlier arbitrary? It is! In practical data analyses, we tend to avoid a strict definition since what is an unusual observation is highly dependent on the context of the data.

⊙ **Guided Practice 2.38** True or false: there is more data between the median and Q_3 than between Q_1 and the median.[26]

● **Example 2.39** Consider the following ordered data set.

$$5 \quad 5 \quad 9 \quad 10 \quad 15 \quad 16 \quad 20 \quad 30 \quad 80$$

Find the 5 number summary and identify how small or large a value would need to be to be considered an outlier. Are there any outliers in this data set?

There are nine numbers in this data set. Because n is odd, the median is the middle number: 15. When finding Q_1, we find the median of the lower half of the data, which in this case includes 4 numbers (we do not include the 15 as belonging to either half of the data set). Q_1 then is the average of 5 and 9, which is $Q_1 = 7$, and Q_3 is the average of 20 and 30, so $Q_3 = 35$. The min is 5 and the max is 80. To see how small a number needs to be to be an outlier on the low end we do:

$$\begin{aligned} Q_1 - 1.5 \times IQR &= Q_1 - 1.5 \times (Q_3 - Q_1) \\ &= 7 - 1.5 \times (35 - 7) \\ &= -35 \end{aligned}$$

On the high end we need:

$$\begin{aligned} Q_3 + 1.5 \times IQR &= Q_3 + 1.5 \times (Q_3 - Q_1) \\ &= 35 + 1.5 \times (35 - 7) \\ &= 77 \end{aligned}$$

There are no numbers less than -41, so there are no outliers on the low end. The observation at 80 is greater than 77, so 80 is an outlier on the high end.

2.2.4 Calculator: summarizing 1-variable statistics

▶ TI-83/84: Entering data

The first step in summarizing data or making a graph is to enter the data set into a list. Use STAT, Edit.

1. Press STAT.

2. Choose 1:Edit.

3. Enter data into L1 or another list.

---

[26]False. Since Q_1 is the 25th percentile and the median is the 50th percentile, 25% of the data fall between Q_1 and the median. Similarly, 25% of the data fall between Q_2 and the median. The distance between the median and Q_3 is larger because that 25% of the data is more spread out.

🎥 Casio fx-9750GII: Entering data

1. Navigate to STAT (MENU button, then hit the 2 button or select STAT).

2. Optional: use the left or right arrows to select a particular list.

3. Enter each numerical value and hit EXE.

🎥 TI-84: Calculating Summary Statistics

Use the STAT, CALC, 1-Var Stats command to find summary statistics such as mean, standard deviation, and quartiles.

1. Enter the data as described previously.

2. Press STAT.

3. Right arrow to CALC.

4. Choose 1:1-Var Stats.

5. Enter L1 (i.e. 2ND 1) for List. If the data is in a list other than L1, type the name of that list.

6. Leave FreqList blank.

7. Choose Calculate and hit ENTER.

TI-83: Do steps 1-4, then type L1 (i.e. 2nd 1) or the list's name and hit ENTER.

Calculating the summary statistics will return the following information. It will be necessary to hit the down arrow to see all of the summary statistics.

$\bar{x}$	Mean	minX	Minimum
Σx	Sum of all the data values	Q_1	First quartile
Σx^2	Sum of all the squared data values	Med	Median
σx	Population standard deviation	maxX	Maximum
n	Sample size or # of data points		

🎥 TI-83/84: Drawing a box plot

1. Enter the data to be graphed as described previously.

2. Hit 2ND Y= (i.e. STAT PLOT).

3. Hit ENTER (to choose the first plot).

4. Hit ENTER to choose ON.

5. Down arrow and then right arrow three times to select box plot with outliers.

6. Down arrow again and make Xlist: L1 and Freq: 1.

7. Choose ZOOM and then 9:ZoomStat to get a good viewing window.

Casio fx-9750GII: Drawing a box plot and 1-variable statistics

1. Navigate to STAT (MENU, then hit 2) and enter the data into a list.

2. Go to GRPH (F1).

3. Next go to SET (F6) to set the graphing parameters.

4. To use the 2nd or 3rd graph instead of GPH1, select F2 or F3.

5. Move down to Graph Type and select the ▷ (F6) option to see more graphing options, then select Box (F2).

6. If XList does not show the list where you entered the data, hit LIST (F1) and enter the correct list number.

7. Leave Frequency at 1.

8. For Outliers, choose On (F1).

9. Hit EXE and then choose the graph where you set the parameters F1 (most common), F2, or F3.

10. If desired, explore 1-variable statistics by selecting 1-Var (F1).

● **Example 2.40** Enter the following 10 data points into the first list on a calculator: 5, 8, 1, 19, 3, 1, 11, 18, 20, 5. Find the summary statistics and make a box plot of the data.

The summary statistics should be $\bar{x} = 9.1$, $Sx = 7.475$, $Q1 = 3$, etc. The box plot should be as follows.

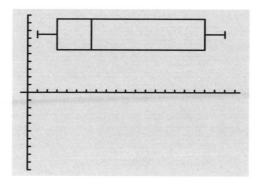

TI-83/84: What to do if you cannot find L1 or another list

Restore lists L1-L6 using the following steps:

1. Press STAT.

2. Choose 5:SetUpEditor.

3. Hit ENTER.

> **▶ Casio fx-9750GII: Deleting a data list**
>
> 1. Navigate to STAT (MENU, then hit 2).
>
> 2. Use the arrow buttons to navigate to the list you would like to delete.
>
> 3. Select ▷ (F6) to see more options.
>
> 4. Select DEL-A (F4) and then F1 to confirm.

2.2.5 Outliers and robust statistics

> **Rules of thumb for identifying outliers**
>
> There are two rules of thumb for identifying outliers:
>
> - More than 1.5× IQR below Q_1 or above Q_3
> - More than 2 standard deviations above or below the mean.
>
> Both are important for the AP exam. In practice, consider these to be only rough
> guidelines.

⊙ **Guided Practice 2.41**   For the email50 data set,$Q_1 = 2,536$ and $Q_3 = 15,411$. $\bar{x}$ $= 11,600$ and $s = 13,130$. What values would be considered an outlier on the low end using each rule?[27]

⊙ **Guided Practice 2.42**   Because there are no negative values in this data set, there can be no outliers on the low end. What does the fact that there are outliers on the high end but not on the low end suggestion?[28]

How are the sample statistics of the num_char data set affected by the observation, 64,401? What would have happened if this email wasn't observed? What would happen to these summary statistics if the observation at 64,401 had been even larger, say 150,000? These scenarios are plotted alongside the original data in Figure 2.18, and sample statistics are computed under each scenario in Table 2.19.

⊙ **Guided Practice 2.43**   (a) Which is more affected by extreme observations, the mean or median? Table 2.19 may be helpful. (b) Is the standard deviation or IQR more affected by extreme observations?[29]

---

[27] $Q_1 - 1.5 \times IQR = 2536 - 1.5 \times (15411 - 2536) = -16,749.5$, so values less than -16,749.5 would be considered an outlier using the first rule of thumb. Using the second rule of thumb, a value less than $\bar{x} - 2 \times s = 11,600 - 2 \times 13,130 = -14,660$ would be considered an outlier. Note tht these are just rules of thumb and yield different values.

[28] It suggests that the distribution has a right hand tail, that is, that it is right skewed.

[29] (a) Mean is affected more. (b) Standard deviation is affected more. Complete explanations are provided in the material following Guided Practice 2.43.

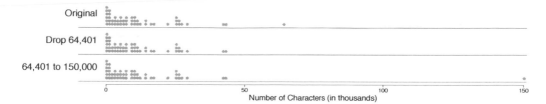

Figure 2.18: Dot plots of the original character count data and two modified data sets.

scenario	robust		not robust	
	median	IQR	$\bar{x}$	s
original `num_char` data	6,890	12,875	11,600	13,130
drop 66,924 observation	6,768	11,702	10,521	10,798
move 66,924 to 150,000	6,890	12,875	13,310	22,434

Table 2.19: A comparison of how the median, IQR, mean ($\bar{x}$), and standard deviation (s) change when extreme observations are present.

The median and IQR are called **robust estimates** because extreme observations have little effect on their values. The mean and standard deviation are much more affected by changes in extreme observations.

● **Example 2.44** The median and IQR do not change much under the three scenarios in Table 2.19. Why might this be the case?

Since there are no large gaps between observations around the three quartiles, adding, deleting, or changing one value, no matter how extreme that value, will have little effect on their values.

⊙ **Guided Practice 2.45** The distribution of vehicle prices tends to be right skewed, with a few luxury and sports cars lingering out into the right tail. If you were searching for a new car and cared about price, should you be more interested in the mean or median price of vehicles sold, assuming you are in the market for a regular car?[30]

2.2.6 Linear transformations of data

● **Example 2.46** Begin with the following list: 1, 1, 5, 5. Multiply all of the numbers by 10. What happens to the mean? What happens to the standard deviation? How do these compare to the mean and the standard deviation of the original list?

The original list has a mean of 3 and a standard deviation of 2. The new list: 10, 10, 50, 50 has a mean of 30 with a standard deviation of 20. Because all of the values were multiplied by 10, both the mean and the standard deviation were multiplied by 10. [31]

---

[30]Buyers of a "regular car" should be concerned about the median price. High-end car sales can drastically inflate the mean price while the median will be more robust to the influence of those sales.

[31]Here, the population standard deviation was used in the calculation. These properties can be proven mathematically using properties of sigma (summation).

● **Example 2.47**   Start with the following list: 1, 1, 5, 5. Multiply all of the numbers by -0.5. What happens to the mean? What happens to the standard deviation? How do these compare to the mean and the standard deviation of the original list?

The new list: -0.5, -0.5, -2.5, -2.5 has a mean of -1.5 with a standard deviation of 1. Because all of the values were multiplied by -0.5, the mean was multiplied by -0.5. Multiplying all of the values by a negative flipped the sign of numbers, which affects the location of the center, but not the spread. Multiplying all of the values by -0.5 multiplied the standard deviation by +0.5 since the standard deviation cannot be negative.

● **Example 2.48**   Again, start with the following list: 1, 1, 5, 5. Add 100 to every entry. How do the new mean and standard deviation compare to the original mean and standard deviation?

The new list is: 101, 101, 105, 105. The new mean of 103 is 100 greater than the original mean of 3. The new standard deviation of 2 is the *same* as the original standard deviation of 2. Adding a constant to every entry shifted the values, but did not stretch them.

Suppose that a researcher is looking at a list of 500 temperatures recorded in Celsius (C). The mean of the temperatures listed is given as 27°C with a standard deviation of 3°C. Because she is not familiar with the Celsius scale, she would like to convert these summary statistics into Fahrenheit (F). To convert from Celsius to Fahrenheit, we use the following conversion:

$$x_F = \frac{9}{5}x_C + 32$$

Fortunately, she does not need to convert each of the 500 temperatures to Fahrenheit and then recalculate the mean and the standard deviation. The unit conversion above is a linear transformation of the following form, where $a = 9/5$ and $b = 32$:

$$aX + b$$

Using the examples as a guide, we can solve this temperature-conversion problem. The mean was 27°C and the standard deviation was 3°C. To convert to Fahrenheit, we multiply all of the values by 9/5, which multiplies both the mean and the standard deviation by 9/5. Then we add 32 to all of the values which adds 32 to the mean but does not change the standard deviation further.

$$\bar{x}_F = \frac{9}{5}\bar{x}_C + 32 \qquad\qquad \sigma_F = \frac{9}{5}\sigma_C$$
$$= \frac{5}{9}(27) + 32 \qquad\qquad = \frac{9}{5}(3)$$
$$= 80.6 \qquad\qquad\qquad = 5.4$$

Adding shifts the values, multiplying stretches or contracts them

Adding a constant to every value in a data set shifts the mean but does not affect the standard deviation. Multiplying the values in a data set by a constant will change the mean and the standard deviation by the same multiple, except that the standard deviation will always remain positive.

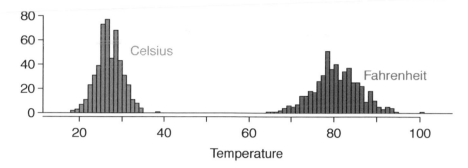

Figure 2.20: 500 temperatures shown in both Celsius and Fahrenheit.

● **Example 2.49** Consider the temperature example. How would converting from Celsuis to Fahrenheit affect the median? The IQR?

The median is affected in the same way as the mean and the IQR is affected in the same way as the standard deviation. To get the new median, multiply the old median by 9/5 and add 32. The IQR is computed by subtracting Q_1 from Q_3. While Q_1 and Q_3 are each affected in the same way as the median, the additional 32 added to each will cancel when we take $Q_3 - Q_1$. That is, the IQR will be increase by a factor of 9/5 but will be unaffected by the addition of 32.

For a more mathematical explanation of the IQR calculation, see the footnote.[32]

2.2.7 Comparing numerical data across groups

Some of the more interesting investigations can be considered by examining numerical data across groups. The methods required here aren't really new. All that is required is to make a numerical plot for each group. To make a direct comparison between two groups, create a pair of dot plots or a pair of histograms drawn using the same scales. It is also common to use back-to-back stem-and-leaf plots, parallel box plots, and hollow histograms, the three of which are explored here.

We will take a look again at the county data set and compare the median household income for counties that gained population from 2000 to 2010 versus counties that had no gain. While we might like to make a causal connection here, remember that these are observational data and so such an interpretation would be unjustified.

There were 2,041 counties where the population increased from 2000 to 2010, and there were 1,099 counties with no gain (all but one were a loss). A random sample of 100 counties from the first group and 50 from the second group are shown in Table 2.21 to give a better sense of some of the raw data, and Figure 2.22 shows a **back-to-back stem-and-leaf plot**.

The **parallel box plot** is a traditional tool for comparing across groups. An example is shown in the left panel of Figure 2.23, where there are two box plots, one for each group, placed into one plotting window and drawn on the same scale.

Another useful plotting method uses **hollow histograms** to compare numerical data across groups. These are just the outlines of histograms of each group put on the same plot, as shown in the right panel of Figure 2.23.

---

[32]new IQR $= \left(\frac{9}{5}Q_3 + 32\right) - \left(\frac{9}{5}Q_1 + 32\right) = \frac{9}{5}\left(Q_3 - Q_1\right) = \frac{9}{5} \times$ (old IQR).

population gain							no gain		
41.2	33.1	30.4	37.3	79.1	34.5		40.3	33.5	34.8
22.9	39.9	31.4	45.1	50.6	59.4		29.5	31.8	41.3
47.9	36.4	42.2	43.2	31.8	36.9		28	39.1	42.8
50.1	27.3	37.5	53.5	26.1	57.2		38.1	39.5	22.3
57.4	42.6	40.6	48.8	28.1	29.4		43.3	37.5	47.1
43.8	26	33.8	35.7	38.5	42.3		43.7	36.7	36
41.3	40.5	68.3	31	46.7	30.5		35.8	38.7	39.8
68.3	48.3	38.7	62	37.6	32.2		46	42.3	48.2
42.6	53.6	50.7	35.1	30.6	56.8		38.6	31.9	31.1
66.4	41.4	34.3	38.9	37.3	41.7		37.6	29.3	30.1
51.9	83.3	46.3	48.4	40.8	42.6		57.5	32.6	31.1
44.5	34	48.7	45.2	34.7	32.2		46.2	26.5	40.1
39.4	38.6	40	57.3	45.2	33.1		38.4	46.7	25.9
43.8	71.7	45.1	32.2	63.3	54.7		36.4	41.5	45.7
71.3	36.3	36.4	41	37	66.7		39.7	37	37.7
50.2	45.8	45.7	60.2	53.1			21.4	29.3	50.1
35.8	40.4	51.5	66.4	36.1			43.6	39.8	

Table 2.21: In this table, median household income (in \$1000s) from a random sample of 100 counties that gained population over 2000-2010 are shown on the left. Median incomes from a random sample of 50 counties that had no population gain are shown on the right.

```
        Population: Gain              Population: No Gain

                           3| 2 |12
                       98766| 2 |66899
             444433222211100| 3 |00112234
      9999888777766666655| 3 |56667788888999
   44433332221111110000| 4 |0000001223344
          9988876665555| 4 |666778
               443221100| 5 |0
                 977775| 5 |8
                    320| 6 |
                  88766| 6 |
                     21| 7 |

Legend: 4 | 5 = 45,000 median income
```

Figure 2.22: Back-to-back stem-and-leaf plot for median income, split by whether the count had a population gain or no gain.

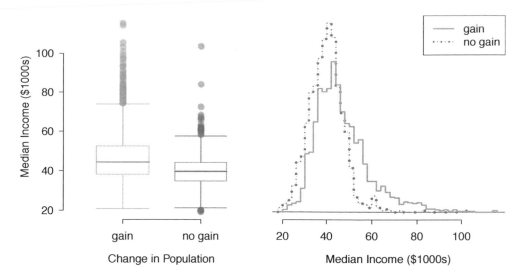

Figure 2.23: Side-by-side box plot (left panel) and hollow histograms (right panel) for med_income, where the counties are split by whether there was a population gain or loss from 2000 to 2010. The income data were collected between 2006 and 2010.

⊙ **Guided Practice 2.50**   Use the plots in Figure 2.23 to compare the incomes for counties across the two groups. What do you notice about the approximate center of each group? What do you notice about the variability between groups? Is the shape relatively consistent between groups? How many *prominent* modes are there for each group?[33]

TIP: Comparing distributions
When comparing distributions, compare them with respect to center, spread, and shape as well as any unusual observations. Such descriptions should be in context.

⊙ **Guided Practice 2.51**   What components of each plot in Figure 2.23 do you find most useful?[34]

⊙ **Guided Practice 2.52**   Do these graphs tell us about any association between income for the two groups?[35]

---

[33]Answers may vary a little. The counties with population gains tend to have higher income (median of about $45,000) versus counties without a gain (median of about $40,000). The variability is also slightly larger for the population gain group. This is evident in the IQR, which is about 50% bigger in the *gain* group. Both distributions show slight to moderate right skew and are unimodal. There is a secondary small bump at about $60,000 for the *no gain* group, visible in the hollow histogram plot, that seems out of place. (Looking into the data set, we would find that 8 of these 15 counties are in Alaska and Texas.) The box plots indicate there are many observations far above the median in each group, though we should anticipate that many observations will fall beyond the whiskers when using such a large data set.

[34]Answers will vary. The parallel box plots are especially useful for comparing centers and spreads, while the hollow histograms are more useful for seeing distribution shape, skew, and groups of anomalies.

[35]No, to see association we require a scatterplot. Moreover, these data are not paired, so the discussion of association does not make sense here.

Looking at an association is different than comparing distributions. When comparing distributions, we are interested in questions such as, "Which distribution has a greater average?" and "How do the shapes of the distribution differ?" The number of elements in each data set need not be the same (e.g. height of women and height of men). When we look at association, we are interested in whether there is a positive, negative, or no association between the variables. This requires two data sets of equal length that are essentially paired (e.g. height and weight of individuals).

TIP: Comparing distributions versus looking at association
We compare two distributions with respect to center, spread, and shape. To compare the distributions visually, we use 2 single-variable graphs, such as two histograms, two dot plots, parallel box plots, or a back-to-back stem-and-leaf. When looking at association, we look for a positive, negative, or no relationship between the variables. To see association visually, we require a scatterplot.

2.2.8   Mapping data (special topic)

The county data set offers many numerical variables that we could plot using dot plots, scatterplots, or box plots, but these miss the true nature of the data. Rather, when we encounter geographic data, we should map it using an **intensity map**, where colors are used to show higher and lower values of a variable. Figures 2.24 and 2.25 shows intensity maps for federal spending per capita (fed_spend), poverty rate in percent (poverty), homeownership rate in percent (homeownership), and median household income (med_income). The color key indicates which colors correspond to which values. Note that the intensity maps are not generally very helpful for getting precise values in any given county, but they are very helpful for seeing geographic trends and generating interesting research questions.

● **Example 2.53**   What interesting features are evident in the fed_spend and poverty intensity maps?

The federal spending intensity map shows substantial spending in the Dakotas and along the central-to-western part of the Canadian border, which may be related to the oil boom in this region. There are several other patches of federal spending, such as a vertical strip in eastern Utah and Arizona and the area where Colorado, Nebraska, and Kansas meet. There are also seemingly random counties with very high federal spending relative to their neighbors. If we did not cap the federal spending range at $18 per capita, we would actually find that some counties have extremely high federal spending while there is almost no federal spending in the neighboring counties. These high-spending counties might contain military bases, companies with large government contracts, or other government facilities with many employees.

Poverty rates are evidently higher in a few locations. Notably, the deep south shows higher poverty rates, as does the southwest border of Texas. The vertical strip of eastern Utah and Arizona, noted above for its higher federal spending, also appears to have higher rates of poverty (though generally little correspondence is seen between the two variables). High poverty rates are evident in the Mississippi flood plains a little north of New Orleans and also in a large section of Kentucky and West Virginia.

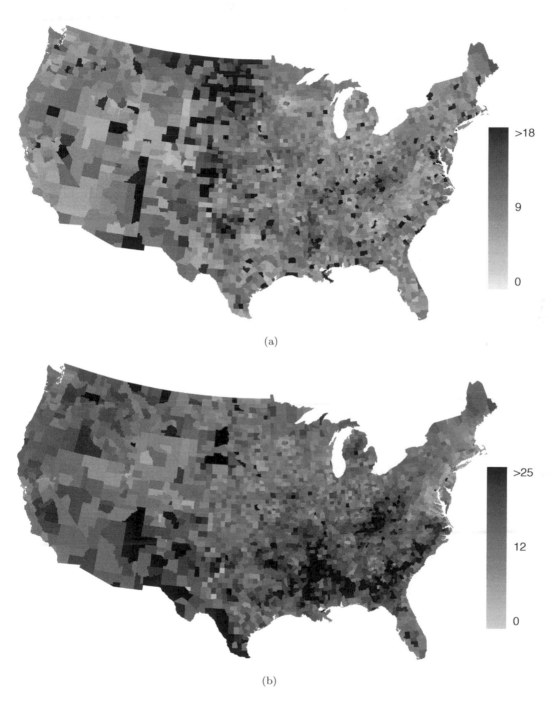

Figure 2.24: (a) Map of federal spending (dollars per capita). (b) Intensity map of poverty rate (percent).

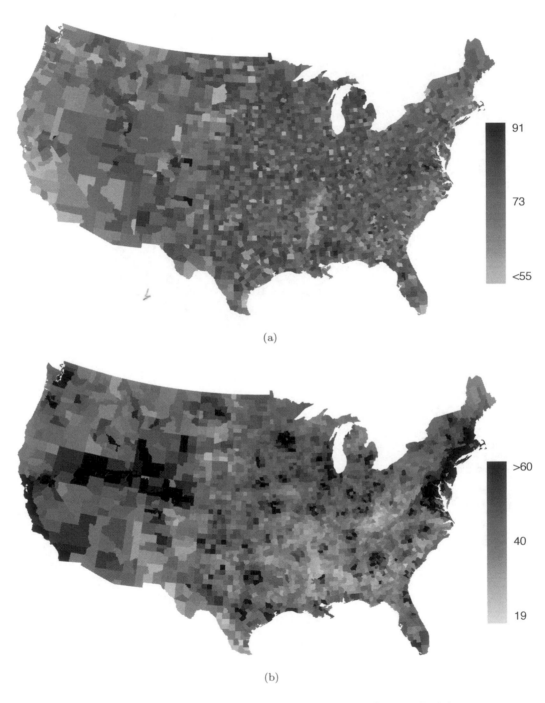

Figure 2.25: (a) Intensity map of homeownership rate (percent). (b) Intensity map of median household income ($1000s).

⊙ **Guided Practice 2.54** What interesting features are evident in the med_income intensity map?[36]

2.3 Considering categorical data

Like numerical data, categorical data can also be organized and analyzed. In this section, we will introduce tables and other basic tools for categorical data that are used throughout this book. The email50 data set represents a sample from a larger email data set called email. This larger data set contains information on 3,921 emails. In this section we will examine whether the presence of numbers, small or large, in an email provides any useful value in classifying email as spam or not spam.

2.3.1 Contingency tables and bar plots

Table 2.26 summarizes two variables: spam and number. Recall that number is a categorical variable that describes whether an email contains no numbers, only small numbers (values under 1 million), or at least one big number (a value of 1 million or more). A table that summarizes data for two categorical variables in this way is called a **contingency table**. Each value in the table represents the number of times a particular combination of variable outcomes occurred. For example, the value 149 corresponds to the number of emails in the data set that are spam *and* had no number listed in the email. Row and column totals are also included. The **row totals** provide the total counts across each row (e.g. $149 + 168 + 50 = 367$), and **column totals** are total counts down each column.

Table 2.27 shows a frequency table for the number variable. If we replaced the counts with percentages or proportions, the table is a **relative frequency table**.

		number			
		none	small	big	Total
spam	spam	149	168	50	367
	not spam	400	2659	495	3554
	Total	549	2827	545	3921

Table 2.26: A contingency table for spam and number.

none	small	big	Total
549	2827	545	3921

Table 2.27: A frequency table for the number variable.

Because the numbers in these tables are counts, not to data points, they cannot be graphed using the methods we applied to numerical data. Instead, another set of graphing methods are needed that are suitable for categorical data.

A bar plot is a common way to display a single categorical variable. The left panel of Figure 2.28 shows a **bar plot** for the number variable. In the right panel, the counts are converted into proportions (e.g. $549/3921 = 0.140$ for none), showing the proportion of observations that are in each level (i.e. in each category).

---

[36]Note: answers will vary. There is a very strong correspondence between high earning and metropolitan areas. You might look for large cities you are familiar with and try to spot them on the map as dark spots.

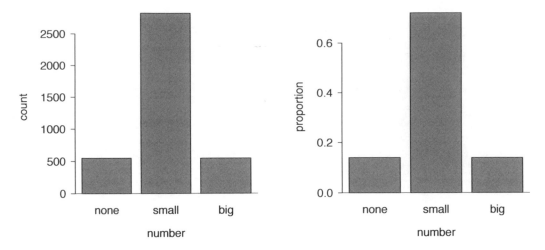

Figure 2.28: Two bar plots of number. The left panel shows the counts, and the right panel shows the proportions in each group.

2.3.2 Row and column proportions

Table 2.29 shows the row proportions for Table 2.26. The **row proportions** are computed as the counts divided by their row totals. The value 149 at the intersection of spam and none is replaced by $149/367 = 0.406$, i.e. 149 divided by its row total, 367. So what does 0.406 represent? It corresponds to the proportion of spam emails in the sample that do not have any numbers.

	none	small	big	Total
spam	$149/367 = 0.406$	$168/367 = 0.458$	$50/367 = 0.136$	1.000
not spam	$400/3554 = 0.113$	$2657/3554 = 0.748$	$495/3554 = 0.139$	1.000
Total	$549/3921 = 0.140$	$2827/3921 = 0.721$	$545/3921 = 0.139$	1.000

Table 2.29: A contingency table with row proportions for the spam and number variables.

A contingency table of the column proportions is computed in a similar way, where each **column proportion** is computed as the count divided by the corresponding column total. Table 2.30 shows such a table, and here the value 0.271 indicates that 27.1% of emails with no numbers were spam. This rate of spam is much higher compared to emails with only small numbers (5.9%) or big numbers (9.2%). Because these spam rates vary between the three levels of number (none, small, big), this provides evidence that the spam and number variables are associated.

	none	small	big	Total
spam	$149/549 = 0.271$	$168/2827 = 0.059$	$50/545 = 0.092$	$367/3921 = 0.094$
not spam	$400/549 = 0.729$	$2659/2827 = 0.941$	$495/545 = 0.908$	$3684/3921 = 0.906$
Total	1.000	1.000	1.000	1.000

Table 2.30: A contingency table with column proportions for the spam and number variables.

We could also have checked for an association between spam and number in Table 2.29 using row proportions. When comparing these row proportions, we would look down columns to see if the fraction of emails with no numbers, small numbers, and big numbers varied from spam to not spam.

⊙ **Guided Practice 2.55** What does 0.458 represent in Table 2.29? What does 0.059 represent in Table 2.30?[37]

⊙ **Guided Practice 2.56** What does 0.139 at the intersection of not spam and big represent in Table 2.29? What does 0.908 represent in the Table 2.30?[38]

● **Example 2.57** Data scientists use statistics to filter spam from incoming email messages. By noting specific characteristics of an email, a data scientist may be able to classify some emails as spam or not spam with high accuracy. One of those characteristics is whether the email contains no numbers, small numbers, or big numbers. Another characteristic is whether or not an email has any HTML content. A contingency table for the spam and format variables from the email data set are shown in Table 2.31. Recall that an HTML email is an email with the capacity for special formatting, e.g. bold text. In Table 2.31, which would be more helpful to someone hoping to classify email as spam or regular email: row or column proportions?

Such a person would be interested in how the proportion of spam changes within each email format. This corresponds to column proportions: the proportion of spam in plain text emails and the proportion of spam in HTML emails.

If we generate the column proportions, we can see that a higher fraction of plain text emails are spam ($209/1195 = 17.5\%$) than compared to HTML emails ($158/2726 = 5.8\%$). This information on its own is insufficient to classify an email as spam or not spam, as over 80% of plain text emails are not spam. Yet, when we carefully combine this information with many other characteristics, such as number and other variables, we stand a reasonable chance of being able to classify some email as spam or not spam.

	text	HTML	Total
spam	209	158	367
not spam	986	2568	3554
Total	1195	2726	3921

Table 2.31: A contingency table for spam and format.

Example 2.57 points out that row and column proportions are not equivalent. Before settling on one form for a table, it is important to consider each to ensure that the most useful table is constructed.

⊙ **Guided Practice 2.58** Look back to Tables 2.29 and 2.30. Which would be more useful to someone hoping to identify spam emails using the number variable?[39]

---

[37]0.458 represents the proportion of spam emails that had a small number. 0.059 represents the fraction of emails with small numbers that are spam.

[38]0.139 represents the fraction of non-spam email that had a big number. 0.908 represents the fraction of emails with big numbers that are non-spam emails.

[39]The column proportions in Table 2.30 will probably be most useful, which makes it easier to see that emails with small numbers are spam about 5.9% of the time (relatively rare). We would also see that about 27.1% of emails with no numbers are spam, and 9.2% of emails with big numbers are spam.

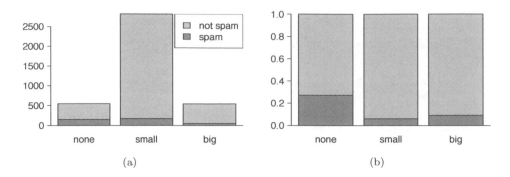

Figure 2.32: (a) Segmented bar plot for numbers found in emails, where the counts have been further broken down by `spam`. (b) Standardized version of Figure (a).

2.3.3 Segmented bar plots

Contingency tables using row or column proportions are especially useful for examining how two categorical variables are related. Segmented bar plots provide a way to visualize the information in these tables.

A **segmented bar plot** is a graphical display of contingency table information. For example, a segmented bar plot representing Table 2.30 is shown in Figure 2.32(a), where we have first created a bar plot using the `number` variable and then separated each group by the levels of `spam`. The column proportions of Table 2.30 have been translated into a standardized segmented bar plot in Figure 2.32(b), which is a helpful visualization of the fraction of spam emails in each level of `number`.

● **Example 2.59** Examine both of the segmented bar plots. Which is more useful?

Figure 2.32(a) contains more information, but Figure 2.32(b) presents the information more clearly. This second plot makes it clear that emails with no number have a relatively high rate of spam email – about 27%! On the other hand, less than 10% of email with small or big numbers are spam.

Since the proportion of spam changes across the groups in Figure 2.32(b), we can conclude the variables are dependent, which is something we were also able to discern using table proportions. Because both the `none` and `big` groups have relatively few observations compared to the `small` group, the association is more difficult to see in Figure 2.32(a).

In some other cases, a segmented bar plot that is not standardized will be more useful in communicating important information. Before settling on a particular segmented bar plot, create standardized and non-standardized forms and decide which is more effective at communicating features of the data.

2.3.4 The only pie chart you will see in this book

While pie charts are well known, they are not typically as useful as other charts in a data analysis. A **pie chart** is shown in Figure 2.33 alongside a bar plot. It is generally more difficult to compare group sizes in a pie chart than in a bar plot, especially when categories have nearly identical counts or proportions. In the case of the `none` and `big` categories, the difference is so slight you may be unable to distinguish any difference in group sizes for either plot!

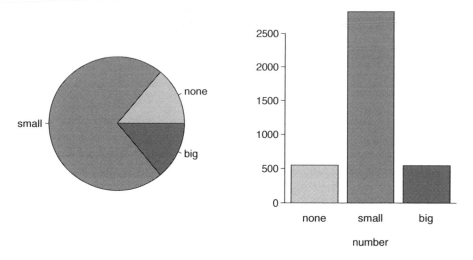

Figure 2.33: A pie chart and bar plot of number for the email data set.

2.4   Case study: gender discrimination (special topic)

⬤ **Example 2.60**   Suppose your professor splits the students in class into two groups: students on the left and students on the right. If $\hat{p}_L$ and $\hat{p}_R$ represent the proportion of students who own an Apple product on the left and right, respectively, would you be surprised if $\hat{p}_L$ did not exactly equal $\hat{p}_R$?

While the proportions would probably be close to each other, it would be unusual for them to be exactly the same. We would probably observe a small difference due to chance.

⊙ **Guided Practice 2.61**   If we don't think the side of the room a person sits on in class is related to whether the person owns an Apple product, what assumption are we making about the relationship between these two variables?[40]

2.4.1   Variability within data

We consider a study investigating gender discrimination in the 1970s, which is set in the context of personnel decisions within a bank.[41]   The research question we hope to answer is, "Are females unfairly discriminated against in promotion decisions made by male managers?"

The participants in this study are 48 male bank supervisors attending a management institute at the University of North Carolina in 1972. They were asked to assume the role of the personnel director of a bank and were given a personnel file to judge whether the person should be promoted to a branch manager position. The files given to the participants were identical, except that half of them indicated the candidate was male and the other half indicated the candidate was female. These files were randomly assigned to the subjects.

---

[40]We would be assuming that these two variables are independent.

[41]Rosen B and Jerdee T. 1974. Influence of sex role stereotypes on personnel decisions. Journal of Applied Psychology 59(1):9-14.

⊙ **Guided Practice 2.62**   Is this an observational study or an experiment? What implications does the study type have on what can be inferred from the results?[42]

For each supervisor we record the gender associated with the assigned file and the promotion decision. Using the results of the study summarized in Table 2.34, we would like to evaluate if females are unfairly discriminated against in promotion decisions. In this study, a smaller proportion of females are promoted than males (0.583 versus 0.875), but it is unclear whether the difference provides *convincing evidence* that females are unfairly discriminated against.

		decision		
		promoted	not promoted	Total
gender	male	21	3	24
	female	14	10	24
	Total	35	13	48

Table 2.34: Summary results for the gender discrimination study.

● **Example 2.63**   Statisticians are sometimes called upon to evaluate the strength of evidence. When looking at the rates of promotion for males and females in this study, what comes to mind as we try to determine whether the data show convincing evidence of a real difference?

The observed promotion rates (58.3% for females versus 87.5% for males) suggest there might be discrimination against women in promotion decisions. However, we cannot be sure if the observed difference represents discrimination or is just from random chance. Generally there is a little bit of fluctuation in sample data, and we wouldn't expect the sample proportions to be *exactly* equal, even if the truth was that the promotion decisions were independent of gender.

Example 2.63 is a reminder that the observed outcomes in the sample may not perfectly reflect the true relationships between variables in the underlying population. Table 2.34 shows there were 7 fewer promotions in the female group than in the male group, a difference in promotion rates of 29.2% $\left(\frac{21}{24} - \frac{14}{24} = 0.292\right)$. This difference is large, but the sample size for the study is small, making it unclear if this observed difference represents discrimination or whether it is simply due to chance. We label these two competing claims, H_0 and H_A:

H_0: **Independence model.** The variables gender and decision are independent. They have no relationship, and the observed difference between the proportion of males and females who were promoted, 29.2%, was due to chance.

H_A: **Alternative model.** The variables gender and decision are *not* independent. The difference in promotion rates of 29.2% was not due to chance, and equally qualified females are less likely to be promoted than males.

What would it mean if the independence model, which says the variables gender and decision are unrelated, is true? It would mean each banker was going to decide whether to promote the candidate without regard to the gender indicated on the file. That is,

---

[42]The study is an experiment, as subjects were randomly assigned a male file or a female file. Since this is an experiment, the results can be used to evaluate a causal relationship between gender of a candidate and the promotion decision.

the difference in the promotion percentages was due to the way the files were randomly divided to the bankers, and the randomization just happened to give rise to a relatively large difference of 29.2%.

Consider the alternative model: bankers were influenced by which gender was listed on the personnel file. If this was true, and especially if this influence was substantial, we would expect to see some difference in the promotion rates of male and female candidates. If this gender bias was against females, we would expect a smaller fraction of promotion decisions for female personnel files relative to the male files.

We choose between these two competing claims by assessing if the data conflict so much with H_0 that the independence model cannot be deemed reasonable. If this is the case, and the data support H_A, then we will reject the notion of independence and conclude there was discrimination.

2.4.2 Simulating the study

Table 2.34 shows that 35 bank supervisors recommended promotion and 13 did not. Now, suppose the bankers' decisions were independent of gender. Then, if we conducted the experiment again with a different random arrangement of files, differences in promotion rates would be based only on random fluctuation. We can actually perform this **randomization**, which simulates what would have happened if the bankers' decisions had been independent of gender but we had distributed the files differently.

In this **simulation**, we thoroughly shuffle 48 personnel files, 24 labeled `male_sim` and 24 labeled `female_sim`, and deal these files into two stacks. We will deal 35 files into the first stack, which will represent the 35 supervisors who recommended promotion. The second stack will have 13 files, and it will represent the 13 supervisors who recommended against promotion. Then, as we did with the original data, we tabulate the results and determine the fraction of `male_sim` and `female_sim` who were promoted. The randomization of files in this simulation is independent of the promotion decisions, which means any difference in the two fractions is entirely due to chance. Table 2.35 show the results of such a simulation.

		decision		
		promoted	not promoted	Total
gender_sim	male_sim	18	6	24
	female_sim	17	7	24
	Total	35	13	48

Table 2.35: Simulation results, where any difference in promotion rates between `male_sim` and `female_sim` is purely due to chance.

⊙ **Guided Practice 2.64** What is the difference in promotion rates between the two simulated groups in Table 2.35? How does this compare to the observed 29.2% in the actual groups?[43]

---

[43] $18/24 - 17/24 = 0.042$ or about 4.2% in favor of the men. This difference due to chance is much smaller than the difference observed in the actual groups.

2.4.3   Checking for independence

We computed one possible difference under the independence model in Guided Practice 2.64, which represents one difference due to chance. While in this first simulation, we physically dealt out files, it is more efficient to perform this simulation using a computer. Repeating the simulation on a computer, we get another difference due to chance: -0.042. And another: 0.208. And so on until we repeat the simulation enough times that we have a good idea of what represents the *distribution of differences from chance alone.* Figure 2.36 shows a plot of the differences found from 100 simulations, where each dot represents a simulated difference between the proportions of male and female files that were recommended for promotion.

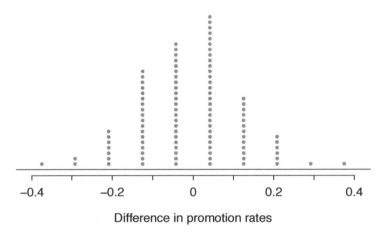

Difference in promotion rates

Figure 2.36: A stacked dot plot of differences from 100 simulations produced under the independence model, H_0, where gender_sim and decision are independent. Two of the 100 simulations had a difference of at least 29.2%, the difference observed in the study.

Note that the distribution of these simulated differences is centered around 0. We simulated these differences assuming that the independence model was true, and under this condition, we expect the difference to be zero with some random fluctuation. We would generally be surprised to see a difference of *exactly* 0: sometimes, just by chance, the difference is higher than 0, and other times it is lower than zero.

● **Example 2.65**   How often would you observe a difference of at least 29.2% (0.292) according to Figure 2.36? Often, sometimes, rarely, or never?

It appears that a difference of at least 29.2% due to chance alone would only happen about 2% of the time according to Figure 2.36. Such a low probability indicates a rare event.

The difference of 29.2% being a rare event suggests two possible interpretations of the results of the study:

H_0 **Independence model.** Gender has no effect on promotion decision, and we observed a difference that would only happen rarely.

H_A **Alternative model.** Gender has an effect on promotion decision, and what we observed was actually due to equally qualified women being discriminated against in promotion decisions, which explains the large difference of 29.2%.

Based on the simulations, we have two options. (1) We conclude that the study results do not provide strong evidence against the independence model. That is, we do not have sufficiently strong evidence to conclude there was gender discrimination. (2) We conclude the evidence is sufficiently strong to reject H_0 and assert that there was gender discrimination. When we conduct formal studies, usually we reject the notion that we just happened to observe a rare event.[44] So in this case, we reject the independence model in favor of the alternative. That is, we are concluding the data provide strong evidence of gender discrimination against women by the supervisors.

One field of statistics, statistical inference, is built on evaluating whether such differences are due to chance. In statistical inference, statisticians evaluate which model is most reasonable given the data. Errors do occur, just like rare events, and we might choose the wrong model. While we do not always choose correctly, statistical inference gives us tools to control and evaluate how often these errors occur. In Chapter 5, we give a formal introduction to the problem of model selection. We spend the next two chapters building a foundation of probability and theory necessary to make that discussion rigorous.

---

[44]This reasoning does not generally extend to anecdotal observations. Each of us observes incredibly rare events every day, events we could not possibly hope to predict. However, in the non-rigorous setting of anecdotal evidence, almost anything may appear to be a rare event, so the idea of looking for rare events in day-to-day activities is treacherous. For example, we might look at the lottery: there was only a 1 in 176 million chance that the Mega Millions numbers for the largest jackpot in history (March 30, 2012) would be (2, 4, 23, 38, 46) with a Mega ball of (23), but nonetheless those numbers came up! However, no matter what numbers had turned up, they would have had the same incredibly rare odds. That is, *any set of numbers we could have observed would ultimately be incredibly rare*. This type of situation is typical of our daily lives: each possible event in itself seems incredibly rare, but if we consider every alternative, those outcomes are also incredibly rare. We should be cautious not to misinterpret such anecdotal evidence.

2.5 Exercises

2.5.1 Examining numerical data

2.1 ACS, Part I. Each year, the US Census Bureau surveys about 3.5 million households with The American Community Survey (ACS). Data collected from the ACS have been crucial in government and policy decisions, helping to determine the allocation of federal and state funds each year. Some of the questions asked on the survey are about their income, age (in years), and gender. The table below contains this information for a random sample of 20 respondents to the 2012 ACS.[45]

	Income	Age	Gender		Income	Age	Gender
1	53,000	28	male	11	670	34	female
2	1600	18	female	12	29,000	55	female
3	70,000	54	male	13	44,000	33	female
4	12,800	22	male	14	48,000	41	male
5	1,200	18	female	15	30,000	47	female
6	30,000	34	male	16	60,000	30	male
7	4,500	21	male	17	108,000	61	male
8	20,000	28	female	18	5,800	50	female
9	25,000	29	female	19	50,000	24	female
10	42,000	33	male	20	11,000	19	male

(a) Create a scatterplot of income vs. age, and describe the relationship between these two variables.

(b) Now create two scatterplots: one for income vs. age for males and another for females.

(c) How, if at all, do the relationships between income and age differ for males and females?

2.2 MLB stats. A baseball team's success in a season is usually measured by their number of wins. In order to win, the team has to have scored more points (runs) than their opponent in any given game. As such, number of runs is often a good proxy for the success of the team. The table below shows number of runs, home runs, and batting averages for a random sample of 10 teams in the 2014 Major League Baseball season.[46]

	Team	Runs	Home runs	Batting avg.
1	Baltimore	705	211	0.256
2	Boston	634	123	0.244
3	Cincinnati	595	131	0.238
4	Cleveland	669	142	0.253
5	Detroit	757	155	0.277
6	Houston	629	163	0.242
7	Minnesota	715	128	0.254
8	NY Yankees	633	147	0.245
9	Pittsburgh	682	156	0.259
10	San Francisco	665	132	0.255

(a) Draw a scatterplot of runs vs. home runs.

(b) Draw a scatterplot of runs vs. batting averages.

(c) Are home runs or batting averages more strongly associated with number of runs? Explain your reasoning.

---

[45]United States Census Bureau. Summary File. 2012 American Community Survey. U.S. Census Bureaus American Community Survey Office, 2013. Web.

[46]ESPN: MLB Team Stats - 2014.

2.3 Fiber in your cereal. The Cereal FACTS report provides information on nutrition content of cereals as well as who they are targeted for (adults, children, families). We have selected a random sample of 20 cereals from the data provided in this report. Shown below are the fiber contents (percentage of fiber per gram of cereal) for these cereals.[47]

	Brand	Fiber %		Brand	Fiber %
1	Pebbles Fruity	0.0%	11	Cinnamon Toast Crunch	3.3%
2	Rice Krispies Treats	0.0%	12	Reese's Puffs	3.4%
3	Pebbles Cocoa	0.0%	13	Cheerios Honey Nut	7.1%
4	Pebbles Marshmallow	0.0%	14	Lucky Charms	7.4%
5	Frosted Rice Krispies	0.0%	15	Pebbles Boulders Chocolate PB	7.4%
6	Rice Krispies	3.0%	16	Corn Pops	9.4%
7	Trix	3.1%	17	Frosted Flakes Reduced Sugar	10.0%
8	Honey Comb	3.1%	18	Clifford Crunch	10.0%
9	Rice Krispies Gluten Free	3.3%	19	Apple Jacks	10.7%
10	Frosted Flakes	3.3%	20	Dora the Explorer	11.1%

(a) Create a stem and leaf plot of the distribution of the fiber content of these cereals.

(b) Create a dot plot of the fiber content of these cereals.

(c) Create a histogram and a relative frequency histogram of the fiber content of these cereals.

(d) What percent of cereals contain more than 0.7% fiber?

2.4 Sugar in your cereal. The Cereal FACTS report from Exercise 2.3 also provides information on sugar content of cereals. We have selected a random sample of 20 cereals from the data provided in this report. Shown below are the sugar contents (percentage of sugar per gram of cereal) for these cereals.

	Brand	Sugar %		Brand	Sugar %
1	Rice Krispies Gluten Free	3%	11	Corn Pops	31%
2	Rice Krispies	12%	12	Cheerios Honey Nut	32%
3	Dora the Explorer	22%	13	Reese's Puffs	34%
4	Frosted Flakes Red. Sugar	27%	14	Pebbles Fruity	37%
5	Clifford Crunch	27%	15	Pebbles Cocoa	37%
6	Rice Krispies Treats	30%	16	Lucky Charms	37%
7	Pebbles Boulders Choc. PB	30%	17	Frosted Flakes	37%
8	Cinnamon Toast Crunch	30%	18	Pebbles Marshmallow	37%
9	Trix	31%	19	Frosted Rice Krispies	40%
10	Honey Comb	31%	20	Apple Jacks	43%

(a) Create a stem and leaf plot of the distribution of the sugar content of these cereals.

(b) Create a dot plot of the sugar content of these cereals.

(c) Create a histogram and a relative frequency histogram of the sugar content of these cereals.

(d) What percent of cereals contain more than 30% sugar?

---

[47]JL Harris et al. "Cereal FACTS 2012: Limited progress in the nutrition quality and marketing of children's cereals". In: *Rudd Center for Food Policy & Obesity.* 12 (2012).

2.5 Mammal life spans. Data were collected on life spans (in years) and gestation lengths (in days) for 62 mammals. A scatterplot of life span versus length of gestation is shown below.[48]

(a) What type of an association is apparent between life span and length of gestation?

(b) What type of an association would you expect to see if the axes of the plot were reversed, i.e. if we plotted length of gestation versus life span?

(c) Are life span and length of gestation independent? Explain your reasoning.

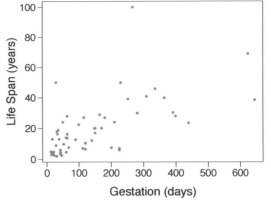

2.6 Associations, Part I. Indicate which of the plots show a

(a) positive association

(b) negative association

(c) no association

Also determine if the positive and negative associations are linear or nonlinear. Each part may refer to more than one plot.

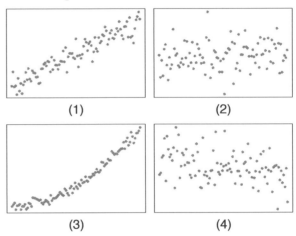

2.7 Office productivity. Office productivity is relatively low when the employees feel no stress about their work or job security. However, high levels of stress can also lead to reduced employee productivity. Sketch a plot to represent the relationship between stress and productivity.

2.8 Reproducing bacteria. Suppose that there is only sufficient space and nutrients to support one million bacterial cells in a petri dish. You place a few bacterial cells in this petri dish, allow them to reproduce freely, and record the number of bacterial cells in the dish over time. Sketch a plot representing the relationship between number of bacterial cells and time.

2.5.2 Numerical summaries and box plots

2.9 Sleeping in college. A recent article in a college newspaper stated that college students get an average of 5.5 hrs of sleep each night. A student who was skeptical about this value decided to conduct a survey by randomly sampling 25 students. On average, the sampled students slept 6.25 hours per night. Identify which value represents the sample mean and which value represents the claimed population mean.

[48]T. Allison and D.V. Cicchetti. "Sleep in mammals: ecological and constitutional correlates". In: *Arch. Hydrobiol* 75 (1975), p. 442.

2.10 Parameters and statistics. Identify which value represents the sample mean and which value represents the claimed population mean.

(a) American households spent an average of about $52 in 2007 on Halloween merchandise such as costumes, decorations and candy. To see if this number had changed, researchers conducted a new survey in 2008 before industry numbers were reported. The survey included 1,500 households and found that average Halloween spending was $58 per household.

(b) The average GPA of students in 2001 at a private university was 3.37. A survey on a sample of 203 students from this university yielded an average GPA of 3.59 in Spring semester of 2012.

2.11 Make-up exam. In a class of 25 students, 24 of them took an exam in class and 1 student took a make-up exam the following day. The professor graded the first batch of 24 exams and found an average score of 74 points with a standard deviation of 8.9 points. The student who took the make-up the following day scored 64 points on the exam.

(a) Does the new student's score increase or decrease the average score?

(b) What is the new average?

(c) Does the new student's score increase or decrease the standard deviation of the scores?

2.12 Days off at a mining plant. Workers at a particular mining site receive an average of 35 days paid vacation, which is lower than the national average. The manager of this plant is under pressure from a local union to increase the amount of paid time off. However, he does not want to give more days off to the workers because that would be costly. Instead he decides he should fire 10 employees in such a way as to raise the average number of days off that are reported by his employees. In order to achieve this goal, should he fire employees who have the most number of days off, least number of days off, or those who have about the average number of days off?

2.13 Smoking habits of UK residents, Part I. A survey was conducted to study the smoking habits of UK residents. The histograms below display the distributions of the number of cigarettes smoked on weekdays and weekends, and they exclude data from people who identified themselves as non-smokers. Describe the two distributions and compare them.[49]

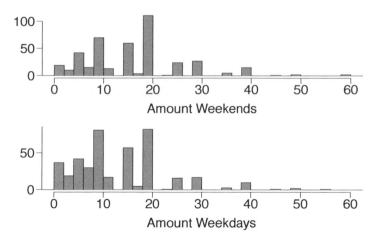

2.14 Stats scores, Part I. Below are the final exam scores of twenty introductory statistics students.

79, 83, 57, 82, 94, 83, 72, 74, 73, 71, 66, 89, 78, 81, 78, 81, 88, 69, 77, 79

Draw a histogram of these data and describe the distribution.

---

[49]Stats4Schools, Smoking.

2.15 Smoking habits of UK residents, Part II. A random sample of 5 smokers from the data set discussed in Exercise 2.13 is provided below.

gender	age	maritalStatus	grossIncome	smoke	amtWeekends	amtWeekdays
Female	51	Married	£2,600 to £5,200	Yes	20 cig/day	20 cig/day
Male	24	Single	£10,400 to £15,600	Yes	20 cig/day	15 cig/day
Female	33	Married	£10,400 to £15,600	Yes	20 cig/day	10 cig/day
Female	17	Single	£5,200 to £10,400	Yes	20 cig/day	15 cig/day
Female	76	Widowed	£5,200 to £10,400	Yes	20 cig/day	20 cig/day

(a) Find the mean amount of cigarettes smoked on weekdays and weekends by these 5 respondents.

(b) Find the standard deviation of the amount of cigarettes smoked on weekdays and on weekends by these 5 respondents. Is the variability higher on weekends or on weekdays?

2.16 Factory defective rate. A factory quality control manager decides to investigate the percentage of defective items produced each day. Within a given work week (Monday through Friday) the percentage of defective items produced was 2%, 1.4%, 4%, 3%, 2.2%.

(a) Calculate the mean for these data.

(b) Calculate the standard deviation for these data, showing each step in detail.

2.17 Medians and IQRs. For each part, compare distributions (1) and (2) based on their medians and IQRs. You do not need to calculate these statistics; simply state how the medians and IQRs compare. Make sure to explain your reasoning.

(a) (1) 3, 5, 6, 7, 9
    (2) 3, 5, 6, 7, 20

(b) (1) 3, 5, 6, 7, 9
    (2) 3, 5, 8, 7, 9

(c) (1) 1, 2, 3, 4, 5
    (2) 6, 7, 8, 9, 10

(d) (1) 0, 10, 50, 60, 100
    (2) 0, 100, 500, 600, 1000

2.18 Means and SDs. For each part, compare distributions (1) and (2) based on their means and standard deviations. You do not need to calculate these statistics; simply state how the means and the standard deviations compare. Make sure to explain your reasoning. *Hint:* It may be useful to sketch dot plots of the distributions.

(a) (1) 3, 5, 5, 5, 8, 11, 11, 11, 13
    (2) 3, 5, 5, 5, 8, 11, 11, 11, 20

(b) (1) -20, 0, 0, 0, 15, 25, 30, 30
    (2) -40, 0, 0, 0, 15, 25, 30, 30

(c) (1) 0, 2, 4, 6, 8, 10
    (2) 20, 22, 24, 26, 28, 30

(d) (1) 100, 200, 300, 400, 500
    (2) 0, 50, 300, 550, 600

2.19 Stats scores, Part II. Create a box plot for the final exam scores of twenty introductory statistics students given in Exercise 2.14. The five number summary provided below may be useful.

Min	Q1	Q2 (Median)	Q3	Max
57	72.5	78.5	82.5	94

2.20 Infant mortality. The infant mortality rate is defined as the number of infant deaths per 1,000 live births. This rate is often used as an indicator of the level of health in a country. The relative frequency histogram below shows the distribution of estimated infant death rates in 2012 for 222 countries.[50]

(a) Estimate Q1, the median, and Q3 from the histogram.

(b) Would you expect the mean of this data set to be smaller or larger than the median? Explain your reasoning.

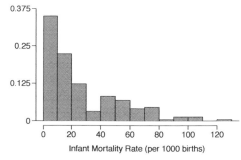

2.21 Matching histograms and box plots. Describe the distribution in the histograms below and match them to the box plots.

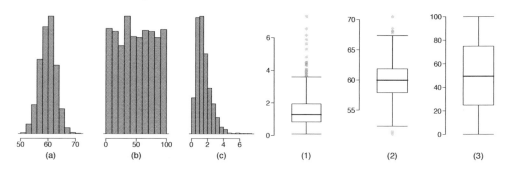

2.22 Air quality. Daily air quality is measured by the air quality index (AQI) reported by the Environmental Protection Agency. This index reports the pollution level and what associated health effects might be a concern. The index is calculated for five major air pollutants regulated by the Clean Air Act and takes values from 0 to 300, where a higher value indicates lower air quality. AQI was reported for a sample of 91 days in 2011 in Durham, NC. The relative frequency histogram below shows the distribution of the AQI values on these days.[51]

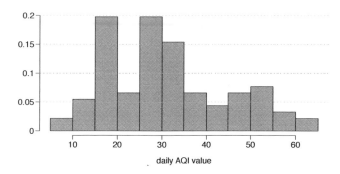

(a) Estimate the median AQI value of this sample.

(b) Would you expect the mean AQI value of this sample to be higher or lower than the median? Explain your reasoning.

(c) Estimate Q1, Q3, and IQR for the distribution.

---

[50]CIA Factbook, Country Comparison: Infant Mortality Rate, 2012.
[51]US Environmental Protection Agency, AirData, 2011.

2.23 Histograms and box plots. Compare the two plots below. What characteristics of the distribution are apparent in the histogram and not in the box plot? What characteristics are apparent in the box plot but not in the histogram?

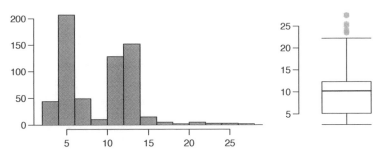

2.24 Marathon winners. The histogram and box plots below show the distribution of finishing times for male and female winners of the New York Marathon between 1970 and 1999.

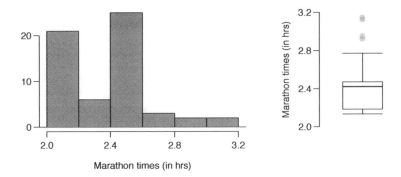

(a) What features of the distribution are apparent in the histogram and not the box plot? What features are apparent in the box plot but not in the histogram?

(b) What may be the reason for the bimodal distribution? Explain.

(c) Compare the distribution of marathon times for men and women based on the box plot shown below.

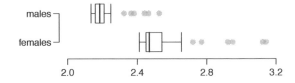

(d) The time series plot shown below is another way to look at these data. Describe what is visible in this plot but not in the others.

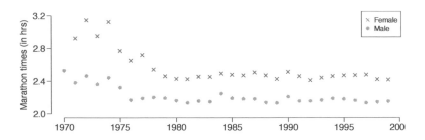

2.25 ACS, Part II. The hollow histograms below show the distribution of incomes of respondents to the American Community Survey introduced in Exercise 2.1.

(a) Compare the distributions of incomes of males and females.

(b) Suggest an alternative visualization for displaying and comparing the distributions of incomes of males and females.

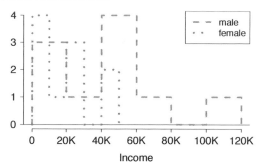

2.26 AP Stats. The table below shows scores (out of 100) of twenty college students on a college level statistical reasoning test given at the beginning of the semester in their introductory statistics course. Ten of these students have taken AP Stats in high school, and the other ten have not taken AP Stats.

Took AP stats: 52.5, 57.5, 60, 65, 70, 70, 72.5, 77.5, 80, 85

Did not take AP stats: 40, 45, 45, 50, 52.5, 57.5, 57.5, 60, 65, 72.5

(a) Create a relative frequency histogram of all students' scores on the statistical reasoning test.

(b) What percent of all students scored above 50 on this test?

(c) Compare the performances of students who did and did not take AP stats. The side-by-side box plots and the hollow histograms shown below might be helpful for this task.

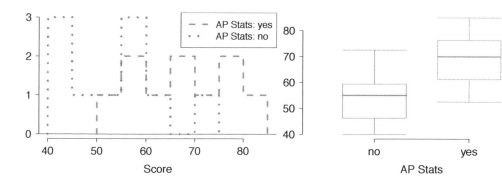

2.27 Distributions and appropriate statistics, Part I. For each of the following, describe whether you expect the distribution to be symmetric, right skewed, or left skewed. Also specify whether the mean or median would best represent a typical observation in the data and whether the variability of observations would be best represented using the standard deviation or IQR.

(a) Number of pets per household.

(b) Distance to work, i.e. number of miles between work and home.

(c) Heights of adult males.

2.28   Distributions and appropriate statistics, Part II. For each of the following, describe whether you expect the distribution to be symmetric, right skewed, or left skewed. Also specify whether the mean or median would best represent a typical observation in the data, and whether the variability of observations would be best represented using the standard deviation or IQR.

(a) Housing prices in a country where 25% of the houses cost below $350,000, 50% of the houses cost below $450,000, 75% of the houses cost below $1,000,000 and there are a meaningful number of houses that cost more than $6,000,000.

(b) Housing prices in a country where 25% of the houses cost below $300,000, 50% of the houses cost below $600,000, 75% of the houses cost below $900,000 and very few houses that cost more than $1,200,000.

(c) Number of alcoholic drinks consumed by college students in a given week. Assume that most of these students don't drink since they are under 21 years old, and only a few drink excessively.

(d) Annual salaries of the employees at a Fortune 500 company where only a few high level executives earn much higher salaries than the all other employees.

2.29   TV watchers. Students in an AP Statistics class were asked how many hours of television they watch per week (including online streaming). This sample yielded an average of 4.71 hours, with a standard deviation of 4.18 hours. Is the distribution of number of hours students watch television weekly symmetric? If not, what shape would you expect this distribution to have? Explain your reasoning.

2.30   Exam scores. The average on a history exam (scored out of 100 points) was 85, with a standard deviation of 15. Is the distribution of the scores on this exam symmetric? If not, what shape would you expect this distribution to have? Explain your reasoning.

2.31   Facebook friends. Facebook data indicate that 50% of Facebook users have 100 or more friends, and that the average friend count of users is 190. What do these findings suggest about the shape of the distribution of number of friends of Facebook users?[52]

2.32   A new statistic. The statistic $\frac{\bar{x}}{median}$ can be used as a measure of skewness. Suppose we have a distribution where all observations are greater than 0, $x_i > 0$. What is the expected shape of the distribution under the following conditions? Explain your reasoning.

(a) $\frac{\bar{x}}{median} = 1$ (b) $\frac{\bar{x}}{median} < 1$ (c) $\frac{\bar{x}}{median} > 1$

2.33   Income at the coffee shop, Part I. The first histogram below shows the distribution of the yearly incomes of 40 patrons at a college coffee shop. Suppose two new people walk into the coffee shop: one making $225,000 and the other $250,000. The second histogram shows the new income distribution. Summary statistics are also provided.

---
[52]Lars Backstrom. "Anatomy of Facebook". In: *Facebook Data Teams Notes* (2011).

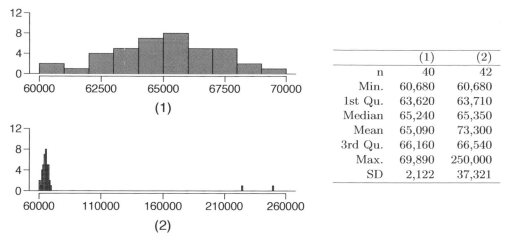

	(1)	(2)
n	40	42
Min.	60,680	60,680
1st Qu.	63,620	63,710
Median	65,240	65,350
Mean	65,090	73,300
3rd Qu.	66,160	66,540
Max.	69,890	250,000
SD	2,122	37,321

(a) Would the mean or the median best represent what we might think of as a typical income for the 42 patrons at this coffee shop? What does this say about the robustness of the two measures?

(b) Would the standard deviation or the IQR best represent the amount of variability in the incomes of the 42 patrons at this coffee shop? What does this say about the robustness of the two measures?

2.34   Midrange. The *midrange* of a distribution is defined as the average of the maximum and the minimum of that distribution. Is this statistic robust to outliers and extreme skew? Explain your reasoning

2.35   Commute times, Part I. The histogram to the right shows the distribution of mean commute times in 3,143 US counties in 2010. Describe the distribution and comment on whether or not a log transformation may be advisable for these data.

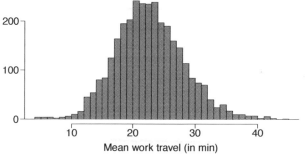

2.36   Hispanic population, Part I. The histogram below shows the distribution of the percentage of the population that is Hispanic in 3,143 counties in the US in 2010. Also shown is a histogram of logs of these values. Describe the distribution and comment on why we might want to use log-transformed values in analyzing or modeling these data.

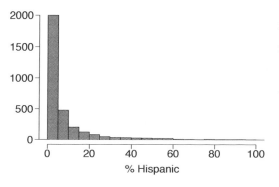

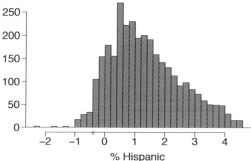

2.37 Income at the coffee shop, Part II. Suppose each of the 40 people in the coffee shop in Exercise 2.33 got a 5% raise. What would the new mean, median, and the standard deviation of their incomes be?

2.38 LA weather. The temperatures in June in Los Angeles have a mean of 77°F, with a standard deviation of 5°F. To convert from Celsius to Fahrenheit, we use the following conversion:

$$x_C = (x_F - 32)\frac{5}{9}$$

(a) What is the mean temperature in June in LA in degrees Celcius?

(b) What is the standard deviation of temperatures in June in LA in degrees Celcius?

2.39 Smoking habits of UK residents, Part III. The UK residents in Exercise 2.15 smoke on average 16 cigarettes per day on weekdays, with a standard deviation of 4.18. Suppose these residents participated in a smoking cessation program and at the end of the first week of the program reduced their weekday smoking by 3 cigarettes / day. Find the new mean and standard deviation of the number of cigarettes they smoke on weekdays.

2.40 Stats scores, Part III. The introductory statistics students in Exercise 2.14 scored on average 77.7 points, with a standard deviation of 8.44. The median score was 78.5. Suppose these students completed an extra credit exercise that earned them additional two points on their exams. Calculate the new mean, median, standard deviation, and IQR of their scores.

2.41 Commute times, Part II. Exercise 2.35 displays histograms of mean commute times in 3,143 US counties in 2010. Describe the spatial distribution of commuting times using the map below.

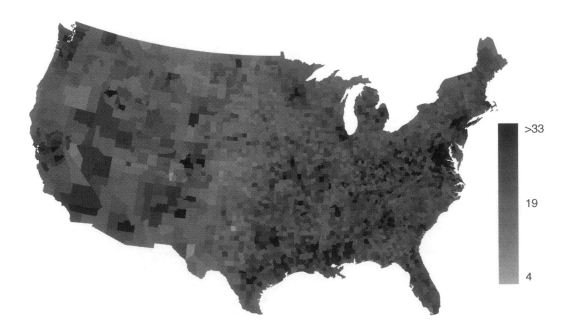

2.42 Hispanic population, Part II. Exercise 2.36 displays histograms of the distribution of the percentage of the population that is Hispanic in 3,143 counties in the US in 2010.

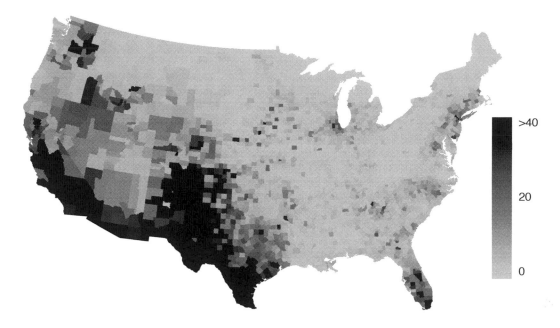

(a) What features of this distribution are apparent in the map but not in the histogram?

(b) What features are apparent in the histogram but not the map?

(c) Is one visualization more appropriate or helpful than the other? Explain your reasoning.

2.5.3 Considering categorical data

2.43 Antibiotic use in children. The bar plot and the pie chart below show the distribution of pre-existing medical conditions of children involved in a study on the optimal duration of antibiotic use in treatment of tracheitis, which is an upper respiratory infection.

(a) What features are apparent in the bar plot but not in the pie chart?

(b) What features are apparent in the pie chart but not in the bar plot?

(c) Which graph would you prefer to use for displaying these categorical data?

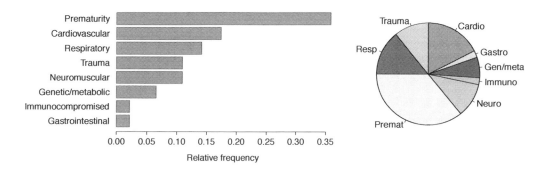

2.44 Views on immigration. 910 randomly sampled registered voters from Tampa, FL were asked if they thought workers who have illegally entered the US should be (i) allowed to keep their jobs and apply for US citizenship, (ii) allowed to keep their jobs as temporary guest workers but not allowed to apply for US citizenship, or (iii) lose their jobs and have to leave the country. The results of the survey by political ideology are shown below.[53]

		Political ideology			
		Conservative	Moderate	Liberal	Total
Response	(i) Apply for citizenship	57	120	101	278
	(ii) Guest worker	121	113	28	262
	(iii) Leave the country	179	126	45	350
	(iv) Not sure	15	4	1	20
	Total	372	363	175	910

(a) What percent of these Tampa, FL voters identify themselves as conservatives?

(b) What percent of these Tampa, FL voters are in favor of the citizenship option?

(c) What percent of these Tampa, FL voters identify themselves as conservatives and are in favor of the citizenship option?

(d) What percent of these Tampa, FL voters who identify themselves as conservatives are also in favor of the citizenship option? What percent of moderates and liberal share this view?

(e) Do political ideology and views on immigration appear to be independent? Explain your reasoning.

2.5.4 Case study: gender discrimination (special topic)

2.45 Side effects of Avandia, Part I. Rosiglitazone is the active ingredient in the controversial type 2 diabetes medicine Avandia and has been linked to an increased risk of serious cardiovascular problems such as stroke, heart failure, and death. A common alternative treatment is pioglitazone, the active ingredient in a diabetes medicine called Actos. In a nationwide retrospective observational study of 227,571 Medicare beneficiaries aged 65 years or older, it was found that 2,593 of the 67,593 patients using rosiglitazone and 5,386 of the 159,978 using pioglitazone had serious cardiovascular problems. These data are summarized in the contingency table below.[54]

		Cardiovascular problems		
		Yes	No	Total
Treatment	Rosiglitazone	2,593	65,000	67,593
	Pioglitazone	5,386	154,592	159,978
	Total	7,979	219,592	227,571

Determine if each of the following statements is true or false. If false, explain why. *Be careful:* The reasoning may be wrong even if the statement's conclusion is correct. In such cases, the statement should be considered false.

(a) Since more patients on pioglitazone had cardiovascular problems (5,386 vs. 2,593), we can conclude that the rate of cardiovascular problems for those on a pioglitazone treatment is higher.

(b) The data suggest that diabetic patients who are taking rosiglitazone are more likely to have cardiovascular problems since the rate of incidence was (2,593 / 67,593 = 0.038) 3.8% for patients on this treatment, while it was only (5,386 / 159,978 = 0.034) 3.4% for patients on pioglitazone. **(Note that parts (c) and (d) for this question are on the next page.)**

[53]SurveyUSA, News Poll #18927, data collected Jan 27-29, 2012.

[54]D.J. Graham et al. "Risk of acute myocardial infarction, stroke, heart failure, and death in elderly Medicare patients treated with rosiglitazone or pioglitazone". In: *JAMA* 304.4 (2010), p. 411. ISSN: 0098-7484.

(c) The fact that the rate of incidence is higher for the rosiglitazone group proves that rosiglitazone causes serious cardiovascular problems.

(d) Based on the information provided so far, we cannot tell if the difference between the rates of incidences is due to a relationship between the two variables or due to chance.

2.46 Heart transplants. The Stanford University Heart Transplant Study was conducted to determine whether an experimental heart transplant program increased lifespan. Each patient entering the program was designated an official heart transplant candidate, meaning that he was gravely ill and would most likely benefit from a new heart. Some patients got a transplant and some did not. The variable `transplant` indicates which group the patients were in; patients in the treatment group got a transplant and those in the control group did not. Another variable called `survived` was used to indicate whether or not the patient was alive at the end of the study. Of the 34 patients in the control group, 4 were alive at the end of the study. Of the 69 patients in the treatment group, 24 were alive. The contingency table below summarizes these results.[55]

		Group		
		Control	Treatment	Total
Outcome	Alive	4	24	28
	Dead	30	45	75
	Total	34	69	103

(a) What proportion of patients in the treatment group and what proportion of patients in the control group died?

(b) One approach for investigating whether or not the treatment is effective is to use a randomization technique.

   i. What are the claims being tested?

  ii. The paragraph below describes the set up for such approach, if we were to do it without using statistical software. Fill in the blanks with a number or phrase, whichever is appropriate.

> We write *alive* on _____ cards representing patients who were alive at the end of the study, and *dead* on _____ cards representing patients who were not. Then, we shuffle these cards and split them into two groups: one group of size _____ representing treatment, and another group of size _____ representing control. We calculate the difference between the proportion of *dead* cards in the treatment and control groups (treatment - control) and record this value. We repeat this 100 times to build a distribution centered at _____. Lastly, we calculate the fraction of simulations where the simulated differences in proportions are _____. If this fraction is low, we conclude that it is unlikely to have observed such an outcome by chance and that the null hypothesis (independence model) should be rejected in favor of the alternative.

  iii. What do the simulation results shown below suggest about the effectiveness of the transplant program?

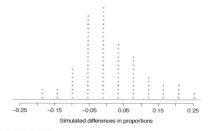

Simulated differences in proportions

---

[55]B. Turnbull et al. "Survivorship of Heart Transplant Data". In: *Journal of the American Statistical Association* 69 (1974), pp. 74–80.

2.47 Side effects of Avandia, Part II. Exercise 2.45 introduces a study that compares the rates of serious cardiovascular problems for diabetic patients on rosiglitazone and pioglitazone treatments. The table below summarizes the results of the study.

| | | *Cardiovascular problems* | | |
		Yes	No	Total
Treatment	Rosiglitazone	2,593	65,000	67,593
	Pioglitazone	5,386	154,592	159,978
	Total	7,979	219,592	227,571

(a) What proportion of all patients had cardiovascular problems?

(b) If the type of treatment and having cardiovascular problems were independent, about how many patients in the rosiglitazone group would we expect to have had cardiovascular problems?

(c) We can investigate the relationship between outcome and treatment in this study using a randomization technique. While in reality we would carry out the simulations required for randomization using statistical software, suppose we actually simulate using index cards. In order to simulate from the independence model, which states that the outcomes were independent of the treatment, we write whether or not each patient had a cardiovascular problem on cards, shuffled all the cards together, then deal them into two groups of size 67,593 and 159,978. We repeat this simulation 1,000 times and each time record the number of people in the rosiglitazone group who had cardiovascular problems. Below is a relative frequency histogram of these counts.

  i. What are the claims being tested?

  ii. Compared to the number calculated in part (b), which would provide more support for the alternative hypothesis, *more* or *fewer* patients with cardiovascular problems in the rosiglitazone group?

  iii. What do the simulation results suggest about the relationship between taking rosiglitazone and having cardiovascular problems in diabetic patients?

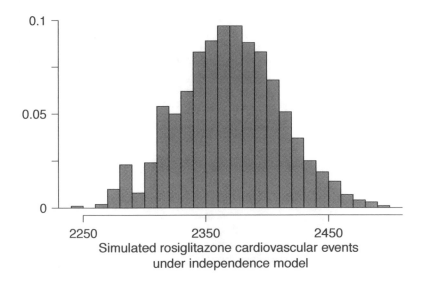

Simulated rosiglitazone cardiovascular events
under independence model

2.48 Sinusitis and antibiotics, Part II. Researchers studying the effect of antibiotic treatment compared to symptomatic treatment for acute sinusitis randomly assigned 166 adults diagnosed with sinusitis into two groups (as discussed in Exercise 1.2). Participants in the antibiotic group received a 10-day course of an antibiotic, and the rest received symptomatic treatments as a placebo. These pills had the same taste and packaging as the antibiotic. At the end of the 10-day period patients were asked if they experienced improvement in symptoms since the beginning of the study. The distribution of responses is summarized below.[56]

| | | *Self reported improvement in symptoms* | | |
		Yes	No	Total
Treatment	Antibiotic	66	19	85
	Placebo	65	16	81
	Total	131	35	166

(a) What type of a study is this?

(b) Does this study make use of blinding?

(c) At first glance, does antibiotic or placebo appear to be more effective for the treatment of sinusitis? Explain your reasoning using appropriate statistics.

(d) There are two competing claims that this study is used to compare: the independence model and the alternative model. Write out these competing claims in easy-to-understand language and in the context of the application. *Hint:* The researchers are studying the effectiveness of antibiotic treatment.

(e) Based on your finding in (c), does the evidence favor the alternative model? If not, then explain why. If so, what would you do to check if whether this is strong evidence?

[56] J.M. Garbutt et al. "Amoxicillin for Acute Rhinosinusitis: A Randomized Controlled Trial". In: *JAMA: The Journal of the American Medical Association* 307.7 (2012), pp. 685–692.

Chapter 3

Probability

Probability forms a foundation for statistics. You might already be familiar with many aspects of probability, however, formalization of the concepts is new for most. This chapter aims to introduce probability on familiar terms using processes most people have seen before.

3.1   Defining probability

● **Example 3.1**   A "die", the singular of dice, is a cube with six faces numbered 1, 2, 3, 4, 5, and 6. What is the chance of getting 1 when rolling a die?

If the die is fair, then the chance of a 1 is as good as the chance of any other number. Since there are six outcomes, the chance must be 1-in-6 or, equivalently, 1/6.

● **Example 3.2**   What is the chance of getting a 1 or 2 in the next roll?

1 and 2 constitute two of the six equally likely possible outcomes, so the chance of getting one of these two outcomes must be $2/6 = 1/3$.

● **Example 3.3**   What is the chance of getting either 1, 2, 3, 4, 5, or 6 on the next roll?

100%. The outcome must be one of these numbers.

● **Example 3.4**   What is the chance of not rolling a 2?

Since the chance of rolling a 2 is 1/6 or $16.\overline{6}\%$, the chance of not rolling a 2 must be $100\% - 16.\overline{6}\% = 83.\overline{3}\%$ or 5/6.

Alternatively, we could have noticed that not rolling a 2 is the same as getting a 1, 3, 4, 5, or 6, which makes up five of the six equally likely outcomes and has probability 5/6.

● **Example 3.5**   Consider rolling two dice. If $1/6^{th}$ of the time the first die is a 1 and $1/6^{th}$ of those times the second die is a 1, what is the chance of getting two 1s?

If $16.\overline{6}\%$ of the time the first die is a 1 and $1/6^{th}$ of *those* times the second die is also a 1, then the chance that both dice are 1 is $(1/6) \times (1/6)$ or 1/36.

3.1.1 Probability

We use probability to build tools to describe and understand apparent randomness. We often frame probability in terms of a **random process** giving rise to an **outcome**.

$$\begin{array}{lcl}
\text{Roll a die} & \rightarrow & 1, 2, 3, 4, 5, \text{ or } 6 \\
\text{Flip a coin} & \rightarrow & \texttt{H} \text{ or } \texttt{T}
\end{array}$$

Rolling a die or flipping a coin is a seemingly random process and each gives rise to an outcome.

Probability

The **probability** of an outcome is the proportion of times the outcome would occur if we observed the random process an infinite number of times.

Probability is defined as a proportion, and it always takes values between 0 and 1 (inclusively). It may also be displayed as a percentage between 0% and 100%.

Probability can be illustrated by rolling a die many times. Consider the event "roll a 1". The **relative frequency** of an event is the proportion of times the event occurs out of the number of trials. Let $\hat{p}_n$ be the proportion of outcomes that are 1 after the first n rolls. As the number of rolls increases, $\hat{p}_n$ (the relative frequency of rolls) will converge to the probability of rolling a 1, $p = 1/6$. Figure 3.1 shows this convergence for 100,000 die rolls. The tendency of $\hat{p}_n$ to stabilize around p, that is, the tendency of the relative frequency to stabilize around the true probability, is described by the **Law of Large Numbers**.

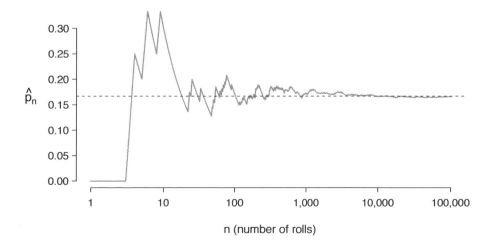

Figure 3.1: The fraction of die rolls that are 1 at each stage in a simulation. The relative frequency tends to get closer to the probability $1/6 \approx 0.167$ as the number of rolls increases.

Law of Large Numbers

As more observations are collected, the observed proportion $\hat{p}_n$ of occurrences with a particular outcome after n trials converges to the true probability p of that outcome.

Occasionally the proportion will veer off from the probability and appear to defy the Law of Large Numbers, as $\hat{p}_n$ does many times in Figure 3.1. However, these deviations become smaller as the number of rolls increases.

Above we write p as the probability of rolling a 1. We can also write this probability as

$P(A)$

Probability of outcome A

$$P(\text{rolling a 1})$$

As we become more comfortable with this notation, we will abbreviate it further. For instance, if it is clear that the process is "rolling a die", we could abbreviate $P(\text{rolling a 1})$ as $P(1)$.

⊙ **Guided Practice 3.6** Random processes include rolling a die and flipping a coin. (a) Think of another random process. (b) Describe all the possible outcomes of that process. For instance, rolling a die is a random process with potential outcomes 1, 2, ..., 6. [1]

What we think of as random processes are not necessarily random, but they may just be too difficult to understand exactly. The fourth example in the footnote solution to Guided Practice 3.6 suggests a roommate's behavior is a random process. However, even if a roommate's behavior is not truly random, modeling her behavior as a random process can still be useful.

TIP: Modeling a process as random
It can be helpful to model a process as random even if it is not truly random.

3.1.2 Disjoint or mutually exclusive outcomes

Two outcomes are called **disjoint** or **mutually exclusive** if they cannot both happen in the same trial. For instance, if we roll a die, the outcomes 1 and 2 are disjoint since they cannot both occur on a single roll. On the other hand, the outcomes 1 and "rolling an odd number" are not disjoint since both occur if the outcome of the roll is a 1. The terms *disjoint* and *mutually exclusive* are equivalent and interchangeable.

Calculating the probability of disjoint outcomes is easy. When rolling a die, the outcomes 1 and 2 are disjoint, and we compute the probability that one of these outcomes

---

[1]Here are four examples. (i) Whether someone gets sick in the next month or not is an apparently random process with outcomes sick and not. (ii) We can *generate* a random process by randomly picking a person and measuring that person's height. The outcome of this process will be a positive number. (iii) Whether the stock market goes up or down next week is a seemingly random process with possible outcomes up, down, and no_change. Alternatively, we could have used the percent change in the stock market as a numerical outcome. (iv) Whether your roommate cleans her dishes tonight probably seems like a random process with possible outcomes cleans_dishes and leaves_dishes.

will occur by adding their separate probabilities:

$$P(1 \text{ or } 2) = P(1) + P(2) = 1/6 + 1/6 = 1/3$$

What about the probability of rolling a 1, 2, 3, 4, 5, or 6? Here again, all of the outcomes are disjoint so we add the probabilities:

$$P(1 \text{ or } 2 \text{ or } 3 \text{ or } 4 \text{ or } 5 \text{ or } 6)$$
$$= P(1) + P(2) + P(3) + P(4) + P(5) + P(6)$$
$$= 1/6 + 1/6 + 1/6 + 1/6 + 1/6 + 1/6 = 1.$$

The **Addition Rule** guarantees the accuracy of this approach when the outcomes are disjoint.

Addition Rule of disjoint outcomes

If A_1 and A_2 represent two disjoint outcomes, then the probability that one of them occurs is given by

$$P(A_1 \text{ or } A_2) = P(A_1) + P(A_2)$$

If there are many disjoint outcomes A_1, ..., A_k, then the probability that one of these outcomes will occur is

$$P(A_1) + P(A_2) + \cdots + P(A_k) \tag{3.7}$$

⊙ **Guided Practice 3.8** We are interested in the probability of rolling a 1, 4, or 5. (a) Explain why the outcomes 1, 4, and 5 are disjoint. (b) Apply the Addition Rule for disjoint outcomes to determine $P(1 \text{ or } 4 \text{ or } 5)$.[2]

⊙ **Guided Practice 3.9** In the `email` data set in Chapter 2, the `number` variable described whether no number (labeled `none`), only one or more small numbers (`small`), or whether at least one big number appeared in an email (`big`). Of the 3,921 emails, 549 had no numbers, 2,827 had only one or more small numbers, and 545 had at least one big number. (a) Are the outcomes `none`, `small`, and `big` disjoint? (b) Determine the proportion of emails with value `small` and `big` separately. (c) Use the Addition Rule for disjoint outcomes to compute the probability a randomly selected email from the data set has a number in it, small or big.[3]

Statisticians rarely work with individual outcomes and instead consider *sets* or *collections* of outcomes. Let A represent the event where a die roll results in 1 or 2 and B represent the event that the die roll is a 4 or a 6. We write A as the set of outcomes $\{1, 2\}$ and $B = \{4, 6\}$. These sets are commonly called **events**. Because A and B have no elements in common, they are disjoint events. A and B are represented in Figure 3.2.

---

[2](a) The random process is a die roll, and at most one of these outcomes can come up. This means they are disjoint outcomes. (b) $P(1 \text{ or } 4 \text{ or } 5) = P(1) + P(4) + P(5) = \frac{1}{6} + \frac{1}{6} + \frac{1}{6} = \frac{3}{6} = \frac{1}{2}$.

[3](a) Yes. Each email is categorized in only one level of `number`. (b) Small: $\frac{2827}{3921} = 0.721$. Big: $\frac{545}{3921} = 0.139$. (c) $P(\text{small or big}) = P(\text{small}) + P(\text{big}) = 0.721 + 0.139 = 0.860$.

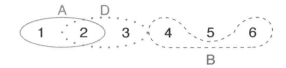

Figure 3.2: Three events, A, B, and D, consist of outcomes from rolling a die. A and B are disjoint since they do not have any outcomes in common.

2♣	3♣	4♣	5♣	6♣	7♣	8♣	9♣	10♣	J♣	Q♣	K♣	A♣
2♢	3♢	4♢	5♢	6♢	7♢	8♢	9♢	10♢	J♢	Q♢	K♢	A♢
2♡	3♡	4♡	5♡	6♡	7♡	8♡	9♡	10♡	J♡	Q♡	K♡	A♡
2♠	3♠	4♠	5♠	6♠	7♠	8♠	9♠	10♠	J♠	Q♠	K♠	A♠

Table 3.3: Representations of the 52 unique cards in a deck.

The Addition Rule applies to both disjoint outcomes and disjoint events. The probability that one of the disjoint events A or B occurs is the sum of the separate probabilities:

$$P(A \text{ or } B) = P(A) + P(B) = 1/3 + 1/3 = 2/3$$

⊙ **Guided Practice 3.10**   (a) Verify the probability of event A, $P(A)$, is $1/3$ using the Addition Rule. (b) Do the same for event B.[4]

⊙ **Guided Practice 3.11**   (a) Using Figure 3.2 as a reference, what outcomes are represented by event D? (b) Are events B and D disjoint? (c) Are events A and D disjoint?[5]

⊙ **Guided Practice 3.12**   In Guided Practice 3.11, you confirmed B and D from Figure 3.2 are disjoint. Compute the probability that either event B or event D occurs.[6]

3.1.3   Probabilities when events are not disjoint

Let's consider calculations for two events that are not disjoint in the context of a regular deck of 52 cards, represented in Table 3.3. If you are unfamiliar with the cards in a regular deck, please see the footnote.[7]

⊙ **Guided Practice 3.13**   (a) What is the probability that a randomly selected card is a diamond? (b) What is the probability that a randomly selected card is a face card?[8]

---

[4](a) $P(A) = P(1 \text{ or } 2) = P(1) + P(2) = \frac{1}{6} + \frac{1}{6} = \frac{2}{6} = \frac{1}{3}$. (b) Similarly, $P(B) = 1/3$.

[5](a) Outcomes 2 and 3. (b) Yes, events B and D are disjoint because they share no outcomes. (c) The events A and D share an outcome in common, 2, and so are not disjoint.

[6]Since B and D are disjoint events, use the Addition Rule: $P(B \text{ or } D) = P(B) + P(D) = \frac{1}{3} + \frac{1}{3} = \frac{2}{3}$.

[7]The 52 cards are split into four **suits**: ♣ (club), ♢ (diamond), ♡ (heart), ♠ (spade). Each suit has its 13 cards labeled: 2, 3, ..., 10, J (jack), Q (queen), K (king), and A (ace). Thus, each card is a unique combination of a suit and a label, e.g. 4♡ and J♣. The 12 cards represented by the jacks, queens, and kings are called **face cards**. The cards that are ♢ or ♡ are typically colored red while the other two suits are typically colored black.

[8](a) There are 52 cards and 13 diamonds. If the cards are thoroughly shuffled, each card has an equal chance of being drawn, so the probability that a randomly selected card is a diamond is $P(♢) = \frac{13}{52} = 0.250$. (b) Likewise, there are 12 face cards, so $P(\text{face card}) = \frac{12}{52} = \frac{3}{13} = 0.231$.

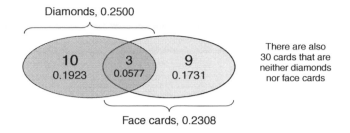

Figure 3.4: A Venn diagram for diamonds and face cards.

Venn diagrams are useful when outcomes can be categorized as "in" or "out" for two or three variables, attributes, or random processes. The Venn diagram in Figure 3.4 uses a circle to represent diamonds and another to represent face cards. If a card is both a diamond and a face card, it falls into the intersection of the circles. If it is a diamond but not a face card, it will be in part of the left circle that is not in the right circle (and so on). The total number of cards that are diamonds is given by the total number of cards in the diamonds circle: $10 + 3 = 13$. The probabilities are also shown (e.g. $10/52 = 0.1923$).

⊙ **Guided Practice 3.14**   Using the Venn diagram, verify $P(\text{face card}) = 12/52 = 3/13$.[9]

Let A represent the event that a randomly selected card is a diamond and B represent the event that it is a face card. How do we compute $P(A \text{ or } B)$? Events A and B are not disjoint – the cards $J\Diamond$, $Q\Diamond$, and $K\Diamond$ fall into both categories – so we cannot use the Addition Rule for disjoint events. Instead we use the Venn diagram. We start by adding the probabilities of the two events:

$$P(A) + P(B) = P(\Diamond) + P(\text{face card}) = 13/52 + 12/52$$

However, the three cards that are in both events were counted twice, once in each probability. We must correct this double counting:

$$
\begin{aligned}
P(A \text{ or } B) &= P(\Diamond) + P(\text{face card}) \\
&= P(\Diamond) + P(\text{face card}) - P(\Diamond \text{ and face card}) \quad\quad (3.15) \\
&= 13/52 + 12/52 - 3/52 \\
&= 22/52 = 11/26
\end{aligned}
$$

Equation (3.15) is an example of the **General Addition Rule**.

---

[9]The Venn diagram shows face cards split up into "face card but not $\Diamond$" and "face card and $\Diamond$". Since these correspond to disjoint events, $P(\text{face card})$ is found by adding the two corresponding probabilities: $\frac{3}{52} + \frac{9}{52} = \frac{12}{52} = \frac{3}{13}$.

General Addition Rule

If A and B are any two events, disjoint or not, then the probability that A or B will occur is

$$P(A \text{ or } B) = P(A) + P(B) - P(A \text{ and } B) \qquad (3.16)$$

where $P(A \text{ and } B)$ is the probability that both events occur.

TIP: Symbolic notation for "and" and "or"
The symbol $\cap$ means intersection and is equivalent to "and".
The symbol $\cup$ means union and is equivalent to "or".
It is common to see the General Addition Rule written as

$$P(A \cup B) = P(A) + P(B) - P(A \cap B) \qquad (3.17)$$

TIP: "or" is inclusive
When we write, "or" in statistics, we mean "and/or" unless we explicitly state otherwise. Thus, A or B occurs means A, B, or both A and B occur. This is equivalent to at least one of A or B occurring.

⊙ **Guided Practice 3.18** (a) If A and B are disjoint, describe why this implies $P(A$ and $B) = 0$. (b) Using part (a), verify that the General Addition Rule simplifies to the simpler Addition Rule for disjoint events if A and B are disjoint.[10]

⊙ **Guided Practice 3.19** In the `email` data set with 3,921 emails, 367 were spam, 2,827 contained some small numbers but no big numbers, and 168 had both characteristics. Create a Venn diagram for this setup.[11]

⊙ **Guided Practice 3.20** (a) Use your Venn diagram from Guided Practice 3.19 to determine the probability a randomly drawn email from the `email` data set is spam and had small numbers (but not big numbers). (b) What is the probability that the email had either of these attributes?[12]

---

[10](a) If A and B are disjoint, A and B can never occur simultaneously. (b) If A and B are disjoint, then the last term of Equation (3.16) is 0 (see part (a)) and we are left with the Addition Rule for disjoint events.

[11]Both the counts and corresponding probabilities (e.g. 2659/3921 = 0.678) are shown. Notice that the number of emails represented in the left circle corresponds to 2659 + 168 = 2827, and the number represented in the right circle is 168 + 199 = 367.

small numbers and no big numbers		spam
2659	168	199
0.678	0.043	0.051

Other emails: 3921–2659–168–199 = 895 (0.228)

[12](a) The solution is represented by the intersection of the two circles: 0.043. (b) This is the sum of the three disjoint probabilities shown in the circles: $0.678 + 0.043 + 0.051 = 0.772$.

3.1.4 Complement of an event

Rolling a die produces a value in the set $\{1, 2, 3, 4, 5, 6\}$. This set of all possible outcomes is called the **sample space** (S) for rolling a die. We often use the sample space to examine the scenario where an event does not occur.

S
Sample space

Let $D = \{2, 3\}$ represent the event that the outcome of a die roll is 2 or 3. Then the **complement** of D represents all outcomes in our sample space that are not in D, which is denoted by $D^c = \{1, 4, 5, 6\}$. That is, D^c is the set of all possible outcomes not already included in D. Figure 3.5 shows the relationship between D, D^c, and the sample space S.

A^c
Complement
of outcome A

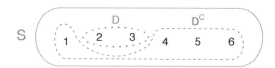

Figure 3.5: Event $D = \{2, 3\}$ and its complement, $D^c = \{1, 4, 5, 6\}$. S represents the sample space, which is the set of all possible events.

⊙ **Guided Practice 3.21** (a) Compute $P(D^c) = P(\text{rolling a 1, 4, 5, or 6})$. (b) What is $P(D) + P(D^c)$?[13]

⊙ **Guided Practice 3.22** Events $A = \{1, 2\}$ and $B = \{4, 6\}$ are shown in Figure 3.2 on page 104. (a) Write out what A^c and B^c represent. (b) Compute $P(A^c)$ and $P(B^c)$. (c) Compute $P(A) + P(A^c)$ and $P(B) + P(B^c)$.[14]

An event A together with its complement A^c comprise the entire sample space. Because of this we can say that $P(A) + P(A^c) = 1$.

Complement

The complement of event A is denoted A^c, and A^c represents all outcomes not in A. A and A^c are mathematically related:

$$P(A) + P(A^c) = 1, \quad \text{i.e.} \quad P(A) = 1 - P(A^c) \qquad (3.23)$$

In simple examples, computing A or A^c is feasible in a few steps. However, using the complement can save a lot of time as problems grow in complexity.

⊙ **Guided Practice 3.24** A die is rolled 10 times. (a) What is the complement of getting at least one 6 in 10 rolls of the die? (b) What is the complement of getting at most three 6's in 10 rolls of the die?[15]

---

[13](a) The outcomes are disjoint and each has probability 1/6, so the total probability is $4/6 = 2/3$. (b) We can also see that $P(D) = \frac{1}{6} + \frac{1}{6} = 1/3$. Since D and D^c are disjoint, $P(D) + P(D^c) = 1$.

[14]Brief solutions: (a) $A^c = \{3, 4, 5, 6\}$ and $B^c = \{1, 2, 3, 5\}$. (b) Noting that each outcome is disjoint, add the individual outcome probabilities to get $P(A^c) = 2/3$ and $P(B^c) = 2/3$. (c) A and A^c are disjoint, and the same is true of B and B^c. Therefore, $P(A) + P(A^c) = 1$ and $P(B) + P(B^c) = 1$.

[15](a) The complement of getting at least one 6 in ten rolls of a die is getting zero 6's in the 10 rolls. (b) The complement of getting at most three 6's in 10 rolls is getting four, five, ..., nine, or ten 6's in 10 rolls.

3.1.5   Independence

Just as variables and observations can be independent, random processes can be independent, too. Two processes are **independent** if knowing the outcome of one provides no useful information about the outcome of the other. For instance, flipping a coin and rolling a die are two independent processes – knowing the coin was heads does not help determine the outcome of a die roll. On the other hand, stock prices usually move up or down together, so they are not independent.

Example 3.5 provides a basic example of two independent processes: rolling two dice. We want to determine the probability that both will be 1. Suppose one of the dice is red and the other white. If the outcome of the red die is a 1, it provides no information about the outcome of the white die. We first encountered this same question in Example 3.5 (page 100), where we calculated the probability using the following reasoning: $1/6^{th}$ of the time the red die is a 1, and $1/6^{th}$ of *those* times the white die will also be 1. This is illustrated in Figure 3.6. Because the rolls are independent, the probabilities of the corresponding outcomes can be multiplied to get the final answer: $(1/6) \times (1/6) = 1/36$. This can be generalized to many independent processes.

All rolls

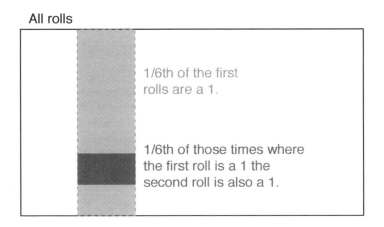

Figure 3.6: $1/6^{th}$ of the time, the first roll is a 1. Then $1/6^{th}$ of *those* times, the second roll will also be a 1.

● **Example 3.25** What if there was also a blue die independent of the other two? What is the probability of rolling the three dice and getting all 1s?

───────────

The same logic applies from Example 3.5. If $1/36^{th}$ of the time the white and red dice are both 1, then $1/6^{th}$ of *those* times the blue die will also be 1, so multiply:

$$P(white = 1 \text{ and } red = 1 \text{ and } blue = 1) = P(white = 1) \times P(red = 1) \times P(blue = 1)$$
$$= (1/6) \times (1/6) \times (1/6) = 1/216$$

Examples 3.5 and 3.25 illustrate what is called the Multiplication Rule for independent processes.

Multiplication Rule for independent processes

If A and B represent events from two different and independent processes, then the probability that both A and B occur can be calculated as the product of their separate probabilities:

$$P(A \text{ and } B) = P(A) \times P(B) \qquad (3.26)$$

Similarly, if there are k events A_1, ..., A_k from k independent processes, then the probability they all occur is

$$P(A_1) \times P(A_2) \times \cdots \times P(A_k)$$

⊙ **Guided Practice 3.27** About 9% of people are left-handed. Suppose 2 people are selected at random from the U.S. population. Because the sample size of 2 is very small relative to the population, it is reasonable to assume these two people are independent. (a) What is the probability that both are left-handed? (b) What is the probability that both are right-handed?[16]

⊙ **Guided Practice 3.28** Suppose 5 people are selected at random.[17]

  (a) What is the probability that all are right-handed?

  (b) What is the probability that all are left-handed?

  (c) What is the probability that not all of the people are right-handed?

Suppose the variables **handedness** and **gender** are independent, i.e. knowing someone's **gender** provides no useful information about their **handedness** and vice-versa. Then we can compute whether a randomly selected person is right-handed and female[18] using the Multiplication Rule:

$$
\begin{aligned}
P(\text{right-handed and female}) &= P(\text{right-handed}) \times P(\text{female}) \\
&= 0.91 \times 0.50 = 0.455
\end{aligned}
$$

---

[16](a) The probability the first person is left-handed is 0.09, which is the same for the second person. We apply the Multiplication Rule for independent processes to determine the probability that both will be left-handed: $0.09 \times 0.09 = 0.0081$.

(b) It is reasonable to assume the proportion of people who are ambidextrous (both right and left handed) is nearly 0, which results in $P(\text{right-handed}) = 1 - 0.09 = 0.91$. Using the same reasoning as in part (a), the probability that both will be right-handed is $0.91 \times 0.91 = 0.8281$.

[17](a) The abbreviations RH and LH are used for right-handed and left-handed, respectively. Since each are independent, we apply the Multiplication Rule for independent processes:

$$P(\text{all five are RH}) = P(\text{first} = \text{RH}, \text{second} = \text{RH}, ..., \text{fifth} = \text{RH})$$
$$= P(\text{first} = \text{RH}) \times P(\text{second} = \text{RH}) \times \cdots \times P(\text{fifth} = \text{RH})$$
$$= 0.91 \times 0.91 \times 0.91 \times 0.91 \times 0.91 = 0.624$$

(b) Using the same reasoning as in (a), $0.09 \times 0.09 \times 0.09 \times 0.09 \times 0.09 = 0.0000059$

(c) Use the complement, $P(\text{all five are RH})$, to answer this question:

$$P(\text{not all RH}) = 1 - P(\text{all RH}) = 1 - 0.624 = 0.376$$

[18]The actual proportion of the U.S. population that is **female** is about 50%, and so we use 0.5 for the probability of sampling a woman. However, this probability does differ in other countries.

⊙ **Guided Practice 3.29**   Three people are selected at random.[19]

(a) What is the probability that the first person is male and right-handed?

(b) What is the probability that the first two people are male and right-handed?.

(c) What is the probability that the third person is female and left-handed?

(d) What is the probability that the first two people are male and right-handed and the third person is female and left-handed?

Sometimes we wonder if one outcome provides useful information about another outcome. The question we are asking is, are the occurrences of the two events independent? We say that two events A and B are independent if they satisfy Equation (3.26).

● **Example 3.30**   If we shuffle up a deck of cards and draw one, is the event that the card is a heart independent of the event that the card is an ace?

The probability the card is a heart is 1/4 and the probability that it is an ace is 1/13. The probability the card is the ace of hearts is 1/52. We check whether Equation 3.26 is satisfied:

$$P(\heartsuit) \times P(\text{ace}) = \frac{1}{4} \times \frac{1}{13} = \frac{1}{52} = P(\heartsuit \text{ and ace})$$

Because the equation holds, the event that the card is a heart and the event that the card is an ace are independent events.

3.2   Conditional probability

The `family_college` data set contains a sample of 792 cases with two variables, `teen` and `parents`, and is summarized in Table 3.7.[20] The `teen` variable is either `college` or `not`, where the teenager is labeled as `college` if she went to college immediately after high school. The `parents` variable takes the value `degree` if at least one parent of the teenager competed a college degree.

		parents		
		degree	not	Total
teen	college	231	214	445
	not	49	298	347
	Total	280	512	792

Table 3.7: Contingency table summarizing the `family_college` data set.

● **Example 3.31**   If at least one parent completed a college degree, what is the chance their teenager attended college right after high school?

We can estimate this probability using the data. Of the 280 cases in this data set where `parents` takes value `degree`, 231 represent cases where the `teen` variable takes value `college`:

$$P(\text{teen college given parents degree}) = \frac{231}{280} = 0.825$$

---

[19]Brief answers are provided. (a) This can be written in probability notation as P(a randomly selected person is male and right-handed) = 0.455. (b) 0.207. (c) 0.045. (d) 0.0093.

[20]A simulated data set based on real population summaries at nces.ed.gov/pubs2001/2001126.pdf.

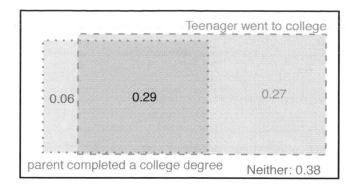

Figure 3.8: A Venn diagram using boxes for the `family_college` data set.

● **Example 3.32** A teenager is randomly selected from the sample and she did not attend college right after high school. What is the probability that at least one of her parents has a college degree?

If the teenager did not attend, then she is one of the 347 teens in the second row. Of these 347 teens, 49 had at least one parent who got a college degree:

$$P(\texttt{parents degree given teen not}) = \frac{49}{347} = 0.141$$

3.2.1 Marginal and joint probabilities

Table 3.7 includes row and column totals for each variable separately in the `family_college` data set. These totals represent **marginal probabilities** for the sample, which are the probabilities based on a single variable without conditioning on any other variables. For instance, a probability based solely on the `teen` variable is a marginal probability:

$$P(\texttt{teen college}) = \frac{445}{792} = 0.56$$

A probability of outcomes for two or more variables or processes is called a **joint probability**:

$$P(\texttt{teen college and parents not}) = \frac{214}{792} = 0.27$$

It is common to substitute a comma for "and" in a joint probability, although either is acceptable. That is,

$$P(\texttt{teen college, parents not})$$

means the same thing as

$$P(\texttt{teen college and parents not})$$

Marginal and joint probabilities

If a probability is based on a single variable, it is a *marginal probability*. The probability of outcomes for two or more variables or processes is called a *joint probability*.

We use **table proportions** to summarize joint probabilities for the `family_college` sample. These proportions are computed by dividing each count in Table 3.7 by the table's total, 792, to obtain the proportions in Table 3.9. The joint probability distribution of the `parents` and `teen` variables is shown in Table 3.10.

	parents: degree	parents: not	Total
teen: college	0.29	0.27	0.56
teen: not	0.06	0.38	0.44
Total	0.35	0.65	1.00

Table 3.9: Probability table summarizing whether at least one parent had a college degree and the teenager attended college.

Joint outcome	Probability
parents degree and teen college	0.29
parents degree and teen not	0.06
parents not and teen college	0.27
parents not and teen not	0.38
Total	1.00

Table 3.10: Joint probability distribution for the `family_college` data set.

⊙ **Guided Practice 3.33** Verify Table 3.10 represents a probability distribution: events are disjoint, all probabilities are non-negative, and the probabilities sum to 1.[21]

We can compute marginal probabilities using joint probabilities in simple cases. For example, the probability a random teenager from the study went to college is found by summing the outcomes from Table 3.10 where `teen` takes value `college`:

$$P(\text{teen college}) = P(\text{parents degree and } \underline{\text{teen college}})$$
$$+ P(\text{parents not and } \underline{\text{teen college}})$$
$$= 0.28 + 0.27$$
$$= 0.56$$

3.2.2 Defining conditional probability

There is some connection between education level of parents and of the teenager: a college degree by a parent is associated with college attendance of the teenager. In this section, we discuss how to use information about associations between two variables to improve probability estimation.

The probability that a random teenager from the study attended college is 0.56. Could we update this probability if we knew that one of the teen's parents has a college degree? Absolutely. To do so, we limit our view to only those 280 cases where a parent has a college degree and look at the fraction where the teenager attended college:

$$P(\text{teen college given parents degree}) = \frac{231}{280} = 0.825$$

---

[21]Each of the four outcome combination are disjoint, all probabilities are indeed non-negative, and the sum of the probabilities is $0.29 + 0.06 + 0.27 + 0.38 = 1.00$.

We call this a **conditional probability** because we computed the probability under a condition: a parent has a college degree. There are two parts to a conditional probability, **the outcome of interest** and the **condition**. It is useful to think of the condition as information we know to be true, and this information usually can be described as a known outcome or event.

We separate the text inside our probability notation into the outcome of interest and the condition:

$$P(\texttt{teen college given parents degree})$$
$$= P(\texttt{teen college} \mid \texttt{parents degree}) = \frac{231}{280} = 0.825 \qquad (3.34)$$

$P(A|B)$

Probability of outcome A given B

The vertical bar "|" is read as *given.*

In Equation (3.34), we computed the probability a teen attended college based on the condition that at least one parent has a college degree as a fraction:

$$P(\texttt{teen college} \mid \texttt{parents degree})$$
$$= \frac{\#\text{ cases where }\texttt{teen college}\text{ and }\texttt{parents degree}}{\#\text{ cases where }\texttt{parents degree}} \qquad (3.35)$$
$$= \frac{231}{280} = 0.825$$

We considered only those cases that met the condition, **parents degree**, and then we computed the ratio of those cases that satisfied our outcome of interest, the teenager attended college.

Frequently, marginal and joint probabilities are provided instead of count data. For example, disease rates are commonly listed in percentages rather than in a count format. We would like to be able to compute conditional probabilities even when no counts are available, and we use Equation (3.35) as an example demonstrating this technique.

We considered only those cases that satisfied the condition, **parents degree**. Of these cases, the conditional probability was the fraction who represented the outcome of interest, **teen college**. Suppose we were provided only the information in Table 3.9, i.e. only probability data. Then if we took a sample of 1000 people, we would anticipate about 35% or $0.35 \times 1000 = 350$ would meet the information criterion (**parents degree**). Similarly, we would expect about 29% or $0.29 \times 1000 = 290$ to meet both the information criteria and represent our outcome of interest. Thus, the conditional probability could be computed:

$$P(\texttt{teen college} \mid \texttt{parents degree})$$
$$= \frac{\#\ (\texttt{teen college}\text{ and }\texttt{parents degree})}{\#\ (\texttt{parents degree})}$$
$$= \frac{290}{350} = \frac{0.29}{0.35} = 0.829 \quad \text{(different from 0.825 due to rounding error)} \qquad (3.36)$$

In Equation (3.36), we examine exactly the fraction of two probabilities, 0.29 and 0.35, which we can write as

$$P(\texttt{teen college}\text{ and }\texttt{parents degree}) \quad \text{and} \quad P(\texttt{parents degree}).$$

The fraction of these probabilities is an example of the general formula for conditional probability.

Conditional probability

The conditional probability of the outcome of interest A given condition B is computed as the following:

$$P(A|B) = \frac{P(A \text{ and } B)}{P(B)} \tag{3.37}$$

⊙ **Guided Practice 3.38** (a) Write out the following statement in conditional probability notation: *"The probability a random case where neither parent has a college degree if it is known that the teenager didn't attend college right after high school"*. Notice that the condition is now based on the teenager, not the parent.

(b) Determine the probability from part (a). Table 3.9 on page 112 may be helpful.[22]

⊙ **Guided Practice 3.39** (a) Determine the probability that one of the parents has a college degree if it is known the teenager did not attend college.

(b) Using the answers from part (a) and Guided Practice 3.38(b), compute

$$P(\texttt{parents degree} \mid \texttt{teen not})$$
$$+ P(\texttt{parents not} \mid \texttt{teen not})$$

(c) Provide an intuitive argument to explain why the sum in (b) is 1.[23]

⊙ **Guided Practice 3.40** The data indicate there is an association between parents having a college degree and their teenager attending college. Does this mean the parents' college degree(s) *caused* the teenager to go to college?[24]

3.2.3 Smallpox in Boston, 1721

The `smallpox` data set provides a sample of 6,224 individuals from the year 1721 who were exposed to smallpox in Boston.[25] Doctors at the time believed that inoculation, which involves exposing a person to the disease in a controlled form, could reduce the likelihood of death.

Each case represents one person with two variables: `inoculated` and `result`. The variable `inoculated` takes two levels: `yes` or `no`, indicating whether the person was inoculated or not. The variable `result` has outcomes `lived` or `died`. These data are summarized in Tables 3.11 and 3.12.

---

[22](a) $P(\texttt{parents not} \mid \texttt{teen not})$. (b) Equation (3.37) for conditional probability indicates we should first find $P(\texttt{parents not and teen not}) = 0.38$ and $P(\texttt{teen not}) = 0.44$. Then the ratio represents the conditional probability: $0.38/0.44 = 0.864$.

[23](a) This probability is $\frac{P(\texttt{parents degree, teen not})}{P(\texttt{teen not})} = \frac{0.06}{0.44} = 0.136$. (b) The total equals 1. (c) Under the condition the teenager didn't attend college, the parents must either have a college degree or not. The complement still works, provided the probabilities condition on the same information.

[24]No. While there is an association, the data are observational. Two potential confounding variables include `income` and `region`. Can you think of others?

[25]Fenner F. 1988. *Smallpox and Its Eradication (History of International Public Health, No. 6).* Geneva: World Health Organization. ISBN 92-4-156110-6.

		inoculated		
		yes	no	Total
result	lived	238	5136	5374
	died	6	844	850
	Total	244	5980	6224

Table 3.11: Contingency table for the `smallpox` data set.

		inoculated		
		yes	no	Total
result	lived	0.0382	0.8252	0.8634
	died	0.0010	0.1356	0.1366
	Total	0.0392	0.9608	1.0000

Table 3.12: Table proportions for the `smallpox` data, computed by dividing each count by the table total, 6224.

⊙ **Guided Practice 3.41** Write out, in formal notation, the probability a randomly selected person who was not inoculated died from smallpox, and find this probability.[26]

⊙ **Guided Practice 3.42** Determine the probability that an inoculated person died from smallpox. How does this result compare with the result of Guided Practice 3.41?[27]

⊙ **Guided Practice 3.43** The people of Boston self-selected whether or not to be inoculated. (a) Is this study observational or was this an experiment? (b) Can we infer any causal connection using these data? (c) What are some potential confounding variables that might influence whether someone lived or died and also affect whether that person was inoculated?[28]

3.2.4 General multiplication rule

Section 3.1.5 introduced the Multiplication Rule for independent processes. Here we provide the **General Multiplication Rule** for events that might not be independent.

General Multiplication Rule

If A and B represent two outcomes or events, then

$$P(A \text{ and } B) = P(A|B) \times P(B)$$

For the term $P(A|B)$, it is useful to think of A as the outcome of interest and B as the condition.

---

[26]$P(\texttt{result = died | not inoculated}) = \frac{P(\texttt{result = died and not inoculated})}{P(\texttt{not inoculated})} = \frac{0.1356}{0.9608} = 0.1411.$

[27]$P(\texttt{died | inoculated}) = \frac{P(\texttt{died and inoculated})}{P(\texttt{inoculated})} = \frac{0.0010}{0.0392} = 0.0255.$ The death rate for individuals who were inoculated is only about 1 in 40 while the death rate is about 1 in 7 for those who were not inoculated.

[28]Brief answers: (a) Observational. (b) No, we cannot infer causation from this observational study. (c) Accessibility to the latest and best medical care, so income may play a role. There are other valid answers for part (c).

This General Multiplication Rule is simply a rearrangement of the definition for conditional probability in Equation (3.37) on page 114.

● **Example 3.44** Consider the `smallpox` data set. Suppose we are given only two pieces of information: 96.08% of residents were not inoculated, and 85.88% of the residents who were not inoculated ended up surviving. How could we compute the probability that a resident was not inoculated and lived?

We will compute our answer using the General Multiplication Rule and then verify it using Table 3.12. We want to determine

$$P(\text{lived and not inoculated})$$

and we are given that

$$P(\text{lived} \mid \text{not inoculated}) = 0.8588$$
$$P(\text{not inoculated}) = 0.9608$$

Among the 96.08% of people who were not inoculated, 85.88% survived:

$$P(\text{lived and not inoculated}) = 0.8588 \times 0.9608 = 0.8251$$

This is equivalent to the General Multiplication Rule. We can confirm this probability in Table 3.12 at the intersection of `no` and `lived` (with a small rounding error).

⊙ **Guided Practice 3.45** Use $P(\text{inoculated}) = 0.0392$ and $P(\text{lived} \mid \text{inoculated}) = 0.9754$ to determine the probability that a person was both inoculated and lived.[29]

⊙ **Guided Practice 3.46** If 97.45% of the people who were inoculated lived, what proportion of inoculated people must have died?[30]

⊙ **Guided Practice 3.47** Based on the probabilities computed above, does it appear that inoculation is effective at reducing the risk of death from smallpox?[31]

---

[29]The answer is 0.0382, which can be verified using Table 3.12.

[30]There were only two possible outcomes: `lived` or `died`. This means that 100% - 97.45% = 2.55% of the people who were inoculated died.

[31]The samples are large relative to the difference in death rates for the "inoculated" and "not inoculated" groups, so it seems there is an association between `inoculated` and `outcome`. However, as noted in the solution to Guided Practice 3.43, this is an observational study and we cannot be sure if there is a causal connection. (Further research has shown that inoculation is effective at reducing death rates.)

3.2.5 Sampling from a small population

● **Example 3.48** Professors sometimes select a student at random to answer a question. If each student has an equal chance of being selected and there are 15 people in your class, what is the chance that she will pick you for the next question?

If there are 15 people to ask and none are skipping class, then the probability is 1/15, or about 0.067.

● **Example 3.49** If the professor asks 3 questions, what is the probability that you will not be selected? Assume that she will not pick the same person twice in a given lecture.

For the first question, she will pick someone else with probability 14/15. When she asks the second question, she only has 14 people who have not yet been asked. Thus, if you were not picked on the first question, the probability you are again not picked is 13/14. Similarly, the probability you are again not picked on the third question is 12/13, and the probability of not being picked for any of the three questions is

$$P(\text{not picked in 3 questions})$$
$$= P(\texttt{Q1 = not\_picked, Q2 = not\_picked, Q3 = not\_picked.})$$
$$= \frac{14}{15} \times \frac{13}{14} \times \frac{12}{13} = \frac{12}{15} = 0.80$$

⊙ **Guided Practice 3.50** What rule permitted us to multiply the probabilities in Example 3.49?[32]

● **Example 3.51** Suppose the professor randomly picks without regard to who she already selected, i.e. students can be picked more than once. What is the probability that you will not be picked for any of the three questions?

Each pick is independent, and the probability of not being picked for any individual question is 14/15. Thus, we can use the Multiplication Rule for independent processes.

$$P(\text{not picked in 3 questions})$$
$$= P(\texttt{Q1 = not\_picked, Q2 = not\_picked, Q3 = not\_picked.})$$
$$= \frac{14}{15} \times \frac{14}{15} \times \frac{14}{15} = 0.813$$

You have a slightly higher chance of not being picked compared to when she picked a new person for each question. However, you now may be picked more than once.

⊙ **Guided Practice 3.52** Under the setup of Example 3.51, what is the probability of being picked to answer all three questions?[33]

---

[32]The three probabilities we computed were actually one marginal probability, $P(\texttt{Q1=not\_picked})$, and two conditional probabilities:

$P(\texttt{Q2 = not\_picked | Q1 = not\_picked})$ $P(\texttt{Q3 = not\_picked | Q1 = not\_picked, Q2 = not\_picked})$

Using the General Multiplication Rule, the product of these three probabilities is the probability of not being picked in 3 questions.

[33]$P(\text{being picked to answer all three questions}) = \left(\frac{1}{15}\right)^3 = 0.00030.$

If we sample from a small population **without replacement**, we no longer have independence between our observations. In Example 3.49, the probability of not being picked for the second question was conditioned on the event that you were not picked for the first question. In Example 3.51, the professor sampled her students **with replacement**: she repeatedly sampled the entire class without regard to who she already picked.

⊙ **Guided Practice 3.53**   Your department is holding a raffle. They sell 30 tickets and offer seven prizes. (a) They place the tickets in a hat and draw one for each prize. The tickets are sampled without replacement, i.e. the selected tickets are not placed back in the hat. What is the probability of winning a prize if you buy one ticket? (b) What if the tickets are sampled with replacement?[34]

⊙ **Guided Practice 3.54**   Compare your answers in Guided Practice 3.53. How much influence does the sampling method have on your chances of winning a prize?[35]

Had we repeated Guided Practice 3.53 with 300 tickets instead of 30, we would have found something interesting: the results would be nearly identical. The probability would be 0.0233 without replacement and 0.0231 with replacement. When the sample size is only a small fraction of the population (under 10%), observations are nearly independent even when sampling without replacement.

3.2.6   Independence considerations in conditional probability

If two processes are independent, then knowing the outcome of one should provide no information about the other. We can show this is mathematically true using conditional probabilities.

⊙ **Guided Practice 3.55**   Let X and Y represent the outcomes of rolling two dice. (a) What is the probability that the first die, X, is 1? (b) What is the probability that both X and Y are 1? (c) Use the formula for conditional probability to compute $P(Y = 1 \,|X = 1)$. (d) What is $P(Y = 1)$? Is this different from the answer from part (c)? Explain.[36]

---

[34](a) First determine the probability of not winning. The tickets are sampled without replacement, which means the probability you do not win on the first draw is 29/30, 28/29 for the second, ..., and 23/24 for the seventh. The probability you win no prize is the product of these separate probabilities: 23/30. That is, the probability of winning a prize is $1 - 23/30 = 7/30 = 0.233$. (b) When the tickets are sampled with replacement, there are seven independent draws. Again we first find the probability of not winning a prize: $(29/30)^7 = 0.789$. Thus, the probability of winning (at least) one prize when drawing with replacement is 0.211.

[35]There is about a 10% larger chance of winning a prize when using sampling without replacement. However, at most one prize may be won under this sampling procedure.

[36]Brief solutions: (a) 1/6. (b) 1/36. (c) $\frac{P(Y= 1 \text{ and } X= 1)}{P(X= 1)} = \frac{1/36}{1/6} = 1/6$. (d) The probability is the same as in part (c): $P(Y = 1) = 1/6$. The probability that $Y = 1$ was unchanged by knowledge about X, which makes sense as X and Y are independent.

We can show in Guided Practice 3.55(c) that the conditioning information has no influence by using the Multiplication Rule for independence processes:

$$
\begin{aligned}
P(Y = 1 | X = 1) &= \frac{P(Y = 1 \text{ and} X = 1)}{P(X = 1)} \\
&= \frac{P(Y = 1) \times P(X = 1)}{P(X = 1)} \\
&= P(Y = 1)
\end{aligned}
$$

⊙ **Guided Practice 3.56** Ron is watching a roulette table in a casino and notices that the last five outcomes were `black`. He figures that the chances of getting `black` six times in a row is very small (about 1/64) and puts his paycheck on red. What is wrong with his reasoning?[37]

3.2.7 Checking for independent and mutually exclusive events

If A and B are independent events, then the probability of A being true is unchanged if B is true. Mathematically, this is written as

$$P(A|B) = P(A)$$

The General Multiplication Rule states that $P(A \text{ and } B)$ equals $P(A|B) \times P(B)$. If A and B are independent events, we can replace $P(A|B)$ with $P(A)$ and the following multiplication rule applies:

$$P(A \text{ and } B) = P(A) \times P(B)$$

TIP: Checking whether two events are independent
When checking whether two events A and B are independent, verify one of the following equations holds (there is no need to check both equations):

$$P(A|B) = P(A) \qquad\qquad P(A \text{ and } B) = P(A) \times P(B)$$

If the equation that is checked holds true (the left and right sides are equal), A and B are independent. If the equation does not hold, then A and B are dependent.

---

[37]He has forgotten that the next roulette spin is independent of the previous spins. Casinos do employ this practice; they post the last several outcomes of many betting games to trick unsuspecting gamblers into believing the odds are in their favor. This is called the **gambler's fallacy**.

● **Example 3.57** Are teenager college attendance and parent college degrees independent or dependent? Table 3.13 may be helpful.

We'll use the first equation above to check for independence. If the `teen` and `parents` variables are independent, it must be true that

$$P(\text{teen college} \mid \text{parent degree}) = P(\text{teen college})$$

Using Table 3.13, we check whether equality holds in this equation.

$$P(\text{teen college} \mid \text{parent degree}) \overset{?}{=} P(\text{teen college})$$
$$0.83 \neq 0.56$$

The value 0.83 came from a probability calculation using Table 3.13: $\frac{231}{280} \approx 0.83$. Because the sides are not equal, teenager college attendance and parent degree are dependent. That is, we estimate the probability a teenager attended college to be higher if we know that one of the teen's parents has a college degree.

		parents		
		degree	not	Total
teen	college	231	214	445
	not	49	298	347
	Total	280	512	792

Table 3.13: Contingency table summarizing the `family_college` data set.

⊙ **Guided Practice 3.58** Use the second equation in the box above to show that teenager college attendance and parent college degrees are dependent.[38]

If A and B are mutually exclusive events, then A and B cannot occur at the same time. Mathematically, this is written as

$$P(A \text{ and } B) = 0$$

The General Addition Rule states that $P(A \text{ or } B)$ equals $P(A) + P(B) - P(A \text{ and } B)$. If A and B are mutually exclusive events, we can replace $P(A \text{ and } B)$ with 0 and the following addition rule applies:

$$P(A \text{ or } B) = P(A) + P(B)$$

---

[38]We check for equality in the following equation:

$$P(\text{teen college, parent degree}) \overset{?}{=} P(\text{teen college}) \times P(\text{parent degree})$$
$$\frac{231}{792} = 0.292 \neq \frac{445}{792} \times \frac{280}{792} = 0.199$$

These terms are not equal, which confirms what we learned in Example 3.57: teenager college attendance and parent college degrees are dependent.

TIP: Checking whether two events are mutually exclusive (disjoint)

If A and B are mutually exclusive events, then they cannot occur at the same time. If asked to determine if events A and B are mutually exclusive, verify one of the following equations holds (there is no need to check both equations):

$$P(A \text{ and } B) = 0 \qquad\qquad P(A \text{ or } B) = P(A) + P(B)$$

If the equation that is checked holds true (the left and right sides are equal), A and B are mutually exclusive. If the equation does not hold, then A and B are not mutually exclusive.

● **Example 3.59** Are teen college attendance and parent college degrees mutually exclusive?

Looking in the table, we see that there are 231 instances where both the teenager attended college and parents have a degree, indicating the probability of both events occurring is greater than 0. Since we have found an example where both of these events happen together, these two events are not mutually exclusive. We could more formally show this by computing the probability both events occur at the same time:

$$P(\texttt{teen college, parent degree}) = \frac{231}{792} \neq 0$$

Since this probability is not zero, teenager college attendance and parent college degrees are not mutually exclusive.

TIP: Mutually exclusive and independent are different

If two events are mutually exclusive, then if one is true, the other cannot be true. This implies the two events are in some way connected, meaning they must be dependent.

If two events are independent, then if one occurs, it is still possible for the other to occur, meaning the events are not mutually exclusive.

Caution: Dependent events need not be mutually exclusive.

If two events are dependent, we cannot simply conclude they are mutually exclusive. For example, the college attendance of teenagers and a college degree by one of their parents are dependent, but those events are not mutually exclusive.

3.2.8   Tree diagrams

Tree diagrams are a tool to organize outcomes and probabilities around the structure of the data. They are most useful when two or more processes occur in a sequence and each process is conditioned on its predecessors.

The `smallpox` data fit this description. We see the population as split by `inoculation`: `yes` and `no`. Following this split, survival rates were observed for each group. This structure is reflected in the tree diagram shown in Figure 3.14. The first branch for `inoculation` is said to be the **primary** branch while the other branches are **secondary**.

Figure 3.14: A tree diagram of the `smallpox` data set.

Tree diagrams are annotated with marginal and conditional probabilities, as shown in Figure 3.14. This tree diagram splits the smallpox data by `inoculation` into the `yes` and `no` groups with respective marginal probabilities 0.0392 and 0.9608. The secondary branches are conditioned on the first, so we assign conditional probabilities to these branches. For example, the top branch in Figure 3.14 is the probability that `lived` conditioned on the information that `inoculated`. We may (and usually do) construct joint probabilities at the end of each branch in our tree by multiplying the numbers we come across as we move from left to right. These joint probabilities are computed using the General Multiplication Rule:

$$P(\texttt{inoculated and lived}) = P(\texttt{inoculated}) \times P(\texttt{lived} \mid \texttt{inoculated})$$
$$= 0.0392 \times 0.9754$$
$$= 0.0382$$

● **Example 3.60**   What is the probability that a randomly selected person who was inoculated died?

This is equivalent to $P(\texttt{died} \mid \texttt{inoculated})$. This conditional probability can be found in the second branch as 0.0246.

● **Example 3.61** What is the probability that a randomly selected person lived?

There are two ways that a person could have lived: be inoculated *and* live OR not be inoculated *and* live. To find this probability, we sum the two disjoint probabilities:

$$P(\texttt{lived}) = 0.0392 \times 0.9745 + 0.9608 \times 0.8589 = 0.03824 + 0.82523 = 0.86347$$

⊙ **Guided Practice 3.62** After an introductory statistics course, 78% of students can successfully construct tree diagrams. Of those who can construct tree diagrams, 97% passed, while only 57% of those students who could not construct tree diagrams passed. (a) Organize this information into a tree diagram. (b) What is the probability that a student who was able to construct tree diagrams did not pass? (c) What is the probability that a randomly selected student was able to successfully construct tree diagrams and passed? (d) What is the probability that a randomly selected student passed? [39]

3.2.9 Bayes' Theorem

In many instances, we are given a conditional probability of the form

$$P(\text{statement about variable 1} \mid \text{statement about variable 2})$$

but we would really like to know the inverted conditional probability:

$$P(\text{statement about variable 2} \mid \text{statement about variable 1})$$

For example, instead of wanting to know $P(\text{lived} \mid \text{inoculated})$, we might want to know $P(\text{inoculated} \mid \text{lived})$. This is more challenging because it cannot be read directly from the tree diagram. In these instances we use **Bayes' Theorem**. Let's begin by looking at a new example.

---

[39](a) The tree diagram is shown to the right. (b) $P(\text{not pass} \mid \text{able to construct tree diagram}) = 0.03$. (c) $P(\text{able to construct tree diagrams and passed}) = P(\text{able to construct tree diagrams}) \times P(\text{passed} \mid \text{able to construct tree diagrams}) = 0.78 \times 0.97 = 0.7566$. (d) $P(\text{passed}) = 0.7566 + 0.1254 = 0.8820$.

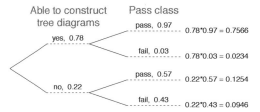

● **Example 3.63** In Canada, about 0.35% of women over 40 will develop breast cancer in any given year. A common screening test for cancer is the mammogram, but this test is not perfect. In about 11% of patients with breast cancer, the test gives a **false negative**: it indicates a woman does not have breast cancer when she does have breast cancer. Similarly, the test gives a **false positive** in 7% of patients who do not have breast cancer: it indicates these patients have breast cancer when they actually do not.[40] If we tested a random woman over 40 for breast cancer using a mammogram and the test came back positive – that is, the test suggested the patient has cancer – what is the probability that the patient actually has breast cancer?

We are given sufficient information to quickly compute the probability of testing positive if a woman has breast cancer ($1.00 - 0.11 = 0.89$). However, we seek the inverted probability of cancer given a positive test result:

$$P(\text{has BC} \mid \text{mammogram}^+)$$

Here, "has BC" is an abbreviation for the patient actually having breast cancer, and "mammogram$^+$" means the mammogram screening was positive, which in this case means the test suggests the patient has breast cancer. (Watch out for the non-intuitive medical language: a *positive* test result suggests the possible presence of cancer in a mammogram screening.) We can use the conditional probability formula from the previous section: $P(A|B) = \frac{P(A \text{ and } B)}{P(B)}$. Our conditional probability can be found as follows:

$$P(\text{has BC} \mid \text{mammogram}^+) = \frac{P(\text{has BC and mammogram}^+)}{P(\text{mammogram}^+)}$$

The probability that a mammogram is positive is as follows.

$$P(\text{mammogram}^+) = P(\text{has BC and mammogram}^+) + P(\text{no BC and mammogram}^+)$$

A tree diagram is useful for identifying each probability and is shown in Figure 3.15. Using the tree diagram, we find that

$$P(\text{has BC} \mid \text{mammogram}^+)$$
$$= \frac{P(\text{has BC and mammogram}^+)}{P(\text{has BC and mammogram}^+) + P(\text{no BC and mammogram}^+)}$$
$$= \frac{0.0035(0.89)}{0.0035(0.89) + 0.9965(0.07)}$$
$$= \frac{0.00312}{0.07288} \approx 0.0428$$

That is, even if a patient has a positive mammogram screening, there is still only a 4% chance that she has breast cancer.

Example 3.63 highlights why doctors often run more tests regardless of a first positive test result. When a medical condition is rare, a single positive test isn't generally definitive.

Consider again the last equation of Example 3.63. Using the tree diagram, we can see that the numerator (the top of the fraction) is equal to the following product:

$$P(\text{has BC and mammogram}^+) = P(\text{mammogram}^+ \mid \text{has BC})P(\text{has BC})$$

---

[40]The probabilities reported here were obtained using studies reported at www.breastcancer.org and www.ncbi.nlm.nih.gov/pmc/articles/PMC1173421.

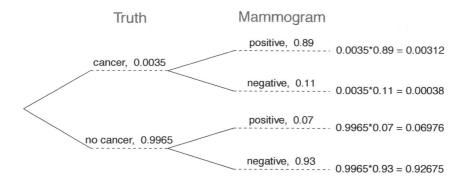

Figure 3.15: Tree diagram for Example 3.63, computing the probability a random patient who tests positive on a mammogram actually has breast cancer.

The denominator – the probability the screening was positive – is equal to the sum of probabilities for each positive screening scenario:

$$P(\text{mammogram}^+) = P(\underline{\text{mammogram}}^+ \text{ and no BC}) + P(\underline{\text{mammogram}}^+ \text{ and has BC})$$

In the example, each of the probabilities on the right side was broken down into a product of a conditional probability and marginal probability using the tree diagram.

$$
\begin{aligned}
P(\text{mammogram}^+) &= P(\text{mammogram}^+ \text{ and no BC}) + P(\text{mammogram}^+ \text{ and has BC}) \\
&= P(\text{mammogram}^+ |\text{ no BC})P(\text{no BC}) \\
&\quad + P(\text{mammogram}^+ |\text{ has BC})P(\text{has BC})
\end{aligned}
$$

We can see an application of Bayes' Theorem by substituting the resulting probability expressions into the numerator and denominator of the original conditional probability.

$$
\begin{aligned}
&P(\text{has BC} |\text{ mammogram}^+) \\
&= \frac{P(\text{mammogram}^+ |\text{ has BC})P(\text{has BC})}{P(\text{mammogram}^+ |\text{ no BC})P(\text{no BC}) + P(\text{mammogram}^+ |\text{ has BC})P(\text{has BC})}
\end{aligned}
$$

Bayes' Theorem: inverting probabilities

Consider the following conditional probability for variable 1 and variable 2:

$$P(\text{outcome } A_1 \text{ of variable 1} |\text{ outcome } B \text{ of variable 2})$$

Bayes' Theorem states that this conditional probability can be identified as the following fraction:

$$\frac{P(B|A_1)P(A_1)}{P(B|A_1)P(A_1) + P(B|A_2)P(A_2) + \cdots + P(B|A_k)P(A_k)} \tag{3.64}$$

where A_2, A_3, ..., and A_k represent all other possible outcomes of the first variable.

Bayes' Theorem is just a generalization of what we have done using tree diagrams. The formula need not be memorized, since it can always be derived using a tree diagram:

- The numerator identifies the probability of getting both A_1 and B.

- The denominator is the overall probability of getting B. Traverse each branch of the tree diagram that ends with event B. Add up the required products.

⊙ **Guided Practice 3.65**   Jose visits campus every Thursday evening. However, some days the parking garage is full, often due to college events. There are academic events on 35% of evenings, sporting events on 20% of evenings, and no events on 45% of evenings. When there is an academic event, the garage fills up about 25% of the time, and it fills up 70% of evenings with sporting events. On evenings when there are no events, it only fills up about 5% of the time. If Jose comes to campus and finds the garage full, what is the probability that there is a sporting event? Use a tree diagram to solve this problem.[41]

The last several exercises offered a way to update our belief about whether there is a sporting event, academic event, or no event going on at the school based on the information that the parking lot was full. This strategy of *updating beliefs* using Bayes' Theorem is actually the foundation of an entire section of statistics called **Bayesian statistics**. While Bayesian statistics is very important and useful, we will not have time to cover it in this book.

3.3   The binomial formula

● **Example 3.66**   Suppose we randomly selected four individuals to participate in the "shock" study. What is the chance exactly one of them will be a success? Let's call the four people Allen (A), Brittany (B), Caroline (C), and Damian (D) for convenience. Also, suppose 35% of people are successes as in the previous version of this example.

Let's consider a scenario where one person refuses:

$$P(A = \textsf{refuse}, \ B = \textsf{shock}, \ C = \textsf{shock}, \ D = \textsf{shock})$$
$$= P(A = \textsf{refuse}) \ P(B = \textsf{shock}) \ P(C = \textsf{shock}) \ P(D = \textsf{shock})$$
$$= (0.35)(0.65)(0.65)(0.65) = (0.35)^1(0.65)^3 = 0.096$$

But there are three other scenarios: Brittany, Caroline, or Damian could have been the one to refuse. In each of these cases, the probability is again $(0.35)^1(0.65)^3$. These four scenarios exhaust all the possible ways that exactly one of these four

---

[41]The tree diagram, with three primary branches, is shown to the right. We want

$P(\text{sporting event}|\text{garage full})$

$= \dfrac{P(\text{sporting event and garage full})}{P(\text{garage full})}$

$= \dfrac{0.14}{0.0875 + 0.14 + 0.0225} = 0.56.$

If the garage is full, there is a 56% probability that there is a sporting event.

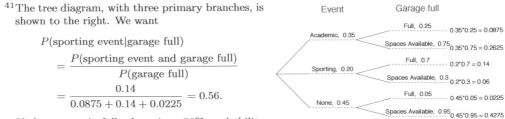

people could refuse to administer the most severe shock, so the total probability is $4 \times (0.35)^1 (0.65)^3 = 0.38$.

⊙ **Guided Practice 3.67**   Verify that the scenario where Brittany is the only one to refuse to give the most severe shock has probability $(0.35)^1 (0.65)^3$. [42]

3.3.1   Understanding the formula

To solve the scenario outlined in Example 3.66 we use what is called the **Binomal Formula**. The binomal formula gives the probability of having k successes in n independent trials where probability of an individual success in one trial is p (in Example 3.66, $n = 4$, $k = 1$, $p = 0.35$). In order to develop this formula, we reexamine each part of the example.

There were four individuals who could have been the one to refuse, and each of these four scenarios had the same probability. Thus, we could identify the final probability as

$$[\# \text{ of scenarios}] \times P(\text{single scenario}) \tag{3.68}$$

The first component of this equation is the number of ways to arrange the $k = 1$ successes among the $n = 4$ trials. The second component is the probability of any of the four (equally probable) scenarios.

Consider $P(\text{single scenario})$ under the general case of k successes and $n - k$ failures in the n trials. In any such scenario, we apply the Multiplication Rule for independent events:

$$p^k (1 - p)^{n-k}$$

This is our general formula for $P(\text{single scenario})$.

Secondly, we introduce a general formula for the number of ways to choose k successes in n trials, i.e. arrange k successes and $n - k$ failures:

$$\binom{n}{k} = \frac{n!}{k!(n-k)!}$$

The quantity $\binom{n}{k}$ is read **n choose k**.[43] The exclamation point notation (e.g. $k!$) denotes a **factorial** expression.

$$0! = 1$$
$$1! = 1$$
$$2! = 2 \times 1 = 2$$
$$3! = 3 \times 2 \times 1 = 6$$
$$4! = 4 \times 3 \times 2 \times 1 = 24$$
$$\vdots$$
$$n! = n \times (n-1) \times \ldots \times 3 \times 2 \times 1$$

Using the formula, we can compute the number of ways to choose $k = 1$ successes in $n = 4$ trials:

$$\binom{4}{1} = \frac{4!}{1!(4-1)!} = \frac{4!}{1!3!} = \frac{4 \times 3 \times 2 \times 1}{(1)(3 \times 2 \times 1)} = 4$$

---

[42]$P(A = \texttt{shock}, \ B = \texttt{refuse}, \ C = \texttt{shock}, \ D = \texttt{shock}) = (0.65)(0.35)(0.65)(0.65) = (0.35)^1 (0.65)^3$.
[43]Other notation for n choose k includes ${}_nC_k$, C_n^k, and $C(n, k)$.

This result is exactly what we found by carefully thinking of each possible scenario in Example 3.66.

Substituting n choose k for the number of scenarios and $p^k(1-p)^{n-k}$ for the single scenario probability in Equation (3.68) yields the general binomial formula.

Binomial formula

Suppose the probability of a single trial being a success is p. Then the probability of observing exactly k successes in n independent trials is given by

$$\binom{n}{k}p^k(1-p)^{n-k} = \frac{n!}{k!(n-k)!}p^k(1-p)^{n-k} \qquad (3.69)$$

3.3.2 When and how to apply the formula

TIP: Is it binomial? Four conditions to check.
(1) The trials are independent.
(2) The number of trials, n, is fixed.
(3) Each trial outcome can be classified as a *success* or *failure*.
(4) The probability of a success, p, is the same for each trial.

● **Example 3.70** What is the probability that 3 of 8 randomly selected students will refuse to administer the worst shock, i.e. 5 of 8 will?

We would like to apply the binomial model, so we check our conditions. The number of trials is fixed ($n = 8$) (condition 2) and each trial outcome can be classified as a success or failure (condition 3). Because the sample is random, the trials are independent (condition 1) and the probability of a success is the same for each trial (condition 4).

In the outcome of interest, there are $k = 3$ successes in $n = 8$ trials, and the probability of a success is $p = 0.35$. So the probability that 3 of 8 will refuse is given by

$$\binom{8}{3}(0.35)^3(1-0.35)^{8-3} = \frac{8!}{3!(8-3)!}(0.35)^3(1-0.35)^{8-3}$$
$$= \frac{8!}{3!5!}(0.35)^3(0.65)^5$$

Dealing with the factorial part:

$$\frac{8!}{3!5!} = \frac{8 \times 7 \times 6 \times 5 \times 4 \times 3 \times 2 \times 1}{(3 \times 2 \times 1)(5 \times 4 \times 3 \times 2 \times 1)} = \frac{8 \times 7 \times 6}{3 \times 2 \times 1} = 56$$

Using $(0.35)^3(0.65)^5 \approx 0.005$, the final probability is about $56 * 0.005 = 0.28$.

> **TIP: computing binomial probabilities**
> The first step in using the binomial model is to check that the model is appropriate.
> If it is, the next step is to identify n, p, and k. The final step is to apply the formulas
> and interpret the results.

⊙ **Guided Practice 3.71**   The probability that a random smoker will develop a severe lung condition in his or her lifetime is about 0.3. If you have 4 friends who smoke, are the conditions for the binomial model satisfied?[44]

⊙ **Guided Practice 3.72**   Suppose these four friends do not know each other and we can treat them as if they were a random sample from the population. Is the binomial model appropriate? What is the probability that (a) none of them will develop a severe lung condition? (b) One will develop a severe lung condition? (c) That no more than one will develop a severe lung condition?[45]

⊙ **Guided Practice 3.73**   What is the probability that at least 2 of your 4 smoking friends will develop a severe lung condition in their lifetimes?[46]

⊙ **Guided Practice 3.74**   Suppose you have 7 friends who are smokers and they can be treated as a random sample of smokers. What is the probability that at most 2 of your 7 friends will develop a severe lung condition.[47]

Below we consider the first term in the binomial probability, n choose k under some special scenarios.

⊙ **Guided Practice 3.75**   Why is it true that $\binom{n}{0} = 1$ and $\binom{n}{n} = 1$ for any number n?[48]

⊙ **Guided Practice 3.76**   How many ways can you arrange one success and $n-1$ failures in n trials? How many ways can you arrange $n-1$ successes and one failure in n trials?[49]

---

[44]One possible answer: if the friends know each other, then the independence assumption is probably not satisfied. For example, acquaintances may have similar smoking habits.

[45]To check if the binomial model is appropriate, we must verify the conditions. (i) Since we are supposing we can treat the friends as a random sample, they are independent. (ii) We have a fixed number of trials ($n = 4$). (iii) Each outcome is a success or failure. (iv) The probability of a success is the same for each trials since the individuals are like a random sample ($p = 0.3$ if we say a "success" is someone getting a lung condition, a morbid choice). Compute parts (a) and (b) from the binomial formula in Equation (3.69): $P(0) = \binom{4}{0}(0.3)^0(0.7)^4 = 1 \times 1 \times 0.7^4 = 0.2401$, $P(1) = \binom{4}{1}(0.3)^1(0.7)^3 = 0.4116$. Note: $0! = 1$, as shown on page 127. Part (c) can be computed as the sum of parts (a) and (b): $P(0)+P(1) = 0.2401+0.4116 = 0.6517$. That is, there is about a 65% chance that no more than one of your four smoking friends will develop a severe lung condition.

[46]The complement (no more than one will develop a severe lung condition) as computed in Guided Practice 3.72 as 0.6517, so we compute one minus this value: 0.3483.

[47]$P(0, 1, \text{or } 2 \text{ develop severe lung condition}) = P(k = 0) + P(k = 1) + P(k = 2) = 0.6471$.

[48]Frame these expressions into words. How many different ways are there to arrange 0 successes and n failures in n trials? (1 way.) How many different ways are there to arrange n successes and 0 failures in n trials? (1 way.)

[49]One success and $n-1$ failures: there are exactly n unique places we can put the success, so there are n ways to arrange one success and $n-1$ failures. A similar argument is used for the second question.

● **Example 3.77**   There are 13 marbles in a bag. 4 are blue and 9 are red. Randomly draw 5 marbles *without replacement*. Find the probability you get exactly 3 blue marbles.

Because the probability of success p is not the same for each trial, we cannot use the binomial formula. However, we can use the same logic to arrive at the following answer.

$$P(x = 3) = (\# \text{ of combinations with 3 blue}) \times P(3 \text{ blue and 2 red in a } specific \text{ order})$$

$$= \binom{5}{3} \times P(\text{RRRBB})$$

$$= \binom{5}{3} \left(\frac{4}{13} \times \frac{3}{12} \times \frac{2}{11} \times \frac{9}{10} \times \frac{8}{9} \right)$$

$$= 0.1112$$

3.3.3   Calculator: binomial probabilities

TI-83/84: Computing the binomial coefficient, $\binom{n}{k}$

Use MATH, PRB, nCr to evaluate n choose r. Here r and k are different letters for the same quantity.

1. Type the value of n.
2. Select MATH.
3. Right arrow to PRB.
4. Choose 3:nCr.
5. Type the value of k.
6. Hit ENTER.

Example: 5 nCr 3 means *5 choose 3*.

Casio fx-9750GII: Computing the binomial coefficient, $\binom{n}{k}$

1. Navigate to the RUN-MAT section (hit MENU, then hit 1).
2. Enter a value for n.
3. Go to CATALOG (hit buttons SHIFT and then 7).
4. Type C (hit the ln button), then navigate down to the bolded C and hit EXE.
5. Enter the value of k. Example of what it should look like: 7C3.
6. Hit EXE.

Mathematically, we show these results by verifying the following two equations:

$$\binom{n}{1} = n, \qquad \binom{n}{n-1} = n$$

TI-84: Computing the binomial formula, $P(X = k) = \binom{n}{k}p^k(1-p)^{n-k}$

Use 2ND VARS, `binompdf` to evaluate the probability of *exactly* k occurrences out of n independent trials of an event with probability p.

1. Select 2ND VARS (i.e. DISTR)
2. Choose A:`binompdf` (use the down arrow).
3. Let `trials` be n.
4. Let `p` be p
5. Let `x` value be k.
6. Select Paste and hit ENTER.

TI-83: Do steps 1-2, then enter n, p, and k separated by commas: `binompdf(n, p, k)`. Then hit ENTER.

TI-84: Computing $P(X \le k) = \binom{n}{0}p^0(1-p)^{n-0} + ... + \binom{n}{k}p^k(1-p)^{n-k}$

Use 2ND VARS, `binomcdf` to evaluate the cumulative probability of *at most* k occurrences out of n independent trials of an event with probability p.

1. Select 2ND VARS (i.e. DISTR)
2. Choose B:`binomcdf` (use the down arrow).
3. Let `trials` be n.
4. Let `p` be p
5. Let `x` value be k.
6. Select Paste and hit ENTER.

TI-83: Do steps 1-2, then enter the values for n, p, and k separated by commas as follows: `binomcdf(n, p, k)`. Then hit ENTER.

Casio fx-9750GII: Binomial calculations

1. Navigate to STAT (MENU, then hit 2).

2. Select DIST (F5), and then BINM (F5).

3. Choose whether to calculate the binomial distribution for a specific number of successes, $P(X = k)$, or for a range $P(X \leq k)$ of values (0 successes, 1 success, ..., k successes).

   - For a specific number of successes, choose Bpd (F1).
   - To consider the range 0, 1, ..., k successes, choose Bcd(F1).

4. If needed, set Data to Variable (Var option, which is F2).

5. Enter the value for x (k), Numtrial (n), and p (probability of a success).

6. Hit EXE.

⊙ **Guided Practice 3.78** Find the number of ways of arranging 3 blue marbles and 2 red marbles.[50]

⊙ **Guided Practice 3.79** There are 13 marbles in a bag. 4 are blue and 9 are red. Randomly draw 5 marbles *with replacement*. Find the probability you get exactly 3 blue marbles.[51]

⊙ **Guided Practice 3.80** There are 13 marbles in a bag. 4 are blue and 9 are red. Randomly draw 5 marbles *with replacement*. Find the probability you get *at most* 3 blue marbles (i.e. less than or equal to 3 blue marbles).[52]

3.4   Simulations

In the previous section we saw how to apply the binomial formula to find the probability of exactly k successes in n independent trials when a success has probability p. Sometimes we have a problem we want to solve but we don't know the appropriate formula, or even worse, a formula may not exist. In this case, one common approach is to estimate the probability using **simulations**.

You may already be familiar with simulations. Want to know the probability of rolling a sum of 7 with a pair of dice? Roll a pair of dice many, many, many times and see what proportion of times the sum was 7. The more times you roll the pair of dice, the better the estimate will tend to be. Of course, such experiments can be time consuming or even infeasible.

In this section, we consider simulations using **random numbers**. Random numbers (or technically, *psuedo-random numbers*) can be produced using a calculator or computer. Random digits are produced such that each digit, *0-9*, is equally likely to come up in each spot. You'll find that occasionally we may have the same number in a row – sometimes multiple times – but in the long run, each digit should appear 1/10th of the time.

---

[50]Use $n = 5$ and $k = 3$ to get 10.
[51]Use $n = 5$, $p = 4/13$, and x (k) = 3 to get 0.1396.
[52]Use $n = 5$, $p = 4/13$, and $x = 3$ to get 0.9662.

	Column			
Row	1-5	6-10	11-15	16-20
1	43087	41864	51009	39689
2	63432	72132	40269	56103
3	19025	83056	62511	52598
4	85117	16706	31083	24816
5	16285	56280	01494	90240
6	94342	18473	50845	77757
7	61099	14136	39052	50235
8	37537	58839	56876	02960
9	04510	16172	90838	15210
10	27217	12151	52645	96218

Table 3.16: Random number table. A full page of random numbers may be found in Appendix B.1 on page 447.

● **Example 3.81** Mika's favorite brand of cereal is running a special where 20% of the cereal boxes contain a prize. Mika really wants that prize. If her mother buys 6 boxes of the cereal over the next few months, what is the probability Mika will get a prize?

To solve this problem using simulation, we need to be able to assign digits to outcomes. Each box should have a 20% chance of having a prize and an 80% chance of not having a prize. Therefore, a valid assignment would be:

$$0, 1 \rightarrow \text{prize}$$
$$2\text{-}9 \rightarrow \text{no prize}$$

Of the ten possible digits (*0, 1, 2, ..., 8, 9*), two of them, i.e. 20% of them, correspond to winning a prize, which exactly matches the odds that a cereal box contains a prize.

In Mika's simulation, one trial will consist of 6 boxes of cereal, and therefore a trial will require six digits (each digit will correspond to one box of cereal). We will repeat the simulation for 20 trials. Therefore we will need 20 sets of 6 digits. Let's begin on row 1 of the random digit table, shown in Table 3.16. If a trial consisted of 5 digits, we could use the first 5 digits going across: *43087*. Because here a trial consists of 6 digits, it may be easier to read down the table, rather than read across. We will let trial 1 consist of the first 6 digits in column 1 (*461819*), trial 2 consist of the first 6 digits in column 2 (*339564*), etc. For this simulation, we will end up using the first 6 rows of each of the 20 columns.

In trial 1, there are two *1*'s, so we record that as a success; in this trial there were actually two prizes. In trial 2 there were no *0*'s or *1*'s, therefore we do not record this as a success. In trial 3 there were three prizes, so we record this as a success. The rest of this exercise is left as a Guided Practice problem for you to complete.

◉ **Guided Practice 3.82** Finish the simulation above and report the estimate for the probability that Mika will get a prize if her mother buys 6 boxes of cereal where each one has a 20% chance of containing a prize.[53]

---

[53] The trials that contain at least one 0 or 1 and therefore are successes are trials: 1, 3, 4, 6, 7, 8, 9, 10, 11, 12, 13, 14, 15, 17, 18, 19, and 20. There were 17 successes among the 20 trials, so our estimate of the probability based on this simulation is $17/20 = 0.85$.

⊙ **Guided Practice 3.83** In the previous example, the probability that a box of cereal contains a prize is 20%. The question presented is equivalent to asking, what is the probability of getting at least one prize in six randomly selected boxes of cereal. This probability question can be solved explicitly using the method of complements. Find this probability. How does the estimate arrived at by simulation compare to this probability?[54]

We can also use simulations to estimate quantities other than probabilities. Consider the following example.

● **Example 3.84** Let's say that instead of buying exactly 6 boxes of cereal, Mika's mother agrees to buy boxes of this cereal *until* she finds one with a prize. On average, how many boxes of cereal would one have to buy until one gets a prize?

For this question, we can use the same digit assignment. However, our stopping rule is different. Each trial may require a different number of digits. For each trial, the stopping rule is: look at digits until we encounter a *0* or a *1*. Then, record how many digits/boxes of cereal it took. Repeat the simulation for 20 trials, and then average the numbers from each trial.

Let's begin again at row 1. We can read across or down, depending upon what is most convenient. Since there are 20 columns and we want 20 trials, we will read down the columns. Starting at column 1, we count how many digits (boxes of cereal) we encounter until we reach a *0* or *1* (which represent a prize). For trial 1 we see *461*, so we record 3. For trial 2 we see *3395641*, so we record 7. For trial 3, we see *0*, so we record 1. The rest of this exercise is left as a Guided Practice problem for you to complete.

⊙ **Guided Practice 3.85** Finish the simulation above and report your estimate for the average number of boxes of cereal one would have to buy until encountering a prize, where the probability of a prize in each box is 20%.[55]

● **Example 3.86** Now, consider a case where the probability of interest is not 20%, but rather 28%. Which digits should correspond to success and which to failure?

This example is more complicated because with only 10 digits, there is no way to select exactly 28% of them. Therefore, each observation will have to consist of *two* digits. We can use two digits at a time and assign pairs of digits as follows:

$$00\text{-}27 \to \text{success}$$
$$28\text{-}99 \to \text{failure}$$

⊙ **Guided Practice 3.87** Assume the probability of winning a particular casino game is 45%. We want to carry out a simulation to estimate the probability that we will win at least 5 times in 10 plays. We will use 30 trials of the simulation. Assign digits to outcomes. Also, how many total digits will we require to run this simulation?[56]

---

[54]The true probability is given by $1 - P(\text{no prizes in six boxes}) = 1 - 0.8^6 = 0.74$. The estimate arrived at by simulation was 11% too high. Note: We only repeated the simulation 20 times. If we had repeated it 1000 times, we would (very likely) have gotten an estimate closer to the true probability.

[55]For the 20 trials, the number of digits we see until we encounter a *0* or *1* is: 3,7,1,4,9, 4,1,2,4,5, 5,1,1,1,3, 8,5,2,2,6. Now we take the average of these 20 numbers to get $74/20 = 3.7$.

[56]One possible assignment is: *00-44* → win and *45-99* → lose. Each trial requires 10 pairs of digits, so we will need 30 sets of 10 pairs of digits for a total of $30 \times 10 \times 2 = 600$ digits.

⊙ **Guided Practice 3.88** Assume carnival spinner has 7 slots. We want to carry out a simulation to estimate the probability that we will win at least 10 times in 60 plays. Repeat 100 trials of the simulation. Assign digits to outcomes. Also, how many total digits will we require to run this simulation?[57]

Does anyone perform simulations like this? Sort of. Simulations are used a lot in statistics, and these often require the same principles covered in this section to properly set up those simulations. The difference is in implementation after the setup. Rather than use a random number table, a statistician will write a program that uses a pseudo-random number generator in a computer to run the simulations very quickly – often times millions of trials each second, which provides much more accurate estimates than running a couple dozen trials by hand.

3.5 Random variables

● **Example 3.89** Two books are assigned for a statistics class: a textbook and its corresponding study guide. The university bookstore determined 20% of enrolled students do not buy either book, 55% buy the textbook only, and 25% buy both books, and these percentages are relatively constant from one term to another. If there are 100 students enrolled, how many books should the bookstore expect to sell to this class?

Around 20 students will not buy either book (0 books total), about 55 will buy one book (55 books total), and approximately 25 will buy two books (totaling 50 books for these 25 students). The bookstore should expect to sell about 105 books for this class.

⊙ **Guided Practice 3.90** Would you be surprised if the bookstore sold slightly more or less than 105 books?[58]

● **Example 3.91** The textbook costs $137 and the study guide $33. How much revenue should the bookstore expect from this class of 100 students?

About 55 students will just buy a textbook, providing revenue of

$$\$137 \times 55 = \$7,535$$

The roughly 25 students who buy both the textbook and the study guide would pay a total of

$$(\$137 + \$33) \times 25 = \$170 \times 25 = \$4,250$$

Thus, the bookstore should expect to generate about $7,535 + $4,250 = $11,785 from these 100 students for this one class. However, there might be some *sampling variability* so the actual amount may differ by a little bit.

---

[57]Note that $1/7 = 0.142857...$ This makes it tricky to assign digits to outcomes. The best approach here would be to exclude some of the digits from the simulation. We can assign *0* to success and *1-6* to failure. This corresponds to a 1/7 chance of getting a success. If we encounter a *7, 8,* or *9*, we will just skip over it. Because we don't know how many *7, 8,* or *9*'s we will encounter, we do not know how many total digits we will end up using for the simulation. (If you want a challenge, try to estimate the total number of digits you would need.)

[58]If they sell a little more or a little less, this should not be a surprise. Hopefully Chapter 2 helped make clear that there is natural variability in observed data. For example, if we would flip a coin 100 times, it will not usually come up heads exactly half the time, but it will probably be close.

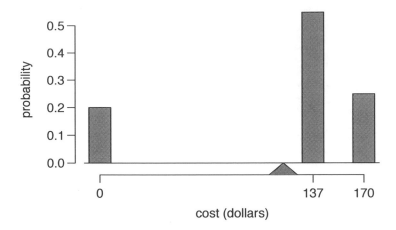

Figure 3.17: Probability distribution for the bookstore's revenue from a single student. The distribution balances on a triangle representing the average revenue per student.

● **Example 3.92** What is the average revenue per student for this course?

The expected total revenue is \$11,785, and there are 100 students. Therefore the expected revenue per student is $\$11,785/100 = \$117.85$.

3.5.1 Probability distributions

A **probability distribution** is a table of all disjoint outcomes and their associated probabilities. Table 3.18 shows the probability distribution for the sum of two dice.

Rules for probability distributions

A probability distribution is a list of the possible outcomes with corresponding probabilities that satisfies three rules:

1. The outcomes listed must be disjoint.

2. Each probability must be between 0 and 1.

3. The probabilities must total 1.

⊙ **Guided Practice 3.93** Table 3.19 suggests three distributions for household income in the United States. Only one is correct. Which one must it be? What is wrong with the other two?[59]

---

[59]The probabilities of (a) do not sum to 1. The second probability in (b) is negative. This leaves (c), which sure enough satisfies the requirements of a distribution. One of the three was said to be the actual distribution of US household incomes, so it must be (c).

Dice sum	2	3	4	5	6	7	8	9	10	11	12
Probability	$\frac{1}{36}$	$\frac{2}{36}$	$\frac{3}{36}$	$\frac{4}{36}$	$\frac{5}{36}$	$\frac{6}{36}$	$\frac{5}{36}$	$\frac{4}{36}$	$\frac{3}{36}$	$\frac{2}{36}$	$\frac{1}{36}$

Table 3.18: Probability distribution for the sum of two dice.

Income range ($1000s)	0-25	25-50	50-100	100+
(a)	0.18	0.39	0.33	0.16
(b)	0.38	-0.27	0.52	0.37
(c)	0.28	0.27	0.29	0.16

Table 3.19: Proposed distributions of US household incomes (Guided Practice 3.93).

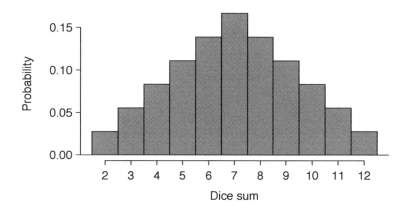

Figure 3.20: A histogram for the probability distribution of the sum of two dice.

Chapter 2 emphasized the importance of plotting data to provide quick summaries. Probability distributions can also be summarized in a histogram or bar plot. The probability distribution for the sum of two dice is shown in Table 3.18 and its histogram is plotted in Figure 3.20. The distribution of US household incomes is shown in Figure 3.21 as a bar plot. The presence of the 100+ category makes it difficult to represent it with a regular histogram.[60]

In these bar plots, the bar heights represent the probabilities of outcomes. If the outcomes are numerical and discrete, it is usually (visually) convenient to make a histogram, as in the case of the sum of two dice. Another example of plotting the bars at their respective locations is shown in Figure 3.17.

---

[60]It is also possible to construct a distribution plot when income is not artificially binned into four groups. Density histograms for *continuous* distributions are considered in Section 3.6.

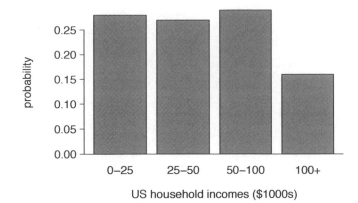

Figure 3.21: A bar graph for the probability distribution of US household income. Because it is artificially separated into four unequal bins, this graph fails to show the shape or skew of the distribution.

3.5.2   Expectation

We call a variable or process with a numerical outcome a **random variable**, and we usually represent this random variable with a capital letter such as X, Y, or Z. The amount of money a single student will spend on her statistics books is a random variable, and we represent it by X.

Random variable

A random process or variable with a numerical outcome.

The possible outcomes of X are labeled with a corresponding lower case letter x and subscripts. For example, we write $x_1 = \$0$, $x_2 = \$137$, and $x_3 = \$170$, which occur with probabilities 0.20, 0.55, and 0.25. The distribution of X is summarized in Figure 3.17 and Table 3.22.

i	1	2	3	Total
x_i	$0	$137	$170	–
p_i	0.20	0.55	0.25	1.00

Table 3.22: The probability distribution for the random variable X, representing the bookstore's revenue from a single student. We use p_i to represent the probability of x_i.

$E(X)$

Expected value of X

We computed the average outcome of X as \$117.85 in Example 3.92. We call this average the **expected value** of X, denoted by $E(X)$. The expected value of a random variable is computed by adding each outcome weighted by its probability:

$$E(X) = 0 \times P(X = 0) + 137 \times P(X = 137) + 170 \times P(X = 170)$$
$$= 0 \times 0.20 + 137 \times 0.55 + 170 \times 0.25 = 117.85$$

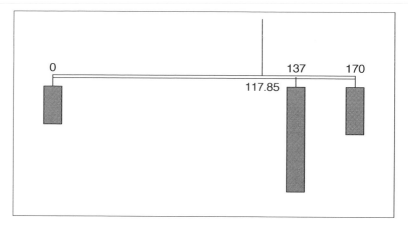

Figure 3.23: A weight system representing the probability distribution for X. The string holds the distribution at the mean to keep the system balanced.

Expected value of a discrete random variable

If X takes outcomes x_1, x_2, ..., x_n with probabilities p_1, p_2, ..., p_n, the expected value of X is the sum of each outcome multiplied by its corresponding probability:

$$E(X) = \mu_x = x_1 \times p_1 + x_2 \times p_2 + \cdots + x_n \times p_n$$

$$= \sum_{i=1}^{n} (x_i \times p_i) \tag{3.94}$$

The expected value for a random variable represents the average outcome. For example, $E(X) = 117.85$ represents the average amount the bookstore expects to make from a single student, which we could also write as $\mu = 117.85$. While the bookstore will make more than this on some students and less than this on other students, the average of many randomly selected students will be near $117.85.

It is also possible to compute the expected value of a continuous random variable (see Section 3.6). However, it requires a little calculus and we save it for a later class.[61]

In physics, the expectation holds the same meaning as the center of gravity. The distribution can be represented by a series of weights at each outcome, and the mean represents the balancing point. This is represented in Figures 3.17 and 3.23. The idea of a center of gravity also expands to continuous probability distributions. Figure 3.24 shows a continuous probability distribution balanced atop a wedge placed at the mean.

---

[61]$\mu_x = \int x f(x) dx$ where $f(x)$ represents a function for the density curve.

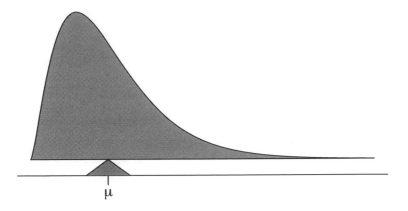

Figure 3.24: A continuous distribution can also be balanced at its mean.

3.5.3   Variability in random variables

Suppose you ran the university bookstore. Besides how much revenue you expect to generate, you might also want to know the volatility (variability) in your revenue.

The variance and standard deviation can be used to describe the variability of a random variable. Section 2.2.2 introduced a method for finding the variance and standard deviation for a data set. We first computed deviations from the mean $(x_i - \mu)$, squared those deviations, and took an average to get the variance. In the case of a random variable, we again compute squared deviations. However, we take their sum weighted by their corresponding probabilities, just like we did for the expectation. This weighted sum of squared deviations equals the variance, and we calculate the standard deviation by taking the square root of the variance, just as we did in Section 2.2.2.

Variance and standard deviation of a discrete random variable

If X takes outcomes x_1, x_2, ..., x_n with probabilities p_1, p_2, ..., p_n and expected value $\mu_x = E(X)$, then to find the standard deviation of X, we first find the variance and then take its square root.

$$Var(X) = \sigma_x^2 = (x_1 - \mu_x)^2 \times p_1 + (x_2 - \mu_x)^2 \times p_2 + \cdots + (x_n - \mu_x)^2 \times p_n$$

$$= \sum_{i=1}^{n}(x_i - \mu_x)^2 \times p_i$$

$$SD(X) = \sigma_x = \sqrt{\sum_{i=1}^{n}(x_i - \mu_x)^2 \times p_i} \tag{3.95}$$

Just as it is possible to compute the mean of a continuous random variable using calculus, we can also use calculus to compute the variance.[62] However, this topic is beyond the scope of the AP exam.

---

[62] $\sigma_x^2 = \int (x - \mu_x)^2 f(x) dx$ where $f(x)$ represents a function for the density curve.

● **Example 3.96** Compute the expected value, variance, and standard deviation of X, the revenue of a single statistics student for the bookstore.

It is useful to construct a table that holds computations for each outcome separately, then add up the results.

i	1	2	3	Total
x_i	\$0	\$137	\$170	
p_i	0.20	0.55	0.25	
$x_i \times p_i$	0	75.35	42.50	117.85

Thus, the expected value is $\mu = 117.85$, which we computed earlier. The variance can be constructed by extending this table:

i	1	2	3	Total
x_i	\$0	\$137	\$170	
p_i	0.20	0.55	0.25	
$x_i \times p_i$	0	75.35	42.50	117.85
$x_i - \mu_X$	-117.85	19.15	52.15	
$(x_i - \mu_X)^2$	13888.62	366.72	2719.62	
$(x_i - \mu_X)^2 \times p_i$	2777.7	201.7	679.9	3659.3

The variance of X is $\sigma_x^2 = 3659.3$, which means the standard deviation is $\sigma_x = \sqrt{3659.3} = \$60.49$.

☉ **Guided Practice 3.97** The bookstore also offers a chemistry textbook for \$159 and a book supplement for \$41. From past experience, they know about 25% of chemistry students just buy the textbook while 60% buy both the textbook and supplement.[63]

(a) What proportion of students don't buy either book? Assume no students buy the supplement without the textbook.

(b) Let Y represent the revenue from a single student. Write out the probability distribution of Y, i.e. a table for each outcome and its associated probability.

(c) Compute the expected revenue from a single chemistry student.

(d) Find the standard deviation to describe the variability associated with the revenue from a single student.

---

[63](a) 100% - 25% - 60% = 15% of students do not buy any books for the class. Part (b) is represented by the first two lines in the table below. The expectation for part (c) is given as the total on the line $y_i \times p_i$. The result of part (d) is the square-root of the variance listed on in the total on the last line: $\sigma_Y = \sqrt{Var(Y)} = \$69.28$.

i (scenario)	1 (noBook)	2 (textbook)	3 (both)	Total
y_i	0.00	159.00	200.00	
p_i	0.15	0.25	0.60	
$y_i \times p_i$	0.00	39.75	120.00	$E(Y) = 159.75$
$y_i - \mu_Y$	-159.75	-0.75	40.25	
$(y_i - \mu_Y)^2$	25520.06	0.56	1620.06	
$(y_i - \mu_Y)^2 \times p_i$	3828.0	0.1	972.0	$Var(Y) \approx 4800$

3.5.4   Linear transformations of a random variable

Let X be a random variable that represents how many books per student a textbook company sells. The probability distribution of X is given in the following table.

x_i	1	2	3
p_i	0.6	0.3	0.1

Using the methods of the previous section we can find that the mean $\mu_x = 1.5$ and the standard deviation $\sigma_x = 0.67$. Suppose that the revenue the textbook company makes per student is \$150 and that each book has a fixed cost of \$30. The profit function, then, is $150X - 30$, where X is the number of books sold. To calculate the mean and standard deviation for the profit of the textbook company, we could define a new variable Y as follows:

$$Y = 150X - 30$$

⊙ **Guided Practice 3.98** Verify that the distribution of Y is given by the table below.[64]

y_i	\$120	\$270	\$420
p_i	0.6	0.3	0.1

Using this new table, we can compute the mean and standard deviation of the textbook company's profit. However, because Y is a linear transformation of X, we can use the properties from Section 2.2.6. Recall that multiplying every X by 150 multiplies both the mean and standard deviation by 150. Subtracting 30 only subtracts 30 from the mean, not the standard deviation. Therefore,

$$\mu_{150X-30} = E(150X - 30) \qquad\qquad \sigma_{150X-30} = SD(150X - 30)$$
$$= 150 \times E(X) - 30 \qquad\qquad\qquad = 150 \times SD(X)$$
$$= 150 \times 1.5 - 30 \qquad\qquad\qquad = 150 \times 0.67$$
$$= 195 \qquad\qquad\qquad\qquad\qquad = 100.5$$

For a randomly selected student, the textbook company can expect to make \$195 dollars, with a standard deviation of \$100.50.

Linear transformations of a random variable

If X is a random variable, then a linear transformation is given by $aX + b$, where a and b are some fixed numbers.

$$E(aX + b) = a \times E(X) + b \qquad\qquad SD(aX + b) = |a| \times SD(X)$$

---

[64] $150 \times 1 - 30 = 120$; $150 \times 2 - 30 = 270$; $150 \times 3 - 30 = 420$

3.5.5 Linear combinations of random variables

So far, we have thought of each variable as being a complete story in and of itself. Sometimes it is more appropriate to use a combination of variables. For instance, the amount of time a person spends commuting to work each week can be broken down into several daily commutes. Similarly, the total gain or loss in a stock portfolio is the sum of the gains and losses in its components.

● **Example 3.99** John travels to work five days a week. We will use X_1 to represent his travel time on Monday, X_2 to represent his travel time on Tuesday, and so on. Write an equation using X_1, ..., X_5 that represents his travel time for the week, denoted by W.

His total weekly travel time is the sum of the five daily values:

$$W = X_1 + X_2 + X_3 + X_4 + X_5$$

Breaking the weekly travel time W into pieces provides a framework for understanding each source of randomness and is useful for modeling W.

● **Example 3.100** It takes John an average of 18 minutes each day to commute to work. What would you expect his average commute time to be for the week?

We were told that the average (i.e. expected value) of the commute time is 18 minutes per day: $E(X_i) = 18$. To get the expected time for the sum of the five days, we can add up the expected time for each individual day:

$$\begin{aligned}
E(W) &= E(X_1 + X_2 + X_3 + X_4 + X_5) \\
&= E(X_1) + E(X_2) + E(X_3) + E(X_4) + E(X_5) \\
&= 18 + 18 + 18 + 18 + 18 = 90 \text{ minutes}
\end{aligned}$$

The expectation of the total time is equal to the sum of the expected individual times. More generally, the expectation of a sum of random variables is always the sum of the expectation for each random variable.

⊙ **Guided Practice 3.101** Elena is selling a TV at a cash auction and also intends to buy a toaster oven in the auction. If X represents the profit for selling the TV and Y represents the cost of the toaster oven, write an equation that represents the net change in Elena's cash.[65]

⊙ **Guided Practice 3.102** Based on past auctions, Elena figures she should expect to make about \$175 on the TV and pay about \$23 for the toaster oven. In total, how much should she expect to make or spend?[66]

⊙ **Guided Practice 3.103** Would you be surprised if John's weekly commute wasn't exactly 90 minutes or if Elena didn't make exactly \$152? Explain.[67]

---

[65]She will make X dollars on the TV but spend Y dollars on the toaster oven: $X - Y$.

[66]$E(X - Y) = E(X) - E(Y) = 175 - 23 = \$152$. She should expect to make about \$152.

[67]No, since there is probably some variability. For example, the traffic will vary from one day to next, and auction prices will vary depending on the quality of the merchandise and the interest of the attendees.

Two important concepts concerning combinations of random variables have so far been introduced. First, a final value can sometimes be described as the sum of its parts in an equation. Second, intuition suggests that putting the individual average values into this equation gives the average value we would expect in total. This second point needs clarification – it is guaranteed to be true in what are called *linear combinations of random variables*.

A **linear combination** of two random variables X and Y is a fancy phrase to describe a combination

$$aX + bY$$

where a and b are some fixed and known numbers. For John's commute time, there were five random variables – one for each work day – and each random variable could be written as having a fixed coefficient of 1:

$$1X_1 + 1X_2 + 1X_3 + 1X_4 + 1X_5$$

For Elena's net gain or loss, the X random variable had a coefficient of $+1$ and the Y random variable had a coefficient of -1.

When considering the average of a linear combination of random variables, it is safe to plug in the mean of each random variable and then compute the final result. For a few examples of nonlinear combinations of random variables – cases where we cannot simply plug in the means – see the footnote.[68]

Linear combinations of random variables and the average result

If X and Y are random variables, then a linear combination of the random variables is given by $aX + bY$, where a and b are some fixed numbers. To compute the average value of a linear combination of random variables, plug in the average of each individual random variable and compute the result:

$$E(aX + bY) = a \times E(X) + b \times E(Y)$$

Recall that the expected value is the same as the mean, i.e. $E(X) = \mu_x$.

● **Example 3.104** Leonard has invested \$6000 in Google Inc. (stock ticker: GOOG) and \$2000 in Exxon Mobil Corp. (XOM). If X represents the change in Google's stock next month and Y represents the change in Exxon Mobil stock next month, write an equation that describes how much money will be made or lost in Leonard's stocks for the month.

For simplicity, we will suppose X and Y are not in percents but are in decimal form (e.g. if Google's stock increases 1%, then $X = 0.01$; or if it loses 1%, then $X = -0.01$). Then we can write an equation for Leonard's gain as

$$\$6000 \times X + \$2000 \times Y$$

If we plug in the change in the stock value for X and Y, this equation gives the change in value of Leonard's stock portfolio for the month. A positive value represents a gain, and a negative value represents a loss.

---

[68]If X and Y are random variables, consider the following combinations: X^{1+Y}, $X \times Y$, X/Y. In such cases, plugging in the average value for each random variable and computing the result will not generally lead to an accurate average value for the end result.

	Mean ($\bar{x}$)	Standard deviation (s)	Variance (s^2)
GOOG	0.0210	0.0846	0.0072
XOM	0.0038	0.0519	0.0027

Table 3.26: The mean, standard deviation, and variance of the GOOG and XOM stocks. These statistics were estimated from historical stock data, so notation used for sample statistics has been used.

⊙ **Guided Practice 3.105** Suppose Google and Exxon Mobil stocks have recently been rising 2.1% and 0.4% per month, respectively. Compute the expected change in Leonard's stock portfolio for next month.[69]

⊙ **Guided Practice 3.106** You should have found that Leonard expects a positive gain in Guided Practice 3.105. However, would you be surprised if he actually had a loss this month?[70]

3.5.6   Variability in linear combinations of random variables

Quantifying the average outcome from a linear combination of random variables is helpful, but it is also important to have some sense of the uncertainty associated with the total outcome of that combination of random variables. The expected net gain or loss of Leonard's stock portfolio was considered in Guided Practice 3.105. However, there was no quantitative discussion of the volatility of this portfolio. For instance, while the average monthly gain might be about \$134 according to the data, that gain is not guaranteed. Figure 3.25 shows the monthly changes in a portfolio like Leonard's during the 36 months from 2009 to 2011. The gains and losses vary widely, and quantifying these fluctuations is important when investing in stocks.

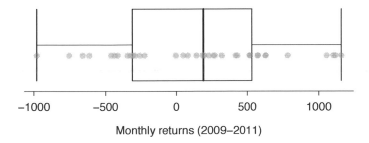

Figure 3.25: The change in a portfolio like Leonard's for the 36 months from 2009 to 2011, where \$6000 is in Google's stock and \$2000 is in Exxon Mobil's.

Just as we have done in many previous cases, we use the variance and standard deviation to describe the uncertainty associated with Leonard's monthly returns. To do so, the standard deviations and variances of each stock's monthly return will be useful, and these are shown in Table 3.26. The stocks' returns are nearly independent.

---

[69] $E(\$6000 \times X + \$2000 \times Y) = \$6000 \times 0.021 + \$2000 \times 0.004 = \$134$.
[70] No. While stocks tend to rise over time, they are often volatile in the short term.

We want to describe the uncertainty of Leonard's monthly returns by finding the standard deviation of the return on his combined portfolio. First, we note that the variance of a sum has a nice property: the variance of a sum is the sum of the variances. That is, if X and Y are independent random variables:

$$Var(X + Y) = Var(X) + Var(Y)$$

Because the standard deviation is the square root of the variance, we can rewrite this equation using standard deviations:

$$(SD_{X+Y})^2 = (SD_X)^2 + (SD_Y)^2$$

This equation might remind you of a theorem from geometry: $c^2 = a^2 + b^2$. The equation for the standard deviation of the sum of two independent random variables looks analogous to the Pythagorean Theorem. Just as the Pythagorean Theorem only holds for right triangles, this equation only holds when X and Y are *independent*.[71]

Standard deviation of the sum and difference of random variables

If X and Y are *independent* random variables:

$$SD_{X+Y} = SD_{X-Y} = \sqrt{(SD_X)^2 + (SD_Y)^2}$$

Because $SD_Y = SD_{-Y}$, the standard deviation of the difference of two variables equals the standard deviation of the sum of two variables. This property holds for more than two variables as well. For example, if X, Y, and Z are independent random variables:

$$SD_{X+Y+Z} = SD_{X-Y-Z} = \sqrt{(SD_X)^2 + (SD_Y)^2 + (SD_Z)^2} \tag{3.107}$$

If we need the standard deviation of a linear combination of independent variables, such as $aX + bY$, we can consider aX and bY as two new variables. Recall that multiplying all of the values of variable by a positive constant multiplies the standard deviation by that constant. Thus, $SD_{aX} = a \times SD_X$ and $SD_{bY} = b \times SD_Y$. It follows that:

$$SD_{aX+bY} = \sqrt{(a \times SD_X)^2 + (b \times SD_Y)^2}$$

This equation can be used to compute the standard deviation of Leonard's monthly return. Recall that Leonard has \$6,000 in Google stock and \$2,000 in Exxon Mobil's stock. From Table 3.26, the standard deviation of Google stock is 0.0846 and the standard deviation of Exxon Mobile stock is 0.0519.

$$
\begin{aligned}
SD_{6000X+2000Y} &= \sqrt{(6000 \times SD_X)^2 + (2000 \times SD_Y)^2} \\
&= \sqrt{(6000 \times 0.0846)^2 + (4000 \times .0519)^2} \\
&= \sqrt{270,000} \\
&= 520
\end{aligned}
$$

---

[71]Another word for independent is orthogonal, meaning right angle! When X and Y are dependent, the equation for SD_{X+Y} becomes analogous to the law of cosines.

The standard deviation of the total is \$520. While an average monthly return of \$134 on an \$8000 investment is nothing to scoff at, the monthly returns are so volatile that Leonard should not expect this income to be very stable.

Standard deviation of linear combinations of random variables

To find the standard deviation of a linear combination of random variables, we first consider aX and bY separately. We find the standard deviation of each, and then we apply the equation for the standard deviation of the sum of two variables:

$$SD_{aX+bY} = \sqrt{(a \times SD_X)^2 + (b \times SD_Y)^2}$$

This equation is valid as long as the random variables X and Y are *independent* of each other.

● **Example 3.108** Suppose John's daily commute has a standard deviation of 4 minutes. What is the uncertainty in his total commute time for the week?

The expression for John's commute time is

$$X_1 + X_2 + X_3 + X_4 + X_5$$

Each coefficient is 1, so the standard deviation of the total weekly commute time is

$$\begin{aligned} SD &= \sqrt{(1 \times 4)^2 + (1 \times 4)^2 + (1 \times 4)^2 + (1 \times 4)^2 + (1 \times 4)^2} \\ &= \sqrt{5 \times (4)^2} \\ &= 8.94 \end{aligned}$$

The standard deviation for John's weekly work commute time is about 9 minutes.

⊙ **Guided Practice 3.109** The computation in Example 3.108 relied on an important assumption: the commute time for each day is independent of the time on other days of that week. Do you think this is valid? Explain.[72]

⊙ **Guided Practice 3.110** Consider Elena's two auctions from Guided Practice 3.101 on page 143. Suppose these auctions are approximately independent and the variability in auction prices associated with the TV and toaster oven can be described using standard deviations of \$25 and \$8. Compute the standard deviation of Elena's net gain.[73]

Consider again Guided Practice 3.110. The negative coefficient for Y in the linear combination was eliminated when we squared the coefficients. This generally holds true: negatives in a linear combination will have no impact on the variability computed for a linear combination, but they do impact the expected value computations.

---

[72]One concern is whether traffic patterns tend to have a weekly cycle (e.g. Fridays may be worse than other days). If that is the case, and John drives, then the assumption is probably not reasonable. However, if John walks to work, then his commute is probably not affected by any weekly traffic cycle.

[73]The equation for Elena can be written as: $(1) \times X + (-1) \times Y$.

To find the SD of this new variable we do:

$$SD_{(1) \times X + (-1) \times Y} = \sqrt{(1 \times SD_X)^2 + (-1 \times SD_Y)^2} = (1 \times 25)^2 + (-1 \times 8)^2 = 26.25$$

The SD is about \$26.25.

3.6   Continuous distributions

● **Example 3.111** Figure 3.27 shows a few different hollow histograms of the variable `height` for 3 million US adults from the mid-90's.[74] How does changing the number of bins allow you to make different interpretations of the data?

Adding more bins provides greater detail. This sample is extremely large, which is why much smaller bins still work well. Usually we do not use so many bins with smaller sample sizes since small counts per bin mean the bin heights are very volatile.

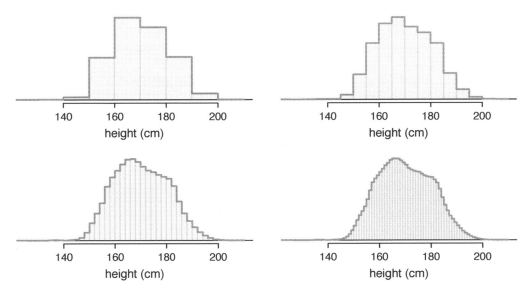

Figure 3.27: Four hollow histograms of US adults heights with varying bin widths.

● **Example 3.112** What proportion of the sample is between 180 cm and 185 cm tall (about 5'11" to 6'1")?

We can add up the heights of the bins in the range 180 cm and 185 and divide by the sample size. For instance, this can be done with the two shaded bins shown in Figure 3.28. The two bins in this region have counts of 195,307 and 156,239 people, resulting in the following estimate of the probability:

$$\frac{195307 + 156239}{3,000,000} = 0.1172$$

This fraction is the same as the proportion of the histogram's area that falls in the range 180 to 185 cm.

---

[74]This sample can be considered a simple random sample from the US population. It relies on the USDA Food Commodity Intake Database.

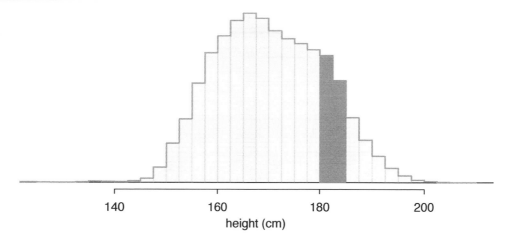

Figure 3.28: A histogram with bin sizes of 2.5 cm. The shaded region represents individuals with heights between 180 and 185 cm.

3.6.1   From histograms to continuous distributions

Examine the transition from a boxy hollow histogram in the top-left of Figure 3.27 to the much smoother plot in the lower-right. In this last plot, the bins are so slim that the hollow histogram is starting to resemble a smooth curve. This suggests the population height as a *continuous* numerical variable might best be explained by a curve that represents the outline of extremely slim bins.

This smooth curve represents a **probability density function** (also called a **density** or **distribution**), and such a curve is shown in Figure 3.29 overlaid on a histogram of the sample. A density has a special property: the total area under the density's curve is 1.

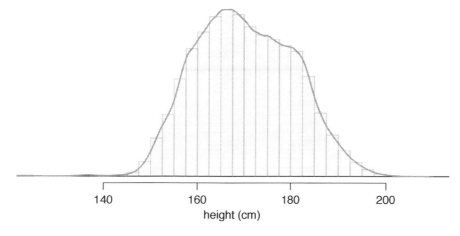

Figure 3.29: The continuous probability distribution of heights for US adults.

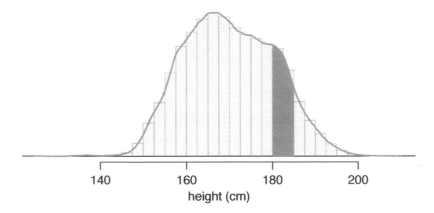

Figure 3.30: Density for heights in the US adult population with the area between 180 and 185 cm shaded. Compare this plot with Figure 3.28.

3.6.2   Probabilities from continuous distributions

We computed the proportion of individuals with heights 180 to 185 cm in Example 3.112 as a fraction:

$$\frac{\text{number of people between 180 and 185}}{\text{total sample size}}$$

We found the number of people with heights between 180 and 185 cm by determining the fraction of the histogram's area in this region. Similarly, we can use the area in the shaded region under the curve to find a probability (with the help of a computer):

$$P(\texttt{height between 180 and 185}) = \text{area between 180 and 185} = 0.1157$$

The probability that a randomly selected person is between 180 and 185 cm is 0.1157. This is very close to the estimate from Example 3.112: 0.1172.

⊙ **Guided Practice 3.113**   Three US adults are randomly selected. The probability a single adult is between 180 and 185 cm is 0.1157.[75]

   (a) What is the probability that all three are between 180 and 185 cm tall?

   (b) What is the probability that none are between 180 and 185 cm?

● **Example 3.114**   What is the probability that a randomly selected person is **exactly** 180 cm? Assume you can measure perfectly.

This probability is zero. A person might be close to 180 cm, but not exactly 180 cm tall. This also makes sense with the definition of probability as area; there is no area captured between 180 cm and 180 cm.

⊙ **Guided Practice 3.115**   Suppose a person's height is rounded to the nearest centimeter. Is there a chance that a random person's **measured** height will be 180 cm?[76]

---

[75]Brief answers: (a) $0.1157 \times 0.1157 \times 0.1157 = 0.0015$. (b) $(1 - 0.1157)^3 = 0.692$

[76]This has positive probability. Anyone between 179.5 cm and 180.5 cm will have a *measured* height of 180 cm. This is probably a more realistic scenario to encounter in practice versus Example 3.114.

3.7 Exercises

3.7.1 Defining probability

3.1 True or false. Determine if the statements below are true or false, and explain your reasoning.

(a) If a fair coin is tossed many times and the last eight tosses are all heads, then the chance that the next toss will be heads is somewhat less than 50%.

(b) Drawing a face card (jack, queen, or king) and drawing a red card from a full deck of playing cards are mutually exclusive events.

(c) Drawing a face card and drawing an ace from a full deck of playing cards are mutually exclusive events.

3.2 Roulette wheel. The game of roulette involves spinning a wheel with 38 slots: 18 red, 18 black, and 2 green. A ball is spun onto the wheel and will eventually land in a slot, where each slot has an equal chance of capturing the ball.

(a) You watch a roulette wheel spin 3 consecutive times and the ball lands on a red slot each time. What is the probability that the ball will land on a red slot on the next spin?

(b) You watch a roulette wheel spin 300 consecutive times and the ball lands on a red slot each time. What is the probability that the ball will land on a red slot on the next spin?

(c) Are you equally confident of your answers to parts (a) and (b)? Why or why not?

Photo by Håkan Dahlström
(http://flic.kr/p/93fEzp)
CC BY 2.0 license

3.3 Four games, one winner. Below are four versions of the same game. Your archnemisis gets to pick the version of the game, and then you get to choose how many times to flip a coin: 10 times or 100 times. Identify how many coin flips you should choose for each version of the game. It costs $1 to play each game. Explain your reasoning.

(a) If the proportion of heads is larger than 0.60, you win $1.

(b) If the proportion of heads is larger than 0.40, you win $1.

(c) If the proportion of heads is between 0.40 and 0.60, you win $1.

(d) If the proportion of heads is smaller than 0.30, you win $1.

3.4 Backgammon. Backgammon is a board game for two players in which the playing pieces are moved according to the roll of two dice. Players win by removing all of their pieces from the board, so it is usually good to roll high numbers. You are playing backgammon with a friend and you roll two 6s in your first roll and two 6s in your second roll. Your friend rolls two 3s in his first roll and again in his second row. Your friend claims that you are cheating, because rolling double 6s twice in a row is very unlikely. Using probability, show that your rolls were just as likely as his.

3.5 Coin flips. If you flip a fair coin 10 times, what is the probability of

(a) getting all tails?   (b) getting all heads?   (c) getting at least one tails?

3.6 Dice rolls. If you roll a pair of fair dice, what is the probability of

(a) getting a sum of 1?   (b) getting a sum of 5?   (c) getting a sum of 12?

3.7 Swing voters. A 2012 Pew Research survey asked 2,373 randomly sampled registered voters their political affiliation (Republican, Democrat, or Independent) and whether or not they identify as swing voters. 35% of respondents identified as Independent, 23% identified as swing voters, and 11% identified as both.[77]

(a) Are being Independent and being a swing voter disjoint, i.e. mutually exclusive?

(b) Draw a Venn diagram summarizing the variables and their associated probabilities.

(c) What percent of voters are Independent but not swing voters?

(d) What percent of voters are Independent or swing voters?

(e) What percent of voters are neither Independent nor swing voters?

(f) Is the event that someone is a swing voter independent of the event that someone is a political Independent?

3.8 Poverty and language. The American Community Survey is an ongoing survey that provides data every year to give communities the current information they need to plan investments and services. The 2010 American Community Survey estimates that 14.6% of Americans live below the poverty line, 20.7% speak a language other that English at home, and 4.2% fall into both categories.[78]

(a) Are living below the poverty line and speaking a language other than English at home disjoint?

(b) Draw a Venn diagram summarizing the variables and their associated probabilities.

(c) What percent of Americans live below the poverty line and only speak English at home?

(d) What percent of Americans live below the poverty line or speak a language other than English at home?

(e) What percent of Americans live above the poverty line and only speak English at home?

(f) Is the event that someone lives below the poverty line independent of the event that the person speaks a language other than English at home?

3.9 Disjoint vs. independent. In parts (a) and (b), identify whether the events are disjoint, independent, or neither (events cannot be both disjoint and independent).

(a) You and a randomly selected student from your class both earn A's in this course.

(b) You and your class study partner both earn A's in this course.

(c) If two events can occur at the same time, must they be dependent?

3.10 Guessing on an exam. In a multiple choice exam, there are 5 questions and 4 choices for each question (a, b, c, d). Nancy has not studied for the exam at all and decides to randomly guess the answers. What is the probability that:

(a) the first question she gets right is the 5th question?

(b) she gets all of the questions right?

(c) she gets at least one question right?

---

[77]Pew Research Center, With Voters Focused on Economy, Obama Lead Narrows, data collected between April 4-15, 2012.

[78]U.S. Census Bureau, 2010 American Community Survey 1-Year Estimates, Characteristics of People by Language Spoken at Home.

3.11 Educational attainment of couples. The table below shows the distribution of education level attained by US residents by gender based on data collected during the 2010 American Community Survey.[79]

		Gender	
		Male	Female
	Less than 9th grade	0.07	0.13
	9th to 12th grade, no diploma	0.10	0.09
Highest	High school graduate, GED, or alternative	0.30	0.20
education	Some college, no degree	0.22	0.24
attained	Associate's degree	0.06	0.08
	Bachelor's degree	0.16	0.17
	Graduate or professional degree	0.09	0.09
	Total	1.00	1.00

(a) What is the probability that a randomly chosen man has at least a Bachelor's degree?

(b) What is the probability that a randomly chosen woman has at least a Bachelor's degree?

(c) What is the probability that a man and a woman getting married both have at least a Bachelor's degree? Note any assumptions you must make to answer this question.

(d) If you made an assumption in part (c), do you think it was reasonable? If you didn't make an assumption, double check your earlier answer and then return to this part.

3.12 School absences. Data collected at elementary schools in DeKalb County, GA suggest that each year roughly 25% of students miss exactly one day of school, 15% miss 2 days, and 28% miss 3 or more days due to sickness.[80]

(a) What is the probability that a student chosen at random doesn't miss any days of school due to sickness this year?

(b) What is the probability that a student chosen at random misses no more than one day?

(c) What is the probability that a student chosen at random misses at least one day?

(d) If a parent has two kids at a DeKalb County elementary school, what is the probability that neither kid will miss any school? Note any assumption you must make to answer this question.

(e) If a parent has two kids at a DeKalb County elementary school, what is the probability that both kids will miss some school, i.e. at least one day? Note any assumption you make.

(f) If you made an assumption in part (d) or (e), do you think it was reasonable? If you didn't make any assumptions, double check your earlier answers.

3.13 Grade distributions. Each row in the table below is a proposed grade distribution for a class. Identify each as a valid or invalid probability distribution, and explain your reasoning.

	Grades				
	A	B	C	D	F
(a)	0.3	0.3	0.3	0.2	0.1
(b)	0	0	1	0	0
(c)	0.3	0.3	0.3	0	0
(d)	0.3	0.5	0.2	0.1	-0.1
(e)	0.2	0.4	0.2	0.1	0.1
(f)	0	-0.1	1.1	0	0

---

[79]U.S. Census Bureau, 2010 American Community Survey 1-Year Estimates, Educational Attainment.
[80]S.S. Mizan et al. "Absence, Extended Absence, and Repeat Tardiness Related to Asthma Status among Elementary School Children". In: *Journal of Asthma* 48.3 (2011), pp. 228–234.

3.14 Health and health coverage, Part I. The Behavioral Risk Factor Surveillance System (BRFSS) is an annual telephone survey designed to identify risk factors in the adult population and report emerging health trends. The following table summarizes two variables for the respondents: health status and health coverage, which describes whether each respondent had health insurance.[81]

		Excellent	Very good	Good	Fair	Poor	Total
		Health Status					
Health	No	459	727	854	385	99	2,524
Coverage	Yes	4,198	6,245	4,821	1,634	578	17,476
	Total	4657	6,972	5,675	2,019	677	20,000

(a) If we draw one individual at random, what is the probability that the respondent has excellent health and doesn't have health coverage?

(b) If we draw one individual at random, what is the probability that the respondent has excellent health or doesn't have health coverage?

3.7.2   Conditional probability

3.15 Joint and conditional probabilities. $P(A) = 0.3$, $P(B) = 0.7$

(a) Can you compute $P(A$ and $B)$ if you only know $P(A)$ and $P(B)$?

(b) Assuming that events A and B arise from independent random processes,

    i. what is $P(A$ and $B)$?

    ii. what is $P(A$ or $B)$?

    iii. what is $P(A|B)$?

(c) If we are given that $P(A$ and $B) = 0.1$, are the random variables giving rise to events A and B independent?

(d) If we are given that $P(A$ and $B) = 0.1$, what is $P(A|B)$?

3.16 PB & J. Suppose 80% of people like peanut butter, 89% like jelly, and 78% like both. Given that a randomly sampled person likes peanut butter, what's the probability that he also likes jelly?

3.17 Global warming. A 2010 Pew Research poll asked 1,306 Americans "From what you've read and heard, is there solid evidence that the average temperature on earth has been getting warmer over the past few decades, or not?". The table below shows the distribution of responses by party and ideology, where the counts have been replaced with relative frequencies.[82]

		Response			
		Earth is warming	Not warming	Don't Know Refuse	Total
	Conservative Republican	0.11	0.20	0.02	0.33
Party and	Mod/Lib Republican	0.06	0.06	0.01	0.13
Ideology	Mod/Cons Democrat	0.25	0.07	0.02	0.34
	Liberal Democrat	0.18	0.01	0.01	0.20
	Total	0.60	0.34	0.06	1.00

---

[81]Office of Surveillance, Epidemiology, and Laboratory Services Behavioral Risk Factor Surveillance System, BRFSS 2010 Survey Data.

[82]Pew Research Center, Majority of Republicans No Longer See Evidence of Global Warming, data collected on October 27, 2010.

(a) Are being the earth is warming and being a liberal Democrat mutually exclusive?

(b) What is the probability that a randomly chosen respondent believes the earth is warming or is a liberal Democrat?

(c) What is the probability that a randomly chosen respondent believes the earth is warming given that he is a liberal Democrat?

(d) What is the probability that a randomly chosen respondent believes the earth is warming given that he is a conservative Republican?

(e) Does it appear that whether or not a respondent believes the earth is warming is independent of their party and ideology? Explain your reasoning.

(f) What is the probability that a randomly chosen respondent is a moderate/liberal Republican given that he does not believe that the earth is warming?

3.18 Health and health coverage, Part II. introduced a contingency table summarizing the relationship between health status and health coverage for a sample of 20,000 Americans. In the table below, the counts have been replaced by relative frequencies (probability estimates).

		Health Status					
		Excellent	Very good	Good	Fair	Poor	Total
Health	No	0.0230	0.0364	0.0427	0.0192	0.0050	0.1262
Coverage	Yes	0.2099	0.3123	0.2410	0.0817	0.0289	0.8738
	Total	0.2329	0.3486	0.2838	0.1009	0.0338	1.0000

(a) Are being in excellent health and having health coverage mutually exclusive?

(b) What is the probability that a randomly chosen individual has excellent health?

(c) What is the probability that a randomly chosen individual has excellent health given that he has health coverage?

(d) What is the probability that a randomly chosen individual has excellent health given that he doesn't have health coverage?

(e) Do having excellent health and having health coverage appear to be independent?

3.19 Burger preferences. A 2010 SurveyUSA poll asked 500 Los Angeles residents, "What is the best hamburger place in Southern California? Five Guys Burgers? In-N-Out Burger? Fat Burger? Tommy's Hamburgers? Umami Burger? Or somewhere else?" The distribution of responses by gender is shown below.[83]

		Gender		
		Male	Female	Total
	Five Guys Burgers	5	6	11
	In-N-Out Burger	162	181	343
Best	Fat Burger	10	12	22
hamburger	Tommy's Hamburgers	27	27	54
place	Umami Burger	5	1	6
	Other	26	20	46
	Not Sure	13	5	18
	Total	248	252	500

(a) Are being female and liking Five Guys Burgers mutually exclusive?

(b) What is the probability that a randomly chosen male likes In-N-Out the best?

(c) What is the probability that a randomly chosen female likes In-N-Out the best?

(d) What is the probability that a man and a woman who are dating both like In-N-Out the best? Note any assumption you make and evaluate whether you think that assumption is reasonable.

(e) What is the probability that a randomly chosen person likes Umami best or that person is female?

---

[83]SurveyUSA, Results of SurveyUSA News Poll #17718, data collected on December 2, 2010.

3.20 Assortative mating. Assortative mating is a nonrandom mating pattern where individuals with similar genotypes and/or phenotypes mate with one another more frequently than what would be expected under a random mating pattern. Researchers studying this topic collected data on eye colors of 204 Scandinavian men and their female partners. The table below summarizes the results. For simplicity, we only include heterosexual relationships in this exercise.[84]

		Partner (female)			
		Blue	Brown	Green	Total
	Blue	78	23	13	114
	Brown	19	23	12	54
Self (male)	Green	11	9	16	36
	Total	108	55	41	204

(a) What is the probability that a randomly chosen male respondent or his partner has blue eyes?

(b) What is the probability that a randomly chosen male respondent with blue eyes has a partner with blue eyes?

(c) What is the probability that a randomly chosen male respondent with brown eyes has a partner with blue eyes? What about the probability of a randomly chosen male respondent with green eyes having a partner with blue eyes?

(d) Does it appear that the eye colors of male respondents and their partners are independent? Explain your reasoning.

3.21 Urns and marbles, Part I. Imagine you have an urn containing 5 red, 3 blue, and 2 orange marbles in it.

(a) What is the probability that the first marble you draw is blue?

(b) Suppose you drew a blue marble in the first draw. If drawing with replacement, what is the probability of drawing a blue marble in the second draw?

(c) Suppose you instead drew an orange marble in the first draw. If drawing with replacement, what is the probability of drawing a blue marble in the second draw?

(d) If drawing with replacement, what is the probability of drawing two blue marbles in a row?

(e) When drawing with replacement, are the draws independent? Explain.

3.22 Socks in a drawer. In your sock drawer you have 4 blue, 5 gray, and 3 black socks. Half asleep one morning you grab 2 socks at random and put them on. Find the probability you end up wearing

(a) 2 blue socks   (c) at least 1 black sock   (e) matching socks

(b) no gray socks   (d) a green sock

3.23 Urns and marbles, Part II. Imagine you have an urn containing 5 red, 3 blue, and 2 orange marbles.

(a) Suppose you draw a marble and it is blue. If drawing without replacement, what is the probability the next is also blue?

(b) Suppose you draw a marble and it is orange, and then you draw a second marble without replacement. What is the probability this second marble is blue?

(c) If drawing without replacement, what is the probability of drawing two blue marbles in a row?

(d) When drawing without replacement, are the draws independent? Explain.

---

[84]B. Laeng et al. "Why do blue-eyed men prefer women with the same eye color?" In: *Behavioral Ecology and Sociobiology* 61.3 (2007), pp. 371–384.

3.24 Books on a bookshelf. The table below shows the distribution of books on a bookcase based on whether they are nonfiction or fiction and hardcover or paperback.

		Format		
		Hardcover	Paperback	Total
Type	Fiction	13	59	72
	Nonfiction	15	8	23
	Total	28	67	95

(a) Find the probability of drawing a hardcover book first then a paperback fiction book second when drawing without replacement.

(b) Determine the probability of drawing a fiction book first and then a hardcover book second, when drawing without replacement.

(c) Calculate the probability of the scenario in part (b), except this time complete the calculations under the scenario where the first book is placed back on the bookcase before randomly drawing the second book.

(d) The final answers to parts (b) and (c) are very similar. Explain why this is the case.

3.25 Student outfits. In a classroom with 24 students, 7 students are wearing jeans, 4 are wearing shorts, 8 are wearing skirts, and the rest are wearing leggings. If we randomly select 3 students without replacement, what is the probability that one of the selected students is wearing leggings and the other two are wearing jeans? Note that these are mutually exclusive clothing options.

3.26 The birthday problem. Suppose we pick three people at random. For each of the following questions, ignore the special case where someone might be born on February 29th, and assume that births are evenly distributed throughout the year.

(a) What is the probability that the first two people share a birthday?

(b) What is the probability that at least two people share a birthday?

3.27 Drawing box plots. After an introductory statistics course, 80% of students can successfully construct box plots. Of those who can construct box plots, 86% passed, while only 65% of those students who could not construct box plots passed.

(a) Construct a tree diagram of this scenario.

(b) Calculate the probability that a student is able to construct a box plot if it is known that he passed.

3.28 Predisposition for thrombosis. A genetic test is used to determine if people have a predisposition for *thrombosis*, which is the formation of a blood clot inside a blood vessel that obstructs the flow of blood through the circulatory system. It is believed that 3% of people actually have this predisposition. The genetic test is 99% accurate if a person actually has the predisposition, meaning that the probability of a positive test result when a person actually has the predisposition is 0.99. The test is 98% accurate if a person does not have the predisposition. What is the probability that a randomly selected person who tests positive for the predisposition by the test actually has the predisposition?

3.29 HIV in Swaziland. Swaziland has the highest HIV prevalence in the world: 25.9% of this country's population is infected with HIV.[85] The ELISA test is one of the first and most accurate tests for HIV. For those who carry HIV, the ELISA test is 99.7% accurate. For those who do not carry HIV, the test is 92.6% accurate. If an individual from Swaziland has tested positive, what is the probability that he carries HIV?

---

[85]Source: CIA Factbook, Country Comparison: HIV/AIDS - Adult Prevalence Rate.

3.30   Exit poll. Edison Research gathered exit poll results from several sources for the Wisconsin recall election of Scott Walker. They found that 53% of the respondents voted in favor of Scott Walker. Additionally, they estimated that of those who did vote in favor for Scott Walker, 37% had a college degree, while 44% of those who voted against Scott Walker had a college degree. Suppose we randomly sampled a person who participated in the exit poll and found that he had a college degree. What is the probability that he voted in favor of Scott Walker?[86]

3.31   It's never lupus. Lupus is a medical phenomenon where antibodies that are supposed to attack foreign cells to prevent infections instead see plasma proteins as foreign bodies, leading to a high risk of blood clotting. It is believed that 2% of the population suffer from this disease. The test is 98% accurate if a person actually has the disease. The test is 74% accurate if a person does not have the disease. There is a line from the Fox television show *House* that is often used after a patient tests positive for lupus: "It's never lupus." Do you think there is truth to this statement? Use appropriate probabilities to support your answer.

3.32   Twins. About 30% of human twins are identical, and the rest are fraternal. Identical twins are necessarily the same sex – half are males and the other half are females. One-quarter of fraternal twins are both male, one-quarter both female, and one-half are mixes: one male, one female. You have just become a parent of twins and are told they are both girls. Given this information, what is the probability that they are identical?

3.7.3   The binomial formula

3.33   Exploring combinations. The formula for the number of ways to arrange n objects is $n! = n \times (n-1) \times \cdots \times 2 \times 1$. This exercise walks you through the derivation of this formula for a couple of special cases.

A small company has five employees: Anna, Ben, Carl, Damian, and Eddy. There are five parking spots in a row at the company, none of which are assigned, and each day the employees pull into a random parking spot. That is, all possible orderings of the cars in the row of spots are equally likely.

(a) On a given day, what is the probability that the employees park in alphabetical order?

(b) If the alphabetical order has an equal chance of occurring relative to all other possible orderings, how many ways must there be to arrange the five cars?

(c) Now consider a sample of 8 employees instead. How many possible ways are there to order these 8 employees' cars?

3.34   Male children. While it is often assumed that the probabilities of having a boy or a girl are the same, the actual probability of having a boy is slightly higher at 0.51. Suppose a couple plans to have 3 kids.

(a) Use the binomial model to calculate the probability that two of them will be boys.

(b) Write out all possible orderings of 3 children, 2 of whom are boys. Use these scenarios to calculate the same probability from part (a) but using the Addition Rule for disjoint events. Confirm that your answers from parts (a) and (b) match.

(c) If we wanted to calculate the probability that a couple who plans to have 8 kids will have 3 boys, briefly describe why the approach from part (b) would be more tedious than the approach from part (a).

---

[86]New York Times, Wisconsin recall exit polls.

3.35 Underage drinking, Part I. The Substance Abuse and Mental Health Services Administration estimated that 70% of 18-20 year olds consumed alcoholic beverages in 2008.[87]

(a) Suppose a random sample of ten 18-20 year olds is taken. Is the use of the binomial distribution appropriate for calculating the probability that exactly six consumed alcoholic beverages? Explain.

(b) Calculate the probability that exactly 6 out of 10 randomly sampled 18-20 year olds consumed an alcoholic drink.

(c) What is the probability that exactly 4 out of the ten 18-20 year olds have *not* consumed an alcoholic beverage?

(d) What is the probability that at most 2 out of 5 randomly sampled 18-20 year olds have consumed alcoholic beverages?

(e) What is the probability that at least 1 out of 5 randomly sampled 18-20 year olds have consumed alcoholic beverages?

3.36 Chickenpox, Part I. The National Vaccine Information Center estimates that 90% of Americans have had chickenpox by the time they reach adulthood.[88]

(a) Suppose we take a random sample of 100 American adults. Is the use of the binomial distribution appropriate for calculating the probability that exactly 97 had chickenpox before they reached adulthood? Explain.

(b) Calculate the probability that exactly 97 out of 100 randomly sampled American adults had chickenpox during childhood.

(c) What is the probability that exactly 3 out of a new sample of 100 American adults have *not* had chickenpox in their childhood?

(d) What is the probability that at least 1 out of 10 randomly sampled American adults have had chickenpox?

(e) What is the probability that at most 3 out of 10 randomly sampled American adults have *not* had chickenpox?

3.7.4 Simulations

3.37 Smog check, Part I. Suppose 16% of cars fail pollution tests (smog checks) in California. We would like to estimate the probability that an entire fleet of seven cars would pass using a simulation. We assume each car is independent. We only want to know if the entire fleet passed, i.e. none of the cars failed. What is wrong with each of the following simulations to represent whether an entire (simulated) fleet passed?

(a) Flip a coin seven times where each toss represents a car. A head means the car passed and a tail means it failed. If all cars passed, we report PASS for the fleet. If at least one car failed, we report FAIL.

(b) Read across a random number table starting at line 5. If a number is a 0 or 1, let it represent a failed car. Otherwise the car passes. We report PASS if all cars passed and FAIL otherwise.

(c) Read across a random number table, looking at two digits for each simulated car. If a pair is in the range [00-16], then the corresponding car failed. If it is in [17-99], the car passed. We report PASS if all cars passed and FAIL otherwise.

---

[87]SAMHSA, Office of Applied Studies, National Survey on Drug Use and Health, 2007 and 2008.
[88]National Vaccine Information Center, Chickenpox, The Disease & The Vaccine Fact Sheet.

3.38   Left-handed. Studies suggest that approximately 10% of the world population is left-handed. Use ten simulations to answer each of the following questions. For each question, describe your simulation scheme clearly.

(a) What is the probability that at least one out of eight people are left-handed?

(b) On average, how many people would you have to sample until the first person who is left-handed?

(c) On average, how many left-handed people would you expect to find among a random sample of six people?

3.39   Smog check, Part II. Consider the fleet of seven cars in Exercise 3.37. Remember that 16% of cars fail pollution tests (smog checks) in California, and that we assume each car is independent.

(a) Write out how to calculate the probability of the fleet failing, i.e. at least one of the cars in the fleet failing, via simulation.

(b) Simulate 5 fleets. Based on these simulations, estimate the probability at least one car will fail in a fleet.

(c) Compute the probability at least one car fails in a fleet of seven.

3.40   To catch a thief. Suppose that at a retail store, $1/5^{th}$ of all employees steal some amount of merchandise. The stores would like to put an end to this practice, and one idea is to use lie detector tests to catch and fire thieves. However, there is a problem: lie detectors are not 100% accurate. Suppose it is known that a lie detector has a failure rate of 25%. A thief will slip by the test 25% of the time and an honest employee will only pass 75% of the time.

(a) Describe how you would simulate whether an employee is honest or is a thief using a random number table. Write your simulation very carefully so someone else can read it and follow the directions exactly.

(b) Using a random number table, simulate 20 employees working at this store and determine if they are honest or not. Make sure to record the random digits assigned to each employee as you will refer back to these in part (c).

(c) Determine the result of the lie detector test for each simulated employee from part (b) using a new simulation scheme.

(d) How many of these employees are "honest and passed" and how many are "honest and failed"?

(e) How many of these employees are "thief and passed" and how many are "thief and failed"?

(f) Suppose the management decided to fire everyone who failed the lie detector test. What percent of fired employees were honest? What percent of not fired employees were thieves?

3.7.5   Random variables

3.41   College smokers. At a university, 13% of students smoke.

(a) Calculate the expected number of smokers in a random sample of 100 students from this university.

(b) The university gym opens at 9am on Saturday mornings. One Saturday morning at 8:55am there are 27 students outside the gym waiting for it to open. Should you use the same approach from part (a) to calculate the expected number of smokers among these 27 students?

3.42   Card game. Consider the following card game with a well-shuffled deck of cards. If you draw a red card, you win nothing. If you get a spade, you win $5. For any club, you win $10 plus an extra $20 for the ace of clubs.

(a) Create a probability model for the amount you win at this game. Also, find the expected winnings for a single game and the standard deviation of the winnings.

(b) What is the maximum amount you would be willing to pay to play this game? Explain.

3.43 Another card game. In a new card game, you start with a well-shuffled full deck and draw 3 cards without replacement. If you draw 3 hearts, you win $50. If you draw 3 black cards, you win $25. For any other draws, you win nothing.

(a) Create a probability model for the amount you win at this game, and find the expected winnings. Also compute the standard deviation of this distribution.

(b) If the game costs $5 to play, what would be the expected value and standard deviation of the net profit (or loss)? *(Hint: profit = winnings − cost; X − 5)*

(c) If the game costs $5 to play, should you play this game? Explain.

3.44 Is it worth it? Andy is always looking for ways to make money fast. Lately, he has been trying to make money by gambling. Here is the game he is considering playing: The game costs $2 to play. He draws a card from a deck. If he gets a number card (2-10), he wins nothing. For any face card (jack, queen or king), he wins $3. For any ace, he wins $5, and he wins an *extra* $20 if he draws the ace of clubs.

(a) Create a probability model and find Andy's expected profit per game.

(b) Would you recommend this game to Andy as a good way to make money? Explain.

3.45 Portfolio return. A portfolio's value increases by 18% during a financial boom and by 9% during normal times. It decreases by 12% during a recession. What is the expected return on this portfolio if each scenario is equally likely?

3.46 A game of roulette, Part I. The game of roulette involves spinning a wheel with 38 slots: 18 red, 18 black, and 2 green. A ball is spun onto the wheel and will eventually land in a slot, where each slot has an equal chance of capturing the ball. Gamblers can place bets on red or black. If the ball lands on their color, they double their money. If it lands on another color, they lose their money. Suppose you bet $1 on red. What's the expected value and standard deviation of your winnings?

3.47 A game of roulette, Part II. Exercise 3.46 describes winnings on a game of roulette.

(a) Suppose you play roulette and bet $3 on a single round. What is the expected value and standard deviation of your total winnings?

(b) Suppose you bet $1 in three different rounds. What is the expected value and standard deviation of your total winnings?

(c) How do your answers to parts (a) and (b) compare? What does this say about the riskiness of the two games?

3.48 Baggage fees. An airline charges the following baggage fees: $25 for the first bag and $35 for the second. Suppose 54% of passengers have no checked luggage, 34% have one piece of checked luggage and 12% have two pieces. We suppose a negligible portion of people check more than two bags.

(a) Build a probability model, compute the average revenue per passenger, and compute the corresponding standard deviation.

(b) About how much revenue should the airline expect for a flight of 120 passengers? With what standard deviation? Note any assumptions you make and if you think they are justified.

3.49 Dodgers vs. Padres. You and your friend decide to bet on the Major League Baseball game happening one evening between the Los Angeles Dodgers and the San Diego Padres. Suppose current statistics indicate that the Dodgers have a 0.46 probability of winning this game against the Padres. If your friend bets you $5 that the Dodgers will win, how much would you need to bet on the Padres to make this a fair game?

3.50 Selling on Ebay. Marcie has been tracking the following two items on Ebay:

- A textbook that sells for an average of $110 with a standard deviation of $4.

- Mario Kart for the Nintendo Wii, which sells for an average of $38 with a standard deviation of $5.

(a) Marcie wants to sell the video game and buy the textbook. How much net money (profits - losses) would she expect to make or spend? Also compute the standard deviation of how much she would make or spend.

(b) Lucy is selling the textbook on Ebay for a friend, and her friend is giving her a 10% commission (Lucy keeps 10% of the revenue). How much money should she expect to make? With what standard deviation?

3.51 Cost of breakfast. Sally gets a cup of coffee and a muffin every day for breakfast from one of the many coffee shops in her neighborhood. She picks a coffee shop each morning at random and independently of previous days. The average price of a cup of coffee is $1.40 with a standard deviation of 30¢($0.30), the average price of a muffin is $2.50 with a standard deviation of 15¢, and the two prices are independent of each other.

(a) What is the mean and standard deviation of the amount she spends on breakfast daily?

(b) What is the mean and standard deviation of the amount she spends on breakfast weekly (7 days)?

3.52 Ice cream. Ice cream usually comes in 1.5 quart boxes (48 fluid ounces), and ice cream scoops hold about 2 ounces. However, there is some variability in the amount of ice cream in a box as well as the amount of ice cream scooped out. We represent the amount of ice cream in the box as X and the amount scooped out as Y. Suppose these random variables have the following means, standard deviations, and variances:

	mean	SD	variance
X	48	1	1
Y	2	0.25	0.0625

(a) An entire box of ice cream, plus 3 scoops from a second box is served at a party. How much ice cream do you expect to have been served at this party? What is the standard deviation of the amount of ice cream served?

(b) How much ice cream would you expect to be left in the box after scooping out one scoop of ice cream? That is, find the expected value of $X - Y$. What is the standard deviation of the amount left in the box?

(c) Using the context of this exercise, explain why we add variances when we subtract one random variable from another.

3.7.6 Continuous distributions

3.53 Cat weights. The histogram shown below represents the weights (in kg) of 47 female and 97 male cats.[89]

(a) What fraction of these cats weigh less than 2.5 kg?

(b) What fraction of these cats weigh between 2.5 and 2.75 kg?

(c) What fraction of these cats weigh between 2.75 and 3.5 kg?

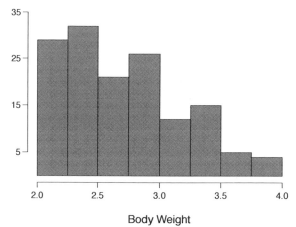

Body Weight

3.54 Income and gender. The relative frequency table below displays the distribution of annual total personal income (in 2009 inflation-adjusted dollars) for a representative sample of 96,420,486 Americans. These data come from the American Community Survey for 2005-2009. This sample is comprised of 59% males and 41% females.[90]

(a) Describe the distribution of total personal income.

(b) What is the probability that a randomly chosen US resident makes less than $50,000 per year?

(c) What is the probability that a randomly chosen US resident makes less than $50,000 per year and is female? Note any assumptions you make.

(d) The same data source indicates that 71.8% of females make less than $50,000 per year. Use this value to determine whether or not the assumption you made in part (c) is valid.

Income	Total
$1 to $9,999 or loss	2.2%
$10,000 to $14,999	4.7%
$15,000 to $24,999	15.8%
$25,000 to $34,999	18.3%
$35,000 to $49,999	21.2%
$50,000 to $64,999	13.9%
$65,000 to $74,999	5.8%
$75,000 to $99,999	8.4%
$100,000 or more	9.7%

[89]W. N. Venables and B. D. Ripley. *Modern Applied Statistics with S.* Fourth Edition. www.stats.ox.ac.uk/pub/MASS4. New York: Springer, 2002.

[90]U.S. Census Bureau, 2005-2009 American Community Survey.

Chapter 4

Distributions of random variables

4.1 Normal distribution

Among all the distributions we see in practice, one is overwhelmingly the most common. The symmetric, unimodal, bell curve is ubiquitous throughout statistics. Indeed it is so common, that people often know it as the **normal curve** or **normal distribution**,[1] shown in Figure 4.1. Variables such as SAT scores and heights of US adult males closely follow the normal distribution.

Normal distribution facts

Many variables are nearly normal, but none are exactly normal. The normal distribution, while never perfect, provides very close approximations for a variety of scenarios. We will use it in data exploration and to solve important problems in statistics.

---

[1]It is also introduced as the Gaussian distribution after Frederic Gauss, the first person to formalize its mathematical expression.

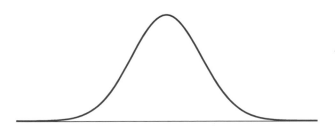

Figure 4.1: A normal curve.

4.1.1 Normal distribution model

The normal distribution model always describes a symmetric, unimodal, bell-shaped curve. However, these curves can look different depending on the details of the model. Specifically, the normal distribution model can be adjusted using two parameters: mean and standard deviation. As you can probably guess, changing the mean shifts the bell curve to the left or right, while changing the standard deviation stretches or constricts the curve. Figure 4.2 shows the normal distribution with mean 0 and standard deviation 1 in the left panel and the normal distributions with mean 19 and standard deviation 4 in the right panel. Figure 4.3 shows these distributions on the same axis.

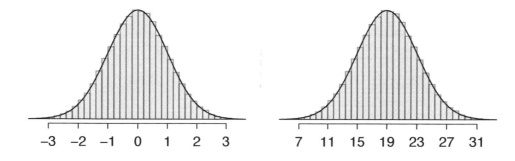

Figure 4.2: Both curves represent the normal distribution, however, they differ in their center and spread. The normal distribution with mean 0 and standard deviation 1 is called the **standard normal distribution**.

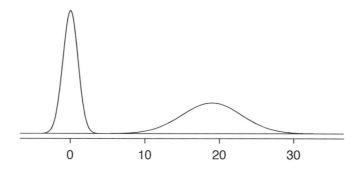

Figure 4.3: The normal models shown in Figure 4.2 but plotted together and on the same scale.

Because the mean and standard deviation describe a normal distribution exactly, they are called the distribution's **parameters**.

	SAT	ACT
Mean	1500	21
SD	300	5

Table 4.4: Mean and standard deviation for the SAT and ACT.

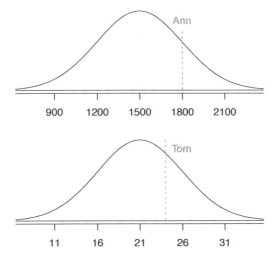

Figure 4.5: Ann's and Tom's scores shown with the distributions of SAT and ACT scores.

4.1.2   Standardizing with Z-scores

● **Example 4.1**   Table 4.4 shows the mean and standard deviation for total scores on the SAT and ACT. The distribution of SAT and ACT scores are both nearly normal. Suppose Ann scored 1800 on her SAT and Tom scored 24 on his ACT. Who performed better?

Since the two distributions are on different scales, we use the standard deviation as a guide. Ann is 1 standard deviation above average on the SAT: $1500 + 300 = 1800$. Tom is 0.6 standard deviations above the mean on the ACT: $21 + 0.6 \times 5 = 24$. In Figure 4.5, we can see that Ann tends to do better with respect to everyone else than Tom did, so her score was better.

Example 4.1 used a standardization technique called a Z-score, a method most commonly employed for nearly normal observations but that may be used with any distribution. The **Z-score** of an observation is defined as the number of standard deviations it falls above or below the mean. If the observation is one standard deviation above the mean, its Z-score is 1. If it is 1.5 standard deviations *below* the mean, then its Z-score is -1.5. If x is an observation from a distribution with mean μ and standard deviation σ, we define the Z-score mathematically as

Z

Z-score, the standardized observation

$$Z = \frac{x - \mu}{\sigma}$$

Using $\mu_{SAT} = 1500$, $\sigma_{SAT} = 300$, and $x_{Ann} = 1800$, we find Ann's Z-score:

$$Z_{Ann} = \frac{x_{Ann} - \mu_{SAT}}{\sigma_{SAT}} = \frac{1800 - 1500}{300} = 1$$

The Z-score

The Z-score of an observation is the number of standard deviations it falls above or below the mean. We compute the Z-score for an observation x that follows a distribution with mean μ and standard deviation σ using

$$Z = \frac{x - \mu}{\sigma}$$

$\odot$ **Guided Practice 4.2** Use Tom's ACT score, 24, along with the ACT mean of 21 and standard deviation of 5 to compute his Z-score.[2]

Observations above the mean always have positive Z-scores while those below the mean have negative Z-scores. If an observation is equal to the mean (e.g. SAT score of 1500), then the Z-score is 0.

$\odot$ **Guided Practice 4.3** Let X represent a random variable from a distribution with $\mu = 3$ and $\sigma = 2$, and suppose we observe $x = 5.19$. (a) Find the Z-score of x. (b) Use the Z-score to determine how many standard deviations above or below the mean x falls.[3]

$\odot$ **Guided Practice 4.4** Head lengths of brushtail possums follow a nearly normal distribution with mean 92.6 mm and standard deviation 3.6 mm. Compute the Z-scores for possums with head lengths of 95.4 mm and 85.8 mm.[4]

We can use Z-scores to roughly identify which observations are more unusual than others. One observation x_1 is said to be more unusual than another observation x_2 if the absolute value of its Z-score is larger than the absolute value of the other observation's Z-score: $|Z_1| > |Z_2|$. This technique is especially insightful when a distribution is symmetric.

$\odot$ **Guided Practice 4.5** Which of the observations in Guided Practice 4.4 is more unusual?[5]

4.1.3   Normal probability table

$\bullet$ **Example 4.6** Ann from Example 4.1 earned a score of 1800 on her SAT with a corresponding $Z = 1$. She would like to know what percentile she falls in among all SAT test-takers.

Ann's **percentile** is the percentage of people who earned a lower SAT score than Ann. We shade the area representing those individuals in Figure 4.6. The total area under the normal curve is always equal to 1, and the proportion of people who scored below Ann on the SAT is equal to the *area* shaded in Figure 4.6: 0.8413. In other words, Ann is in the 84th percentile of SAT takers.

---

[2] $Z_{Tom} = \frac{x_{Tom} - \mu_{ACT}}{\sigma_{ACT}} = \frac{24 - 21}{5} = 0.6$

[3] (a) Its Z-score is given by $Z = \frac{x - \mu}{\sigma} = \frac{5.19 - 3}{2} = 2.19/2 = 1.095$. (b) The observation x is 1.095 standard deviations *above* the mean. We know it must be above the mean since Z is positive.

[4] For $x_1 = 95.4$ mm: $Z_1 = \frac{x_1 - \mu}{\sigma} = \frac{95.4 - 92.6}{3.6} = 0.78$. For $x_2 = 85.8$ mm: $Z_2 = \frac{85.8 - 92.6}{3.6} = -1.89$.

[5] Because the *absolute value* of Z-score for the second observation is larger than that of the first, the second observation has a more unusual head length.

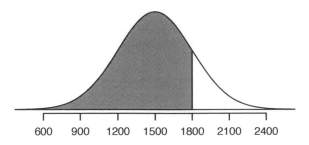

Figure 4.6: The normal model for SAT scores, shading the area of those individuals who scored below Ann.

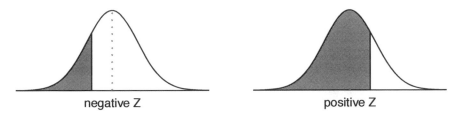

Figure 4.7: The area to the left of Z represents the percentile of the observation.

We can use the normal model to find percentiles. A **normal probability table**, which lists Z-scores and corresponding percentiles, can be used to identify a percentile based on the Z-score (and vice versa). Statistical software can also be used.

A normal probability table is given in Appendix B.2 on page 449 and abbreviated in Table 4.8. We use this table to identify the percentile corresponding to any particular Z-score. For instance, the percentile of $Z = 0.43$ is shown in row 0.4 and column 0.03 in Table 4.8: 0.6664, or the 66.64^{th} percentile. Generally, we round Z to two decimals, identify the proper row in the normal probability table up through the first decimal, and then determine the column representing the second decimal value. The intersection of this row and column is the percentile of the observation.

We can also find the Z-score associated with a percentile. For example, to identify Z for the 80th percentile, we look for the value closest to 0.8000 in the middle portion of the table: 0.7995. We determine the Z-score for the 80th percentile by combining the row and column Z values: 0.84.

⊙ **Guided Practice 4.7**   Determine the proportion of SAT test takers who scored better than Ann on the SAT.[6]

4.1.4   Normal probability examples

Cumulative SAT scores are approximated well by a normal model with mean 1500 and standard deviation 300.

---

[6]If 84% had lower scores than Ann, the number of people who had better scores must be 16%. (Generally ties are ignored when the normal model, or any other continuous distribution, is used.)

| Z | Second decimal place of Z | | | | | | | | | |
	0.00	0.01	0.02	0.03	0.04	0.05	0.06	0.07	0.08	0.09
0.0	0.5000	0.5040	0.5080	0.5120	0.5160	0.5199	0.5239	0.5279	0.5319	0.5359
0.1	0.5398	0.5438	0.5478	0.5517	0.5557	0.5596	0.5636	0.5675	0.5714	0.5753
0.2	0.5793	0.5832	0.5871	0.5910	0.5948	0.5987	0.6026	0.6064	0.6103	0.6141
0.3	0.6179	0.6217	0.6255	0.6293	0.6331	0.6368	0.6406	0.6443	0.6480	0.6517
0.4	0.6554	0.6591	0.6628	0.6664	0.6700	0.6736	0.6772	0.6808	0.6844	0.6879
0.5	0.6915	0.6950	0.6985	0.7019	0.7054	0.7088	0.7123	0.7157	0.7190	0.7224
0.6	0.7257	0.7291	0.7324	0.7357	0.7389	0.7422	0.7454	0.7486	0.7517	0.7549
0.7	0.7580	0.7611	0.7642	0.7673	0.7704	0.7734	0.7764	0.7794	0.7823	0.7852
0.8	0.7881	0.7910	0.7939	0.7967	**0.7995**	0.8023	0.8051	0.8078	0.8106	0.8133
0.9	0.8159	0.8186	0.8212	0.8238	0.8264	0.8289	0.8315	0.8340	0.8365	0.8389
1.0	0.8413	0.8438	0.8461	0.8485	0.8508	0.8531	0.8554	0.8577	0.8599	0.8621
1.1	0.8643	0.8665	0.8686	0.8708	0.8729	0.8749	0.8770	0.8790	0.8810	0.8830
⋮	⋮	⋮	⋮	⋮	⋮	⋮	⋮	⋮	⋮	⋮

Table 4.8: A section of the normal probability table. The percentile for a normal random variable with $Z = 0.43$ has been *highlighted*, and the percentile closest to 0.8000 has also been **highlighted**.

● **Example 4.8** What is the probability that a randomly selected SAT taker scores at least 1630 on the SAT?

The probability that a randomly selected SAT taker scores at least 1630 on the SAT is equivalent to the proportion of all SAT takers that score at least 1630 on the SAT. First, always draw and label a picture of the normal distribution. (Drawings need not be exact to be useful.) We are interested in the probability that a randomly selected score will be above 1630, so we shade this upper tail:

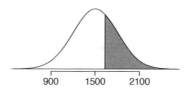

The picture shows the mean and the values at 2 standard deviations above and below the mean. The simplest way to find the shaded area under the curve makes use of the Z-score of the cutoff value. With $\mu = 1500$, $\sigma = 300$, and the cutoff value $x = 1630$, the Z-score is computed as

$$Z = \frac{x - \mu}{\sigma} = \frac{1630 - 1500}{300} = \frac{130}{300} = 0.43$$

We look up the percentile of $Z = 0.43$ in the normal probability table shown in Table 4.8 or in Appendix B.2 on page 449, which yields 0.6664. However, the percentile describes those who had a Z-score *lower* than 0.43. To find the area *above* $Z = 0.43$, we compute one minus the area of the lower tail:

$$1.0000 \quad - \quad 0.6664 \quad = \quad 0.3336$$

The probability that a randomly selected score is at least 1630 on the SAT is 0.3336.

TIP: Always draw a picture first, and find the Z-score second
For any normal probability situation, *always always always* draw and label the normal curve and shade the area of interest first. The picture will provide an estimate of the probability.

After drawing a figure to represent the situation, identify the Z-score for the observation of interest.

⊙ **Guided Practice 4.9**   If the probability that a randomly selected score is at least 1630 is 0.3336, what is the probability that the score is less than 1630? Draw the normal curve representing this exercise, shading the lower region instead of the upper one.[7]

● **Example 4.10**   Edward earned a 1400 on his SAT. What is his percentile?

First, a picture is needed. Edward's percentile is the proportion of people who do not get as high as a 1400. These are the scores to the left of 1400.

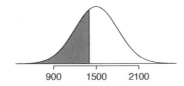

Identifying the mean $\mu = 1500$, the standard deviation $\sigma = 300$, and the cutoff for the tail area $x = 1400$ makes it easy to compute the Z-score:

$$Z = \frac{x - \mu}{\sigma} = \frac{1400 - 1500}{300} = -0.33$$

Using the normal probability table, identify the row of -0.3 and column of 0.03, which corresponds to the probability 0.3707. Edward is at the 37th percentile.

⊙ **Guided Practice 4.11**   Use the results of Example 4.10 to compute the proportion of SAT takers who did better than Edward. Also draw a new picture.[8]

---

[7]We found the probability in Example 4.8: 0.6664. A picture for this exercise is represented by the shaded area below "0.6664" in Example 4.8.

[8]If Edward did better than 37% of SAT takers, then about 63% must have done better than him.

> **TIP: Areas to the right**
> The normal probability table in most books gives the area to the left. If you would like the area to the right, first find the area to the left and then subtract this amount from one.

The last several problems have focused on finding the probability or percentile for a particular observation. It is also possible to identify the value corresponding to a particular percentile.

● **Example 4.12** Carlos believes he can get into his preferred college if he scores at least in the 80th percentile on the SAT. What score should he aim for?

Here, we are given a percentile rather than a Z-score, so we work backwards. As always, first draw the picture.

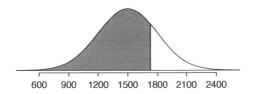

We want to find the observation that corresponds to the 80th percentile. First, we find the Z-score associated with the 80th percentile using the normal probability table. Looking at Table 4.8., we look for the number closest to 0.80 *inside* the table. The closest number we find is 0.7995 (highlighted). 0.7995 falls on row 0.8 and column 0.04, therefore it corresponds to a Z-score of 0.84. In any normal distribution, a value with a Z-score of 0.84 will be at the 80th percentile. Once we have the Z-score, we work backwards to find x.

$$Z = \frac{x - \mu}{\sigma}$$
$$0.84 = \frac{x - 1500}{300}$$
$$0.84 \times 300 + 1500 = x$$
$$x = 1752$$

The 80th percentile on the SAT corresponds to a score of 1752.

⊙ **Guided Practice 4.13** Imani scored at the 72nd percentile on the SAT. What was her SAT score?[9]

> **Caution: If the data are not nearly normal, don't use a normal table**
> Before using the normal table, verify that the data or distribution is approximately normal. If it is not, the normal table will give incorrect results. Also, all answers based on normal approximations are approximations and are not exact.

---

[9]First, draw a picture! The closest percentile in the table to 0.72 is 0.7190, which corresponds to $Z = 0.58$. Next, set up the Z-score formula and solve for x: $0.58 = \frac{x-1500}{300} \rightarrow x = 1674$. Imani scored 1674.

4.1.5   Calculator: finding normal probabilities

▶ TI-84: Finding area under the normal curve

Use 2ND VARS, `normalcdf` to find an area/proportion/probability to the left or right of a Z-score or between two Z-scores.

1. Choose 2ND VARS (i.e. DISTR).
2. Choose 2:`normalcdf`.
3. Enter the Z-scores that correspond to the lower (left) and upper (right) bounds.
4. Leave μ as 0 and σ as 1.
5. Down arrow, choose Paste, and hit ENTER.

TI-83: Do steps 1-2, then enter the lower bound and upper bound separated by a comma, e.g. `normalcdf(2, 5)`, and hit ENTER.

▶ Casio fx-9750GII: Finding area under the normal curve

1. Navigate to STAT (MENU, then hit 2).
2. Select DIST (F5), then NORM (F1), and then Ncd (F2).
3. If needed, set Data to Variable (Var option, which is F2).
4. Enter the Lower Z-score and the Upper Z-score. Set σ to 1 and μ to 0.
   - If finding just a lower tail area, set Lower to -12.
   - For an upper tail area, set Upper to 12.
5. Hit EXE, which will return the area probability (p) along with the Z-scores for the lower and upper bounds.

● **Example 4.14**   Use a calculator to determine what percentile corresponds to a Z-score of 1.5.

Always first sketch a graph:[10]

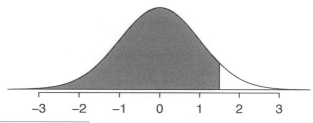

---
[10]`normalcdf` gives the result without drawing the graph. To draw the graph, do 2nd VARS, DRAW, 1:ShadeNorm. However, beware of errors caused by other plots that might interfere with this plot.

To find an area under the normal curve using a calculator, first identify a lower bound and an upper bound. Theoretically, we want all of the area to the left of 1.5, so the left endpoint should be -∞. However, the area under the curve is nearly negligible when Z is smaller than -4, so we will use -5 as the lower bound when not given a lower bound (any other negative number smaller than -5 will also work). Using a lower bound of -5 and an upper bound of 1.5, we get $P(Z < 1.5) = 0.933$.

⊙ **Guided Practice 4.15**   Find the area under the normal curve to right of $Z = 2$. [11]

⊙ **Guided Practice 4.16**   Find the area under the normal curve between -1.5 and 1.5. [12]

▶ TI-84: Find a Z-score that corresponds to a percentile

Use 2ND VARS, invNorm to find the Z-score that corresponds to a given percentile.

1. Choose 2ND VARS (i.e. DISTR).
2. Choose 3:invNorm.
3. Let Area be the percentile as a decimal (the area to the left of desired Z-score).
4. Leave μ as 0 and σ as 1.
5. Down arrow, choose Paste, and hit ENTER.

TI-83: Do steps 1-2, then enter the percentile as a decimal, e.g. invNorm(.40), then hit ENTER.

▶ Casio fx-9750GII: Find a Z-score that corresponds to a percentile

1. Navigate to STAT (MENU, then hit 2).
2. Select DIST (F5), then NORM (F1), and then InvN (F3).
3. If needed, set Data to Variable (Var option, which is F2).
4. Decide which tail area to use (Tail), the tail area (Area), and then enter the σ and μ values.
5. Hit EXE.

● **Example 4.17**   Use a calculator to find the Z-score that corresponds to the 40th percentile.

Letting Area be 0.40, a calculator gives -0.253. This means that $Z = -0.253$ corresponds to the 40th percentile, that is, $P(Z < -0.253) = 0.40$.

---

[11] Now we want to shade to the right. Therefore our lower bound will be 2 and the upper bound will be +5 (or a number bigger than 5) to get $P(Z > 2) = 0.023$.

[12] Here we are given both the lower and the upper bound. Lower bound is -1.5 and upper bound is 1.5. The area under the normal curve between -1.5 and 1.5 = $P(-1.5 < Z < 1.5) = 0.866$.

⊙ **Guided Practice 4.18**   Find the Z-score such that 20 percent of the area is to the right of that Z-score.[13]

● **Example 4.19**   In a large study of birth weight of newborns, the weights of 23,419 newborn boys were recorded.[14] The distribution of weights was approximately normal with a mean of 7.44 lbs (3376 grams) and a standard deviation of 1.33 lbs (603 grams). The government classifies a newborn as having low birth weight if the weight is less than 5.5 pounds. What percent of these newborns had a low birth weight?

We find an area under the normal curve between -5 (or a number smaller than -5, e.g. -10) and a Z-score that we will calculate. There is no need to write calculator commands in a solution. Instead, continue to use standard statistical notation.

$$Z = \frac{5.5 - 7.44}{1.33}$$
$$= -1.49$$
$$P(Z < -1.49) = 0.068$$

Approximately 6.8% of the newborns were of low birth weight.

⊙ **Guided Practice 4.20**   Approximately what percent of these babies weighed greater than 10 pounds?[15]

⊙ **Guided Practice 4.21**   Approximately *how many* of these newborns weighed greater than 10 pounds?[16]

⊙ **Guided Practice 4.22**   How much would a newborn have to weigh in order to be at the 90th percentile among this group?[17]

4.1.6   68-95-99.7 rule

Here, we present a useful rule of thumb for the probability of falling within 1, 2, and 3 standard deviations of the mean in the normal distribution. This will be useful in a wide range of practical settings, especially when trying to make a quick estimate without a calculator or Z table.

⊙ **Guided Practice 4.23**   Use the Z table to confirm that about 68%, 95%, and 99.7% of observations fall within 1, 2, and 3, standard deviations of the mean in the normal distribution, respectively. For instance, first find the area that falls between $Z = -1$ and $Z = 1$, which should have an area of about 0.68. Similarly there should be an area of about 0.95 between $Z = -2$ and $Z = 2$.[18]

---

[13]If 20% of the area is the right, then 80% of the area is to the left. Letting area be 0.80, we get $Z = 0.841$.

[14]www.biomedcentral.com/1471-2393/8/5

[15]$Z = \frac{10-7.44}{1.33} = 1.925$. Using a lower bound of 2 and an upper bound of 5, we get $P(Z > 1.925) = 0.027$. Approximately 2.7% of the newborns weighed over 10 pounds.

[16]Approximately 2.7% of the newborns weighed over 10 pounds. Because there were 23,419 of them, about $0.027 \times 23419 \approx 632$ weighed greater than 10 pounds.

[17]Because we have the percentile, this is the inverse problem. To get the Z-score, use the inverse normal option with 0.90 to get $Z = 1.28$. Then solve for x in $1.28 = \frac{x-7.44}{1.33}$ to get $x = 9.15$. To be at the 90th percentile among this group, a newborn would have to weigh 9.15 pounds.

[18]First draw the pictures. To find the area between $Z = -1$ and $Z = 1$, use the normal probability table to determine the areas below $Z = -1$ and above $Z = 1$. Next verify the area between $Z = -1$ and $Z = 1$ is about 0.68. Repeat this for $Z = -2$ to $Z = 2$ and also for $Z = -3$ to $Z = 3$.

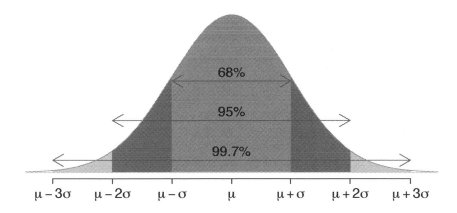

Figure 4.9: Probabilities for falling within 1, 2, and 3 standard deviations of the mean in a normal distribution.

It is possible for a normal random variable to fall 4, 5, or even more standard deviations from the mean. However, these occurrences are very rare if the data are nearly normal. The probability of being further than 4 standard deviations from the mean is about 1-in-30,000. For 5 and 6 standard deviations, it is about 1-in-3.5 million and 1-in-1 billion, respectively.

⊙ **Guided Practice 4.24** SAT scores closely follow the normal model with mean $\mu = 1500$ and standard deviation $\sigma = 300$. (a) About what percent of test takers score 900 to 2100? (b) What percent score between 1500 and 2100?[19]

4.1.7 Evaluating the normal approximation

It is important to remember normality is always an approximation. Testing the appropriateness of the normal assumption is a key step in many data analyses.

The distribution of heights of US males is well approximated by the normal model. We are interested in proceeding under the assumption that the data are normally distributed, but first we must check to see if this is reasonable.

There are two visual methods for checking the assumption of normality that can be implemented and interpreted quickly. The first is a simple histogram with the best fitting normal curve overlaid on the plot, as shown in the left panel of Figure 4.10. The sample mean $\bar{x}$ and standard deviation s are used as the parameters of the best fitting normal curve. The closer this curve fits the histogram, the more reasonable the normal model assumption. Another more common method is examining a **normal probability plot**,[20] shown in the right panel of Figure 4.10. The closer the points are to a perfect straight line, the more confident we can be that the data follow the normal model.

---

[19](a) 900 and 2100 represent two standard deviations above and below the mean, which means about 95% of test takers will score between 900 and 2100. (b) Since the normal model is symmetric, then half of the test takers from part (a) ($\frac{95\%}{2} = 47.5\%$ of all test takers) will score 900 to 1500 while 47.5% score between 1500 and 2100.

[20]Also commonly called a **quantile-quantile plot**.

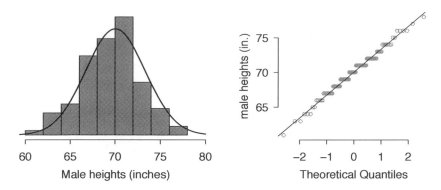

Figure 4.10: A sample of 100 male heights. The observations are rounded to the nearest whole inch, explaining why the points appear to jump in increments in the normal probability plot.

● **Example 4.25** Three data sets of 40, 100, and 400 samples were simulated from a normal distribution, and the histograms and normal probability plots of the data sets are shown in Figure 4.11. These will provide a benchmark for what to look for in plots of real data.

The left panels show the histogram (top) and normal probability plot (bottom) for the simulated data set with 40 observations. The data set is too small to really see clear structure in the histogram. The normal probability plot also reflects this, where there are some deviations from the line. However, these deviations are not strong.

The middle panels show diagnostic plots for the data set with 100 simulated observations. The histogram shows more normality and the normal probability plot shows a better fit. While there is one observation that deviates noticeably from the line, it is not particularly extreme.

The data set with 400 observations has a histogram that greatly resembles the normal distribution, while the normal probability plot is nearly a perfect straight line. Again in the normal probability plot there is one observation (the largest) that deviates slightly from the line. If that observation had deviated 3 times further from the line, it would be of much greater concern in a real data set. Apparent outliers can occur in normally distributed data but they are rare.

Notice the histograms look more normal as the sample size increases, and the normal probability plot becomes straighter and more stable.

● **Example 4.26** Are NBA player heights normally distributed? Consider all 435 NBA players from the 2008-9 season presented in Figure 4.12.[21]

We first create a histogram and normal probability plot of the NBA player heights. The histogram in the left panel is slightly left skewed, which contrasts with the symmetric normal distribution. The points in the normal probability plot do not appear to closely follow a straight line but show what appears to be a "wave". We can compare these characteristics to the sample of 400 normally distributed observations in Example 4.25 and see that they represent much stronger deviations from the normal model. NBA player heights do not appear to come from a normal distribution.

---

[21]These data were collected from www.nba.com.

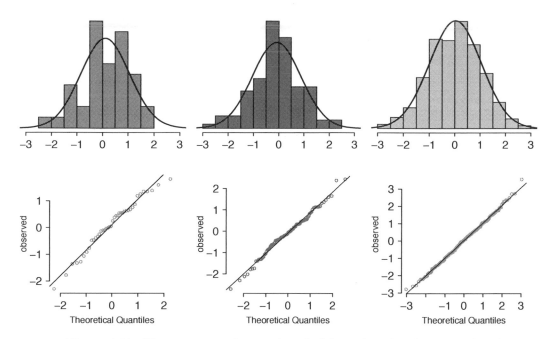

Figure 4.11: Histograms and normal probability plots for three simulated normal data sets; $n = 40$ (left), $n = 100$ (middle), $n = 400$ (right).

● **Example 4.27** Can we approximate poker winnings by a normal distribution? We consider the poker winnings of an individual over 50 days. A histogram and normal probability plot of these data are shown in Figure 4.13.

The data are very strongly right skewed in the histogram, which corresponds to the very strong deviations on the upper right component of the normal probability plot. If we compare these results to the sample of 40 normal observations in Example 4.25, it is apparent that these data show very strong deviations from the normal model.

⊙ **Guided Practice 4.28** Determine which data sets represented in Figure 4.14 plausibly come from a nearly normal distribution. Are you confident in all of your conclusions? There are 100 (top left), 50 (top right), 500 (bottom left), and 15 points (bottom right) in the four plots.[22]

⊙ **Guided Practice 4.29** Figure 4.15 shows normal probability plots for two distributions that are skewed. One distribution is skewed to the low end (left skewed) and the other to the high end (right skewed). Which is which?[23]

---

[22]Answers may vary a little. The top-left plot shows some deviations in the smallest values in the data set; specifically, the left tail of the data set has some outliers we should be wary of. The top-right and bottom-left plots do not show any obvious or extreme deviations from the lines for their respective sample sizes, so a normal model would be reasonable for these data sets. The bottom-right plot has a consistent curvature that suggests it is not from the normal distribution. If we examine just the vertical coordinates of these observations, we see that there is a lot of data between -20 and 0, and then about five observations scattered between 0 and 70. This describes a distribution that has a strong right skew.

[23]Examine where the points fall along the vertical axis. In the first plot, most points are near the low end with fewer observations scattered along the high end; this describes a distribution that is skewed to the high end. The second plot shows the opposite features, and this distribution is skewed to the low end.

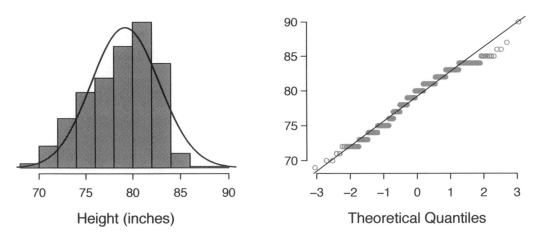

Figure 4.12: Histogram and normal probability plot for the NBA heights from the 2008-9 season.

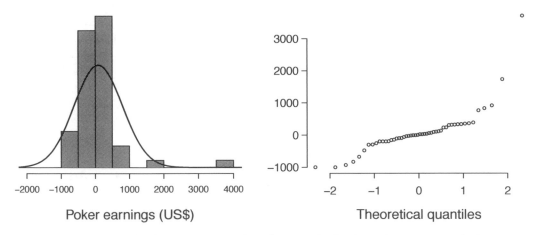

Figure 4.13: A histogram of poker data with the best fitting normal plot and a normal probability plot.

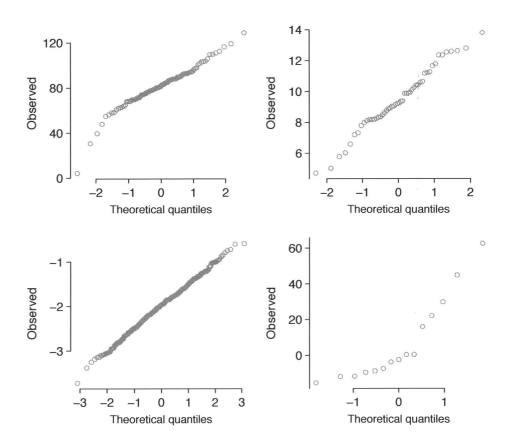

Figure 4.14: Four normal probability plots for Guided Practice 4.28.

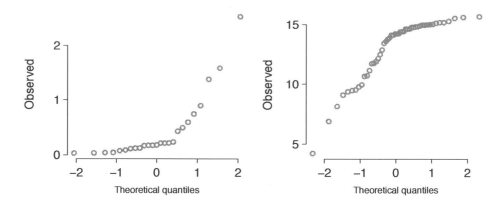

Figure 4.15: Normal probability plots for Guided Practice 4.29.

4.1.8   Normal approximation for sums of random variables

We have seen that many distributions are approximately normal. The sum and the difference of normally distributed variables is also normal. While we cannot prove this here, the usefulness of it is seen in the following example.

● **Example 4.30**   Three friends are playing a cooperative video game in which they have to complete a puzzle as fast as possible. Assume that the individual times of the 3 friends are independent of each other. The individual times of the friends in similar puzzles are approximately normally distributed with the following means and standard deviations.

	Mean	SD
Friend 1	5.6	0.11
Friend 2	5.8	0.13
Friend 3	6.1	0.12

To advance to the next level of the game, the friends' total time must not exceed 17.1 minutes. What is the probability that they will advance to the next level?

Because each friend's time is approximately normally distributed, *the sum of their times is also approximately normally distributed*. We will do a normal approximation, but first we need to find the mean and standard deviation of the *sum*. We learned how to do this in Section 3.5.

Let the three friends be labeled X, Y, Z. We want $P(X + Y + Z < 17.1)$. The mean and standard deviation of the sum of X, Y, and Z is given by:

$$
\begin{aligned}
\mu_{sum} &= E(X + Y + Z) & \sigma_{sum} &= \sqrt{(SD_X)^2 + (SD_Y)^2 + (SD_Z)^2} \\
&= E(X) + E(Y) + E(Z) & &= \sqrt{(0.11)^2 + (0.13)^2 + (0.12)^2} \\
&= 4.6 + 4.8 + 4.5 & &= 0.208 \\
&= 17.5
\end{aligned}
$$

Now we can find the Z-score.

$$Z = \frac{x_{sum} - \mu_{sum}}{\sigma_{sum}}$$
$$= \frac{17.1 - 17.5}{0.208}$$
$$= -1.92$$

Finally, we want the probability that the sum is less than 17.5, so we shade the area to the left of $Z = -1.92$. Using the normal table or a calculator, we get

$$P(Z < -1.92) = 0.027$$

There is a 2.7% chance that the friends will advance to the next level.

⊙ **Guided Practice 4.31** What is the probability that Friend 2 will complete the puzzle with a faster time than Friend 1? Hint: find $P(Y < X)$, or $P(Y - X < 0)$.[24]

4.2 Sampling distribution of a sample mean

4.2.1 The mean and standard deviation of $\bar{x}$

In this section we consider a data set called run10, which represents all 16,924 runners who finished the 2012 Cherry Blossom 10 mile run in Washington, DC.[25] Part of this data set is shown in Table 4.16, and the variables are described in Table 4.17.

ID	time	age	gender	state
1	92.25	38.00	M	MD
2	106.35	33.00	M	DC
3	89.33	55.00	F	VA
4	113.50	24.00	F	VA
⋮	⋮	⋮	⋮	⋮
16923	122.87	37.00	F	VA
16924	93.30	27.00	F	DC

Table 4.16: Six observations from the run10 data set.

variable	description
time	Ten mile run time, in minutes
age	Age, in years
gender	Gender (M for male, F for female)
state	Home state (or country if not from the US)

Table 4.17: Variables and their descriptions for the run10 data set.

---

[24]First find the mean and standard deviation of $Y - X$. The mean of $Y - X$ is $\mu_{Y-X} = 5.8 - 5.6 = 0.2$. The standard deviation is $SD_{Y-X} = \sqrt{(0.13)^2 + (0.11)^2} = 0.170$. Then $Z = \frac{0-0.2}{0.170} = -1.18$ and $P(Z < -1.18) = .119$. There is an 11.9% chance that Friend 2 will complete the puzzle with a faster time than Friend 1.

[25]www.cherryblossom.org

ID	time	age	gender	state
1983	88.31	59	M	MD
8192	100.67	32	M	VA
11020	109.52	33	F	VA
⋮	⋮	⋮	⋮	⋮
1287	89.49	26	M	DC

Table 4.18: Four observations for the `run10Samp` data set, which represents a simple random sample of 100 runners from the 2012 Cherry Blossom Run.

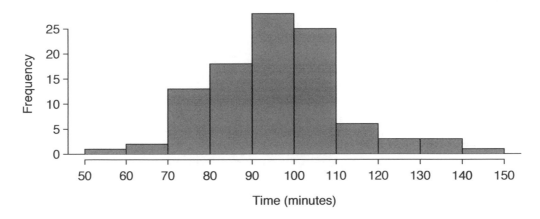

Figure 4.19: Histogram of `time` for a single sample of size 100. The average of the sample is in the mid-90s and the standard deviation of the sample $s \approx 16$ minutes.

These data are special because they include the results for the entire population of runners who finished the 2012 Cherry Blossom Run. We took a simple random sample of this population, which is represented in Table 4.18. A histogram summarizing the time variable in the `run10Samp` data set is shown in Figure 4.19.

From the random sample represented in `run10Samp`, we guessed the average time it takes to run 10 miles is 95.61 minutes. Suppose we take another random sample of 100 individuals and take its mean: 95.30 minutes. Suppose we took another (93.43 minutes) and another (94.16 minutes), and so on. If we do this many many times – which we can do only because we have the entire population data set – we can build up a **sampling distribution** for the sample mean when the sample size is 100, shown in Figure 4.20.

Sampling distribution

The sampling distribution represents the distribution of the point estimates based on samples of a fixed size from a certain population. It is useful to think of a point estimate as being drawn from such a distribution. Understanding the concept of a sampling distribution is central to understanding statistical inference.

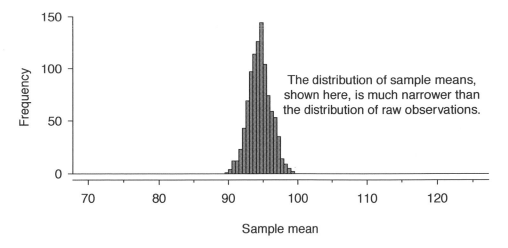

Figure 4.20: A histogram of 1000 sample means for run time, where the samples are of size $n = 100$. This histogram approximates the true sampling distribution of the sample mean, with mean $\mu_{\bar{x}}$ and standard deviation $\sigma_{\bar{x}}$.

The sampling distribution shown in Figure 4.20 is unimodal and approximately symmetric. It is also centered exactly at the true population mean: $\mu = 94.52$. Intuitively, this makes sense. The sample mean should be an unbiased estimator of the population mean. Because we are considering the distribution of the sample mean, we will use $\mu_{\bar{x}} = 94.52$ to describe the true mean of this distribution.

We can see that the sample mean has some variability around the population mean, which can be quantified using the standard deviation of this distribution of sample means. The standard deviation of the sample mean tells us how far the typical estimate is away from the actual population mean, 94.52 minutes. It also describes the typical **error** of a single estimate, and is denoted by the symbol $\sigma_{\bar{x}}$.

Standard deviation of an estimate

The standard deviation associated with an estimate describes the typical error or uncertainty associated with the estimate.

● **Example 4.32** Looking at Figures 4.19 and 4.20, we see that the standard deviation of the sample mean with $n = 100$ is much smaller than the standard deviation of a single sample. Interpret this statement and explain why it is true.

―――――――

The variation from one sample mean to another sample mean is much smaller than the variation from one individual to another individual. This makes sense because when we average over 100 values, the large and small values tend to cancel each other out. While many individuals have a time under 90 minutes, it would be unlikely for the *average* of 100 runners to be less than 90 minutes.

⊙ **Guided Practice 4.33**   (a) Would you rather use a small sample or a large sample when estimating a parameter? Why? (b) Using your reasoning from (a), would you expect a point estimate based on a small sample to have smaller or larger standard deviation than a point estimate based on a larger sample?[26]

When considering how to calculate the standard deviation of a sample mean, there is one problem: there is no obvious way to estimate this from a single sample. However, statistical theory provides a helpful tool to address this issue.

In the sample of 100 runners, the standard deviation of the sample mean is equal to one-tenth of the population standard deviation: $15.93/10 = 1.59$. In other words, the standard deviation of the sample mean based on 100 observations is equal to

$$SD_{\bar{x}} = \sigma_{\bar{x}} = \frac{\sigma_x}{\sqrt{n}} = \frac{15.93}{\sqrt{100}} = 1.59$$

where σ_x is the standard deviation of the individual observations. This is no coincidence. We can show mathematically that this equation is correct when the observations are independent using the probability tools of Section 3.5.

Computing SD for the sample mean

Given n independent observations from a population with standard deviation σ, the standard deviation of the sample mean is equal to

$$SD_{\bar{x}} = \sigma_{\bar{x}} = \frac{\sigma}{\sqrt{n}} \qquad (4.34)$$

A reliable method to ensure sample observations are independent is to conduct a simple random sample consisting of less than 10% of the population.

⊙ **Guided Practice 4.35**   The average of the runners' ages is 35.05 years with a standard deviation of $\sigma = 8.97$. A simple random sample of 100 runners is taken. (a) What is the standard deviation of the sample mean? (b) Would you be surprised to get a sample of size 100 with an average of 36 years?[27]

---

[26](a) Consider two random samples: one of size 10 and one of size 1000. Individual observations in the small sample are highly influential on the estimate while in larger samples these individual observations would more often average each other out. The larger sample would tend to provide a more accurate estimate. (b) If we think an estimate is better, we probably mean it typically has less error. Based on (a), our intuition suggests that a larger sample size corresponds to a smaller standard deviation.

[27](a) Use Equation (4.34) with the population standard deviation to compute the standard deviation of the sample mean: $SD_{\bar{y}} = 8.97/\sqrt{100} = 0.90$ years. (b) It would not be surprising. 36 years is about 1 standard deviation from the true mean of 35.05. Based on the 68, 95 rule, we would get a sample mean at least this far away from the true mean approximately $100\% - 68\% = 32\%$ of the time.

⊙ **Guided Practice 4.36**   (a) Would you be more trusting of a sample that has 100 observations or 400 observations? (b) We want to show mathematically that our estimate tends to be better when the sample size is larger. If the standard deviation of the individual observations is 10, what is our estimate of the standard deviation of the mean when the sample size is 100? What about when it is 400? (c) Explain how your answer to (b) mathematically justifies your intuition in part (a).[28]

4.2.2   Examining the Central Limit Theorem

In Figure 4.20, the sampling distribution of the sample mean looks approximately normally distributed. Will the sampling distribution of a mean always be nearly normal? To address this question, we will investigate three cases to see roughly when the approximation is reasonable.

We consider three data sets: one from a *uniform* distribution, one from an *exponential* distribution, and the other from a *normal* distribution. These distributions are shown in the top panels of Figure 4.21. The uniform distribution is symmetric, and the exponential distribution may be considered as having moderate skew since its right tail is relatively short (few outliers).

The left panel in the $n = 2$ row represents the sampling distribution of $\bar{x}$ if it is the sample mean of two observations from the uniform distribution shown. The dashed line represents the closest approximation of the normal distribution. Similarly, the center and right panels of the $n = 2$ row represent the respective distributions of $\bar{x}$ for data from exponential and log-normal distributions.

⊙ **Guided Practice 4.37**   Examine the distributions in each row of Figure 4.21. What do you notice about the sampling distribution of the mean as the sample size, n, becomes larger?[29]

● **Example 4.38**   In general, would normal approximation for a sample mean be appropriate when the sample size is at least 30?

Yes, the sampling distributions when $n = 30$ all look very much like the normal distribution.

However, the more non-normal a population distribution, the larger a sample size is necessary for the sampling distribution to look nearly normal.

---

[28](a) Extra observations are usually helpful in understanding the population, so a point estimate with 400 observations seems more trustworthy. (b) The standard deviation of the mean when the sample size is 100 is given by $SD_{100} = 10/\sqrt{100} = 1$. For 400: $SD_{400} = 10/\sqrt{400} = 0.5$. The larger sample has a smaller standard deviation of the mean. (c) The standard deviation of the mean of the sample with 400 observations is lower than that of the sample with 100 observations. The standard deviation of $\bar{x}$ describes the typical error, and since it is lower for the larger sample, this mathematically shows the estimate from the larger sample tends to be better – though it does not guarantee that every large sample will provide a better estimate than a particular small sample.

[29]The normal approximation becomes better as larger samples are used. However, in the case when the population is normally distributed, the normal distribution of the sample mean is normal for all sample sizes.

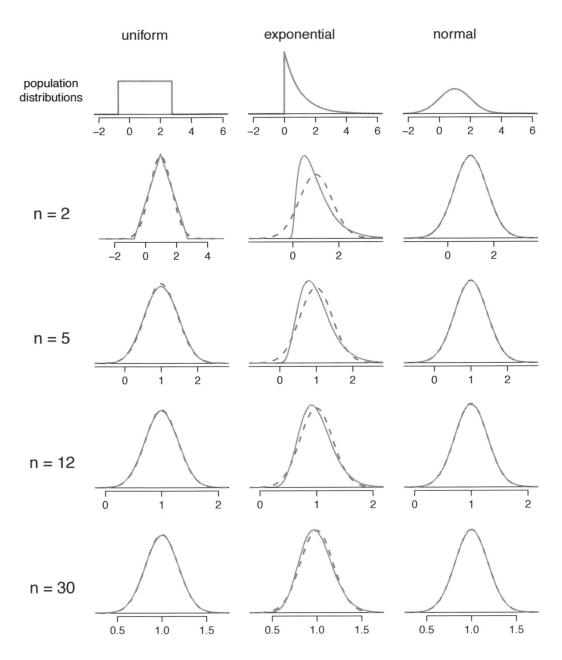

Figure 4.21: Sampling distributions for the mean at different sample sizes and for three different distributions. The dashed red lines show normal distributions.

Determining if the sample mean is normally distributed

If the population is normal, the sampling distribution of $\bar{x}$ will be normal for any sample size.

The less normal the population, the larger n needs to be for the sampling distribution of $\bar{x}$ to be nearly normal. However, a good rule of thumb is that for almost all populations, the sampling distribution of $\bar{x}$ will be approximately normal if $n \geq 30$.

This brings us to the **Central Limit Theorem**, the most fundamental theorem in Statistics.

Central Limit Theorem

When taking a random sample of independent observations from a population with a fixed mean and standard deviation, the distribution of $\bar{x}$ approaches the normal distribution as n increases.

● **Example 4.39** Sometimes we do not know what the population distribution looks like. We have to infer it based on the distribution of a single sample. Figure 4.22 shows a histogram of 20 observations. These represent winnings and losses from 20 consecutive days of a professional poker player. Based on this sample data, can the normal approximation be applied to the distribution of the sample mean?

We should consider each of the required conditions.

(1) These are referred to as **time series data**, because the data arrived in a particular sequence. If the player wins on one day, it may influence how she plays the next. To make the assumption of independence we should perform careful checks on such data.

(2) The sample size is 20, which is smaller than 30.

(3) There are two outliers in the data, both quite extreme, which suggests the population may not be normal and instead may be very strongly skewed or have distant outliers. Outliers can play an important role and affect the distribution of the sample mean and the estimate of the standard deviation of the sample mean.

Since we should be skeptical of the independence of observations and the extreme upper outliers pose a challenge, we should not use the normal model for the sample mean of these 20 observations. If we can obtain a much larger sample, then the concerns about skew and outliers would no longer apply.

Caution: Examine data structure when considering independence
Some data sets are collected in such a way that they have a natural underlying structure between observations, e.g. when observations occur consecutively. Be especially cautious about independence assumptions regarding such data sets.

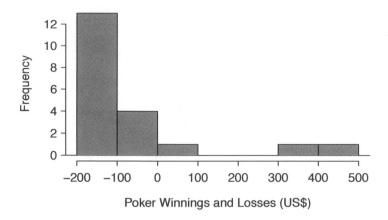

Figure 4.22: Sample distribution of poker winnings. These data include two very clear outliers. These are problematic when considering the normality of the sample mean. For example, outliers are often an indicator of very strong skew.

Caution: Watch out for strong skew and outliers
Strong skew in the population is often identified by the presence of clear outliers in the data. If a data set has prominent outliers, then a larger sample size will be needed for the sampling distribution of $\bar{x}$ to be normal. There are no simple guidelines for what sample size is big enough for each situation. However, we can use the rule of thumb that, in general, an n of at least 30 is sufficient for most cases.

4.2.3   Normal approximation for the sampling distribution of $\bar{x}$

At the beginning of this chapter, we used normal approximation for populations or for data that had an approximately normal distribution. When appropriate conditions are met, we can also use the normal approximation to estimate probabilities about a sample average. We must remember to verify that the conditions are met and use the mean $\mu_{\bar{x}}$ and standard deviation $\sigma_{\bar{x}}$ for the sampling distribution of the sample average.

TIP: Three important facts about the distribution of a sample mean $\bar{x}$
Consider taking a simple random sample from a large population.

1. The mean of a sample mean is denoted by $\mu_{\bar{x}}$, and it is equal to μ.

2. The SD of a sample mean is denoted by $\sigma_{\bar{x}}$, and it is equal to $\frac{\sigma}{\sqrt{n}}$.

3. When the population is normal or when $n \geq 30$, the sample mean closely follows a normal distribution.

● **Example 4.40** In the 2012 Cherry Blossom 10 mile run, the average time for all of the runners is 94.52 minutes with a standard deviation of 8.97 minutes. The distribution of run times is approximately normal. Find the probabiliy that a randomly selected runner completes the run in less than 90 minutes.

Because the distribution of run times is approximately normal, we can use normal approximation.

$$Z = \frac{x - \mu}{\sigma} = \frac{90 - 94.52}{8.97} = -0.504$$
$$P(Z < -0.504) = 0.3072$$

There is a 30.72% probability that a randomly selected runner will complete the run in less than 90 minutes.

● **Example 4.41** Find the probabiliy that the average of 20 runners is less than 90 minutes.

Here, $n = 20 < 30$, but the distribution of the population, that is, the distribution of run times is stated to be approximately normal. Because of this, the sampling distribution will be normal for any sample size.

$$\sigma_{\bar{x}} = \frac{\sigma}{\sqrt{n}} = \frac{8.97}{\sqrt{20}} = 2.01$$
$$Z = \frac{\bar{x} - \mu_{\bar{x}}}{\sigma_{\bar{x}}} = \frac{90 - 94.52}{2.01} = -2.25$$
$$P(Z < -0.504) = 0.0123$$

There is a 1.23% probability that the average run time of 20 randomly selected runners will be less than 90 minutes.

● **Example 4.42** The average of all the runners' ages is 35.05 years with a standard deviation of $\sigma = 8.97$. The distribution of age is somewhat skewed. What is the probability that a randomly selected runner is older than 37 years?

Because the distribution of age is skewed and is not normal, we cannot use normal approximation for this problem. In order to answer this question, we would need to look at all of the data.

⊙ **Guided Practice 4.43** What is the probability that the average of 50 randomly selected runners is greater than 37 years?[30]

TIP: Remember to divide by $\sqrt{n}$
When finding the probability that an *average* or mean is greater or less than a particular value, remember to divide the standard deviation of the population by $\sqrt{n}$ to calculate the correct SD.

---

[30]Because $n = 50 \geq 30$, the sampling distribution of the mean is approximately normal, so we can use normal approximation for this problem. The mean is given as 35.05 years.

$$\sigma_{\bar{x}} = \frac{\sigma}{\sqrt{n}} = \frac{8.97}{\sqrt{50}} = 1.27 \qquad z = \frac{\bar{x} - \mu_{\bar{x}}}{\sigma_{\bar{x}}} = \frac{37 - 35.05}{1.27} = 1.535 \qquad P(Z > 1.535) = 0.062$$

There is a 6.2% chance that the average age of 50 runners will be greater than 37.

4.3   Geometric distribution

How long should we expect to flip a coin until it turns up heads? Or how many times should we expect to roll a die until we get a 1? These questions can be answered using the geometric distribution. We first formalize each trial – such as a single coin flip or die toss – using the Bernoulli distribution, and then we combine these with our tools from probability (Chapter 3) to construct the geometric distribution.

4.3.1   Bernoulli distribution

Stanley Milgram began a series of experiments in 1963 to estimate what proportion of people would willingly obey an authority and give severe shocks to a stranger. Milgram found that about 65% of people would obey the authority and give such shocks. Over the years, additional research suggested this number is approximately consistent across communities and time.[31]

Each person in Milgram's experiment can be thought of as a **trial**. We label a person a **success** if she refuses to administer the worst shock. A person is labeled a **failure** if she administers the worst shock. Because only 35% of individuals refused to administer the most severe shock, we denote the **probability of a success** with $p = 0.35$. The probability of a failure is sometimes denoted with $q = 1 - p$.

Thus, success or failure is recorded for each person in the study. When an individual trial only has two possible outcomes, it is called a **Bernoulli random variable**.

Bernoulli random variable, descriptive

A Bernoulli random variable has exactly two possible outcomes. We typically label one of these outcomes a "success" and the other outcome a "failure". We may also denote a success by 1 and a failure by 0.

TIP: "success" need not be something positive
We chose to label a person who refuses to administer the worst shock a "success" and all others as "failures". However, we could just as easily have reversed these labels. The mathematical framework we will build does not depend on which outcome is labeled a success and which a failure, as long as we are consistent.

Bernoulli random variables are often denoted as 1 for a success and 0 for a failure. In addition to being convenient in entering data, it is also mathematically handy. Suppose we observe ten trials:

$$0\ 1\ 1\ 1\ 1\ 0\ 1\ 1\ 0\ 0$$

Then the **sample proportion**, $\hat{p}$, is the sample mean of these observations:

$$\hat{p} = \frac{\#\text{ of successes}}{\#\text{ of trials}} = \frac{0+1+1+1+1+0+1+1+0+0}{10} = 0.6$$

---
[31]Find further information on Milgram's experiment at
www.cnr.berkeley.edu/ucce50/ag-labor/7article/article35.htm.

This mathematical inquiry of Bernoulli random variables can be extended even further. Because 0 and 1 are numerical outcomes, we can define the mean and standard deviation of a Bernoulli random variable.[32]

Bernoulli random variable, mathematical

If X is a random variable that takes value 1 with probability of success p and 0 with probability $1 - p$, then X is a Bernoulli random variable with mean and standard deviation

$$\mu = p \qquad\qquad \sigma = \sqrt{p(1 - p)}$$

In general, it is useful to think about a Bernoulli random variable as a random process with only two outcomes: a success or failure. Then we build our mathematical framework using the numerical labels 1 and 0 for successes and failures, respectively.

4.3.2 Geometric distribution

● **Example 4.44** Dr. Smith wants to repeat Milgram's experiments but she only wants to sample people until she finds someone who will not inflict the worst shock.[33] If the probability a person will *not* give the most severe shock is still 0.35 and the subjects are independent, what are the chances that she will stop the study after the first person? The second person? The third? What about if it takes her $n - 1$ individuals who will administer the worst shock before finding her first success, i.e. the first success is on the nth person? (If the first success is the fifth person, then we say $n = 5$.)

The probability of stopping after the first person is just the chance the first person will not administer the worst shock: $1 - 0.65 = 0.35$. The probability it will be the second person is

$$P(\text{second person is the first to not administer the worst shock})$$
$$= P(\text{the first will, the second won't}) = (0.65)(0.35) = 0.228$$

Likewise, the probability it will be the third person is $(0.65)(0.65)(0.35) = 0.148$.

If the first success is on the nth person, then there are $n - 1$ failures and finally 1 success, which corresponds to the probability $(0.65)^{n-1}(0.35)$. This is the same as $(1 - 0.35)^{n-1}(0.35)$.

---

[32]If p is the true probability of a success, then the mean of a Bernoulli random variable X is given by

$$\mu = E[X] = P(X = 0) \times 0 + P(X = 1) \times 1$$
$$= (1 - p) \times 0 + p \times 1 = 0 + p = p$$

Similarly, the variance of X can be computed:

$$\sigma^2 = P(X = 0)(0 - p)^2 + P(X = 1)(1 - p)^2$$
$$= (1 - p)p^2 + p(1 - p)^2 = p(1 - p)$$

The standard deviation is $\sigma = \sqrt{p(1 - p)}$.

[33]This is hypothetical since, in reality, this sort of study probably would not be permitted any longer under current ethical standards.

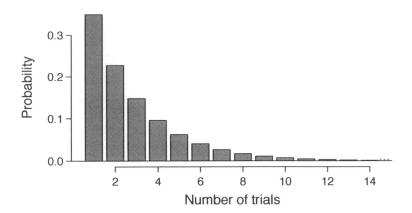

Figure 4.23: The geometric distribution when the probability of success is $p = 0.35$.

Example 4.44 illustrates what is called the geometric distribution, which describes the waiting time until a success for **independent and identically distributed (iid)** Bernoulli random variables. In this case, the *independence* aspect just means the individuals in the example don't affect each other, and *identical* means they each have the same probability of success.

The geometric distribution from Example 4.44 is shown in Figure 4.23. In general, the probabilities for a geometric distribution decrease **exponentially** fast.

While this text will not derive the formulas for the mean (expected) number of trials needed to find the first success or the standard deviation or variance of this distribution, we present general formulas for each.

Geometric distribution

If the probability of a success in one trial is p and the probability of a failure is $1 - p$, then the probability of finding the first success in the nth trial is given by

$$(1 - p)^{n-1}p \qquad (4.45)$$

The mean (i.e. expected value) and standard deviation of this wait time are given by

$$\mu = \frac{1}{p} \qquad\qquad \sigma = \sqrt{\frac{1-p}{p^2}} \qquad (4.46)$$

It is no accident that we use the symbol μ for both the mean and expected value. The mean and the expected value are one and the same.

The left side of Equation (4.46) says that, on average, it takes $1/p$ trials to get a success. This mathematical result is consistent with what we would expect intuitively. If the probability of a success is high (e.g. 0.8), then we don't usually wait very long for a success: $1/0.8 = 1.25$ trials on average. If the probability of a success is low (e.g. 0.1), then we would expect to view many trials before we see a success: $1/0.1 = 10$ trials.

⊙ **Guided Practice 4.47** The probability that an individual would refuse to administer the worst shock is said to be about 0.35. If we were to examine individuals until we found one that did not administer the shock, how many people should we expect to check? The first expression in Equation (4.46) may be useful.[34]

● **Example 4.48** What is the chance that Dr. Smith will find the first success within the first 4 people?

This is the chance it is the first ($n = 1$), second ($n = 2$), third ($n = 3$), or fourth ($n = 4$) person as the first success, which are four disjoint outcomes. Because the individuals in the sample are randomly sampled from a large population, they are independent. We compute the probability of each case and add the separate results:

$$P(n = 1, 2, 3, \text{or} 4)$$
$$= P(n = 1) + P(n = 2) + P(n = 3) + P(n = 4)$$
$$= (0.65)^{1-1}(0.35) + (0.65)^{2-1}(0.35) + (0.65)^{3-1}(0.35) + (0.65)^{4-1}(0.35)$$
$$= 0.82$$

There is an 82% chance that she will end the study within 4 people.

⊙ **Guided Practice 4.49** Determine a more clever way to solve Example 4.48. Show that you get the same result.[35]

● **Example 4.50** Suppose in one region it was found that the proportion of people who would administer the worst shock was "only" 55%. If people were randomly selected from this region, what is the expected number of people who must be checked before one was found that would be deemed a success? What is the standard deviation of this waiting time?

A success is when someone will **not** inflict the worst shock, which has probability $p = 1 - 0.55 = 0.45$ for this region. The expected number of people to be checked is $1/p = 1/0.45 = 2.22$ and the standard deviation is $\sqrt{(1-p)/p^2} = 1.65$.

⊙ **Guided Practice 4.51** Using the results from Example 4.50, $\mu = 2.22$ and $\sigma = 1.65$, would it be appropriate to use the normal model to find what proportion of experiments would end in 3 or fewer trials?[36]

The independence assumption is crucial to the geometric distribution's accurate description of a scenario. Mathematically, we can see that to construct the probability of the success on the nth trial, we had to use the Multiplication Rule for Independent Processes. It is no simple task to generalize the geometric model for dependent trials.

---

[34]We would expect to see about $1/0.35 = 2.86$ individuals to find the first success.

[35]First find the probability of the complement: $P(\text{no success in first 4 trials}) = 0.65^4 = 0.18$. Next, compute one minus this probability: $1 - P(\text{no success in 4 trials}) = 1 - 0.18 = 0.82$.

[36]No. The geometric distribution is always right skewed and can never be well-approximated by the normal model.

4.4 Binomial distribution

4.4.1 An example of a binomial distribution

Take a second look at Guided Practice 3.72 on page 129. We asked many probability questions regarding this scenario that could be solved using the binomial formula. Instead of looking at it piecewise, we could describe the entire *distribution* of possible values and their corresponding probabilities. Since there are 4 smoking friends, there are several possible outcomes for the number who might develop a severe lung condition in their lifetime: 0, 1, 2, 3, 4. We can make a distribution table as we did previously. Recall that the probability that a random smoker will develop a severe lung condition in her lifetime is about 0.3.

x_i	p_i
0	$\binom{4}{0}(0.3)^0(0.7)^4 = 0.058$
1	$\binom{4}{1}(0.3)^1(0.7)^3 = 0.268$
2	$\binom{4}{2}(0.3)^2(0.7)^2 = 0.242$
3	$\binom{4}{3}(0.3)^3(0.7)^1 = 0.075$
4	$\binom{4}{4}(0.3)^4(0.7)^0 = 0.008$

Table 4.24: Probability distribution for the number of 4 smoking friends who will develop a severe lung condition in their lifetime. This is a binomial distribution. Correcting for rounding error, the probabilities add up to 1, as they must for any probability distribution.

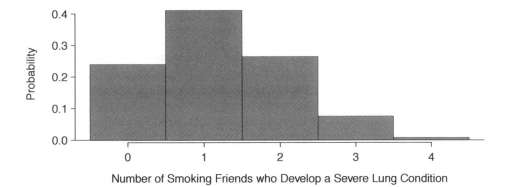

Figure 4.25: Distribution for the number of 4 smoking friends who will develop a severe lung condition.

4.4.2 The mean and standard deviation of a binomial distribution

Since this is a probability distribution we could find the mean and standard deviation of it using the formulas from Chapter 3. Those formulas require a lot of calculations, so it is fortunate that there are shortcuts for the mean and the standard deviation of a binomial random variable.

Mean and standard deviation of the binomial distribution

For a binomial distribution with parameters n and p, where n is the number of trials and p is the probability of a success, the mean and standard deviation of the number of observed successes are

$$\mu_x = np \qquad\qquad \sigma_x = \sqrt{np(1-p)} \qquad (4.52)$$

⬤ **Example 4.53** If the probability that a random smoker will develop a severe lung condition in his or her lifetime is 0.3 and you have 40 smoking friends, about how many would you expect to develop such a condition? What is the standard deviation of the number of people who would develop such a condition? Equation (4.52) may be useful.

We are asked to determine the expected number (the mean) and the standard deviation, both of which can be directly computed from the formulas in Equation (4.52), as shown below. The exact distribution is shown in Figure 4.26.

$$\mu = np = 40 \times 0.3 = 12$$
$$\sigma = \sqrt{np(1-p)} = \sqrt{40 \times 0.3 \times 0.7} = 2.9$$

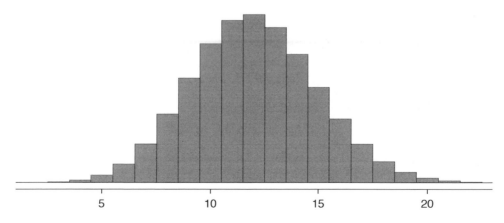

Figure 4.26: Distribution for the number of 40 smoking friends who will develop a severe lung condition, which looks very much like a normal distribution!

4.4.3 Normal approximation to the binomial distribution

The binomial formula is cumbersome when the sample size (n) is large, particularly when we consider a range of observations. Suppose we wanted to find the probability that at least 25 of 40 smoking friends will develop a severe lung condition. We would need to use the binomial formula with $k = 25$, $k = 26$, $k = 27$, ..., $k = 40$. That's a lot of work! In some cases we may use the normal distribution as an easier and faster way to estimate binomial probabilities. While a normal approximation for the distribution in Figure 4.25 would not be appropriate, it would not be too bad for the distribution in Figure 4.26.

● **Example 4.54** Approximately 20% of the US population smokes cigarettes. A local government believed their community had a lower smoker rate and commissioned a survey of 400 randomly selected individuals. The survey found that only 60 of the 400 participants smoke cigarettes. If the true proportion of smokers in the community was really 20%, what is the probability of observing 60 or fewer smokers in a sample of 400 people?

_____

We leave the usual verification that the four conditions for the binomial model are valid as an exercise.

The question posed is equivalent to asking, what is the probability of observing $k = 0$, 1, ..., 59, or 60 smokers in a sample of $n = 400$ when $p = 0.20$? We can compute these 61 different probabilities and add them together to find the answer:

$$P(k = 0 \text{ or } k = 1 \text{ or } \cdots \text{ or } k = 60)$$
$$= P(k = 0) + P(k = 1) + \cdots + P(k = 60)$$
$$= 0.0061$$

If the true proportion of smokers in the community is $p = 0.20$, then the probability of observing 60 or fewer smokers in a sample of $n = 400$ is less than 0.0061.

The computations in Example 4.54 are tedious and long. In general, we should avoid such work if an alternative method exists that is faster, easier, and still accurate. Recall that calculating probabilities of a range of values is much easier in the normal model. We might wonder, is it reasonable to use the normal model in place of the binomial distribution? Surprisingly, yes, if certain conditions are met.

⊙ **Guided Practice 4.55** Here we consider the binomial model when the probability of a success is $p = 0.10$. Figure 4.27 shows four hollow histograms for simulated samples from the binomial distribution using four different sample sizes: $n = 10, 30, 100, 300$. What happens to the shape of the distributions as the sample size increases? What distribution does the last hollow histogram resemble?[37]

_____

[37]The distribution is transformed from a blocky and skewed distribution into one that rather resembles the normal distribution in last hollow histogram

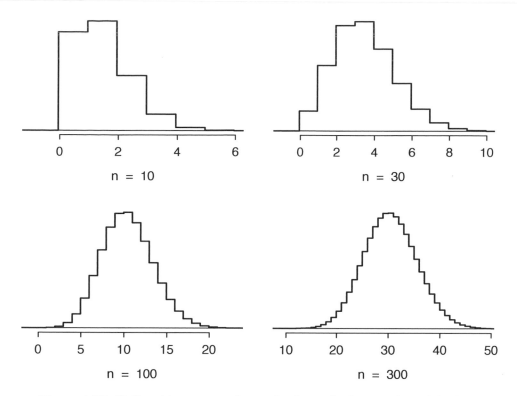

Figure 4.27: Hollow histograms of samples from the binomial model when $p = 0.10$. The sample sizes for the four plots are $n = 10, 30, 100,$ and $300,$ respectively.

Normal approximation of the binomial distribution

The binomial distribution with probability of success p is nearly normal when the sample size n is sufficiently large that $np \geq 10$ and $n(1 - p) \geq 10$. The approximate normal distribution has parameters corresponding to the mean and standard deviation of the binomial distribution:

$$\mu = np \qquad\qquad \sigma = \sqrt{np(1 - p)}$$

The normal approximation may be used when computing the range of many possible successes. For instance, we may apply the normal distribution to the setting described in Example 4.54.

● **Example 4.56** Use the normal approximation to estimate the probability of observing 60 or fewer smokers in a sample of 400, if the true proportion of smokers is $p = 0.20$.

As in Example 4.54, we leave it to the reader to show that the binomial model is reasonable for this context. However, we will verify that both np and $n(1 - p)$ are at least 10 so we can apply the normal model:

$$np = 400(0.20) = 80 \geq 10 \qquad\qquad n(1 - p) = 400(0.8) = 320 \geq 10$$

With these conditions checked, we may use the normal approximation in place of the binomial distribution with the following mean and standard deviation:

$$\mu = np = 400(0.2) = 80 \qquad\qquad \sigma = \sqrt{np(1 - p)} = \sqrt{400(0.2)(0.8)} = 8$$

We want to find the probability of observing 60 or fewer smokers using this model. We know that this probability will be small because 60 is more than 2 standard deviations below the mean:

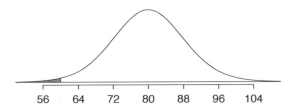

Next, we compute the Z-score as $Z = \frac{60-80}{8} = -2.5$ to find the shaded area in the picture: $P(Z < -2.5) = 0.0062$. This probability of 0.0062 using the normal approximation is remarkably close to the true probability of 0.0061 from the binomial distribution!

4.4.4   The normal approximation breaks down on small intervals (special topic)

Caution: The normal approximation may fail on small intervals
The normal approximation to the binomial distribution tends to perform poorly when estimating the probability of a small range of counts, even when the conditions are met.

Suppose we wanted to compute the probability of observing 69, 70, or 71 smokers in 400 when $p = 0.20$. With such a large sample, we might be tempted to apply the normal approximation and use the range 69 to 71. However, we would find that the binomial solution and the normal approximation notably differ:

Binomial: 0.0703   Normal: 0.0476

We can identify the cause of this discrepancy using Figure 4.28, which shows the areas representing the binomial probability (outlined) and normal approximation (shaded). Notice that the width of the area under the normal distribution is 0.5 units too slim on both sides of the interval. The binomial distribution is a discrete distribution, and the each bar

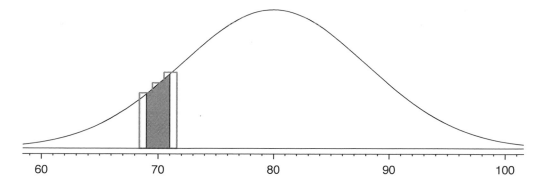

Figure 4.28: A normal curve with the area between 69 and 71 shaded. The outlined area from 68.5 to 71.5 represents the exact binomial probability.

is centered over an integer value. Looking closely at Figure 4.28, we can see that the bar corresponding to 69 begins at 68.5 and ends at 69.5, the bar corresponding to 70 begins at 69.5 and ends at 70.5, etc.

TIP: Improving the accuracy of the normal approximation to the binomial distribution
The normal approximation to the binomial distribution for intervals of values is usually improved if cutoff values for the lower end of a shaded region are reduced by 0.5 and the cutoff value for the upper end are increased by 0.5. This correction is called the continuity correction and accounts for the fact that the binomial distribution is discrete.

● **Example 4.57** Use the method described to find a more accurate estimate for the probability of observing 69, 70, or 71 smokers in 400 randomly selected people when $p = 0.20$.

Instead of standardizing 69 and 71, we will standardize 68.5 and 71.5:

$$Z_{left} = \frac{68.5 - 80}{8} = -1.4375$$
$$Z_{right} = \frac{71.5 - 80}{8} = -1.0625$$
$$P(-1.4375 < Z < -1.0625) = 0.0687$$

The probability 0.0687 is much closer to the true value of 0.0703 than the previous estimate of 0.0476 we calculated using normal approximation without the continuity correction.

It is always possible to apply the continuity correction when finding a normal approximation to the binomial distribution. However, when n is very large or when the interval is wide, the benefit of the modification is limited since the added area becomes negligible compared to the overall area being calculated.

4.5   Sampling distribution of a sample proportion

The binomial distribution shows us the distribution of number of successes in n trials. Often, we are interested in the *proportion* of successes rather than the number of successes. We would like to answer questions such as the following:

1. Approximately 20% of the US population smokes cigarettes. A random sample of size 400 from a particular county found that 15% of the sample smoked. If the smoking rate in this county really is 20%, what is the probability that the sample would contain 15% or fewer smokers?

2. Given a population that is 50% male, what is the probability that a sample of size of 200 people would consist of more than 55% males?

4.5.1   The mean and standard deviation of $\hat{p}$

To answer these questions, we investigate the distribution of the sample proportion $\hat{p}$. In the last section we saw that the *number* of smokers in a sample of size 400 follows a binomial distribution with $p = 0.2$ and $n = 400$ that is centered on 80 and has standard deviation 8. What does the distribution of the *proportion* of smokers in a sample of size 400 look like? To convert from a count to a proportion, we divide the count (i.e. number of yeses) by the sample size, $n = 400$. For example, 60 becomes $60/400 = 0.15$ as a proportion and 61 becomes $61/400 = 0.1525$.

We can find the general formula for the mean (expected value) and standard deviation of a sample proportion $\hat{p}$ using our tools that we've learned so far. To get the sample mean for $\hat{p}$, we divide the binomial mean $\mu_{binomial} = np$ by n:

$$\mu_{\hat{p}} = \frac{\mu_{binomial}}{n} = \frac{np}{n} = p$$

As one might expect, the sample proportion $\hat{p}$ is centered on the true proportion p. Likewise, the standard deviation of $\hat{p}$ is equal to the standard deviation of the binomial distribution divided by n:

$$\sigma_{\hat{p}} = \frac{\sigma_{binomial}}{n} = \frac{\sqrt{np(1-p)}}{n} = \sqrt{\frac{p(1-p)}{n}}$$

Mean and standard deviation of a sample proportion

The mean and standard deviation of the sample proportion describe the center and spread of the distribution of all possible sample proportions $\hat{p}$ from a random sample of size n with true population proportion p.

$$\mu_{\hat{p}} = p \qquad\qquad\qquad \sigma_{\hat{p}} = \sqrt{\frac{p(1-p)}{n}}$$

In analyses, we think of the formula for the standard deviation of a sample proportion, $\sigma_{\hat{p}}$, as describing the uncertainty associated with the estimate $\hat{p}$. That is, $\sigma_{\hat{p}}$ can be thought of as a way to quantify the typical error in our sample estimate $\hat{p}$ of the true proportion p. Understanding the variability of statistics such as $\hat{p}$ is a central component in the study of statistics.

● **Example 4.58** If the rate of smoking in the county is really 20%, find and interpret the mean and standard deviation of the sample proportion for a sample of size 400.

The mean of the sample proportion is the population proportion: 0.20. That is, if we took many, many samples and calculated $\hat{p}$, these values would average out to $p = 0.20$.

The standard deviation of $\hat{p}$ is described by the standard deviation for the proportion:

$$\sigma_{\hat{p}} = \sqrt{\frac{p(1-p)}{n}} = \sqrt{\frac{0.2 \times 0.8}{400}} = .02$$

The sample proportion will typically be about 0.02 or 2% away from the true proportion of $p = 0.20$. We'll become more rigorous about quantifying how close $\hat{p}$ will tend to be to p in Chapter 5.

4.5.2 The Central Limit Theorem revisited

In section 4.2, we saw the Central Limit Theorem, which states that for large enough n, the sample mean $\bar{x}$ is normally distributed.

A natural question is, what does this have to do with sample proportions? In fact, a lot! A sample proportion can be written down as a sample mean. For example, suppose we have 3 successes in 10 trials. If we label each of the 3 success as a 1 and each of the 7 failures as a 0, then the sample proportion is the same as the sample mean:

$$\hat{p} = \frac{1 + 0 + 0 + 1 + 1 + 0 + 0 + 0 + 0 + 0}{10} = \frac{3}{10} = 0.3$$

That is, the distribution of the sample proportion is governed by the Central Limit Theorem, and the Central Limit Theorem is what ties much of the statistical theory we will see together.

TIP: Three important facts about the distribution of a sample proportion $\hat{p}$

Consider taking a simple random sample from a large population.

1. The mean of a sample proportion is p.

2. The SD of a sample proportion is $\sqrt{\frac{p(1-p)}{n}}$.

3. When $np \geq 10$ and $n(1-p) \geq 10$, the sample proportion closely follows a normal distribution.

Using these facts, we can now answer the two questions posed at the beginning of this section.

4.5.3 Normal approximation for the distribution of $\hat{p}$

● **Example 4.59** Find the probability that less than 15% of the sample of 400 people will be smokers if the true proportion is 20%.

In the previous section we verified that np and $n(1-p)$ are at least 10. The mean of the sample proportion is 0.20 and the standard deviation for the sample proportion is given by $\sqrt{\frac{0.2(1-0.2)}{400}} = 0.02$. We can find a Z-score and use our calculator to find the probability:

$$Z = \frac{\hat{p} - \mu_{\hat{p}}}{\sigma_{\hat{p}}} = \frac{0.15 - 0.20}{0.02} = -2.5$$
$$P(Z < 2.5) = 0.0062$$

We leave it to the reader to construct a figure for this example.

● **Example 4.60** The probability 0.0062 is the same probability we calculated when we found the probability of getting 60 or fewer smokers out of 400! Why is this?

Notice that $60/400 = 0.15$. Using the binomial distribution to find the probability of 60 or fewer smokers in the sample is equivalent to using the probability that $\hat{p}$ will be less than or equal to 15%.

⊙ **Guided Practice 4.61** Given a population that is 50% male, what is the probability that a sample of size 200 would have greater than 55% males? Remember to verify that conditions for normal approximation are met.[38]

---

[38]First, verify the conditions: $np = 200 \times 0.5 = 100 \geq 10$ and $n(1-p) = 200 \times 0.5 = 100 \geq 10$, so the normal approximation is reasonable. Next we find the mean and standard deviation of $\hat{p}$:

$$\mu_{\hat{p}} = p = 0.50 \qquad\qquad \sigma_{\hat{p}} = \sqrt{\frac{p(1-p)}{n}} = \sqrt{\frac{0.5 \times 0.5}{200}} = 0.0354$$

Then we find a Z-score and find the upper tail of the normal distribution:

$$Z = \frac{\hat{p} - \mu_{\hat{p}}}{\sigma_{\hat{p}}} = \frac{0.55 - 0.5}{0.0354} = 1.412 \qquad \rightarrow \qquad P(Z > 1.412) = 0.07$$

The probability of getting a sample proportion of 55% or greater is about 0.07.

4.6 Exercises

4.6.1 Normal distribution

4.1 Area under the curve, I. What percent of a standard normal distribution $N(\mu = 0, \sigma = 1)$ is found in each region? Be sure to draw a graph.

(a) $Z < -1.35$ (b) $Z > 1.48$ (c) $-0.4 < Z < 1.5$ (d) $|Z| > 2$

4.2 Area under the curve, II. What percent of a standard normal distribution $N(\mu = 0, \sigma = 1)$ is found in each region? Be sure to draw a graph.

(a) $Z > -1.13$ (b) $Z < 0.18$ (c) $Z > 8$ (d) $|Z| < 0.5$

4.3 GRE scores, Part I. Sophia who took the Graduate Record Examination (GRE) scored 160 on the Verbal Reasoning section and 157 on the Quantitative Reasoning section. The mean score for Verbal Reasoning section for all test takers was 151 with a standard deviation of 7, and the mean score for the Quantitative Reasoning was 153 with a standard deviation of 7.67. Suppose that both distributions are nearly normal.

(a) Write down the short-hand for these two normal distributions.

(b) What is Sophia's Z-score on the Verbal Reasoning section? On the Quantitative Reasoning section? Draw a standard normal distribution curve and mark these two Z-scores.

(c) What do these Z-scores tell you?

(d) Relative to others, which section did she do better on?

(e) Find her percentile scores for the two exams.

(f) What percent of the test takers did better than her on the Verbal Reasoning section? On the Quantitative Reasoning section?

(g) Explain why simply comparing her raw scores from the two sections would lead to the incorrect conclusion that she did better on the Quantitative Reasoning section.

(h) If the distributions of the scores on these exams are not nearly normal, would your answers to parts (b) - (f) change? Explain your reasoning.

4.4 Triathlon times, Part I. In triathlons, it is common for racers to be placed into age and gender groups. Friends Leo and Mary both completed the Hermosa Beach Triathlon, where Leo competed in the *Men, Ages 30 - 34* group while Mary competed in the *Women, Ages 25 - 29* group. Leo completed the race in 1:22:28 (4948 seconds), while Mary completed the race in 1:31:53 (5513 seconds). Obviously Leo finished faster, but they are curious about how they did within their respective groups. Can you help them? Here is some information on the performance of their groups:

- The finishing times of the *Men, Ages 30 - 34* group has a mean of 4313 seconds with a standard deviation of 583 seconds.
- The finishing times of the *Women, Ages 25 - 29* group has a mean of 5261 seconds with a standard deviation of 807 seconds.
- The distributions of finishing times for both groups are approximately Normal.

Remember: a better performance corresponds to a faster finish.

(a) Write down the short-hand for these two normal distributions.

(b) What are the Z-scores for Leo's and Mary's finishing times? What do these Z-scores tell you?

(c) Did Leo or Mary rank better in their respective groups? Explain your reasoning.

(d) What percent of the triathletes did Leo finish faster than in his group?

(e) What percent of the triathletes did Mary finish faster than in her group?

(f) If the distributions of finishing times are not nearly normal, would your answers to parts (b) - (e) change? Explain your reasoning.

4.5 GRE scores, Part II. In Exercise 4.3 we saw two distributions for GRE scores: $N(\mu = 151, \sigma = 7)$ for the verbal part of the exam and $N(\mu = 153, \sigma = 7.67)$ for the quantitative part. Use this information to compute each of the following:

(a) The score of a student who scored in the 80th percentile on the Quantitative Reasoning section.

(b) The score of a student who scored worse than 70% of the test takers in the Verbal Reasoning section.

4.6 Triathlon times, Part II. In Exercise 4.4 we saw two distributions for triathlon times: $N(\mu = 4313, \sigma = 583)$ for *Men, Ages 30 - 34* and $N(\mu = 5261, \sigma = 807)$ for the *Women, Ages 25 - 29* group. Times are listed in seconds. Use this information to compute each of the following:

(a) The cutoff time for the fastest 5% of athletes in the men's group, i.e. those who took the shortest 5% of time to finish.

(b) The cutoff time for the slowest 10% of athletes in the women's group.

4.7 LA weather, Part I. The average daily high temperature in June in LA is 77°F with a standard deviation of 5°F. Suppose that the temperatures in June closely follow a normal distribution.

(a) What is the probability of observing an 83°F temperature or higher in LA during a randomly chosen day in June?

(b) How cold are the coldest 10% of the days during June in LA?

4.8 Portfolio returns. The Capital Asset Pricing Model is a financial model that assumes returns on a portfolio are normally distributed. Suppose a portfolio has an average annual return of 14.7% (i.e. an average gain of 14.7%) with a standard deviation of 33%. A return of 0% means the value of the portfolio doesn't change, a negative return means that the portfolio loses money, and a positive return means that the portfolio gains money.

(a) What percent of years does this portfolio lose money, i.e. have a return less than 0%?

(b) What is the cutoff for the highest 15% of annual returns with this portfolio?

4.9 LA weather, Part II. Exercise 4.7 states that average daily high temperature in June in LA is 77°F with a standard deviation of 5°F, and it can be assumed that they to follow a normal distribution. We use the following equation to convert °F (Fahrenheit) to °C (Celsius):

$$C = (F - 32) \times \frac{5}{9}.$$

(a) Write the probability model for the distribution of temperature in °C in June in LA.

(b) What is the probability of observing a 28°C (which roughly corresponds to 83°F) temperature or higher in June in LA? Calculate using the °C model from part (a).

(c) Did you get the same answer or different answers in part (b) of this question and part (a) of Exercise 4.7? Are you surprised? Explain.

(d) Estimate the IQR of the temperatures (in °C) in June in LA?

4.10 Heights of 10 year olds. Heights of 10 year olds, regardless of gender, closely follow a normal distribution with mean 55 inches and standard deviation 6 inches.

(a) What is the probability that a randomly chosen 10 year old is shorter than 48 inches?

(b) What is the probability that a randomly chosen 10 year old is between 60 and 65 inches?

(c) If the tallest 10% of the class is considered "very tall", what is the height cutoff for "very tall"?

(d) The height requirement for *Batman the Ride* at Six Flags Magic Mountain is 54 inches. What percent of 10 year olds cannot go on this ride?

4.11 Auto insurance premiums. Suppose a newspaper article states that the distribution of auto insurance premiums for residents of California is approximately normal with a mean of $1,650. The article also states that 25% of California residents pay more than $1,800.

(a) What is the Z-score that corresponds to the top 25% (or the 75th percentile) of the standard normal distribution?

(b) What is the mean insurance cost? What is the cutoff for the 75th percentile?

(c) Identify the standard deviation of insurance premiums in LA.

4.12 Speeding on the I-5, Part I. The distribution of passenger vehicle speeds traveling on the Interstate 5 Freeway (I-5) in California is nearly normal with a mean of 72.6 miles/hour and a standard deviation of 4.78 miles/hour.[39]

(a) What percent of passenger vehicles travel slower than 80 miles/hour?

(b) What percent of passenger vehicles travel between 60 and 80 miles/hour?

(c) How fast do the fastest 5% of passenger vehicles travel?

(d) The speed limit on this stretch of the I-5 is 70 miles/hour. Approximate what percentage of the passenger vehicles travel above the speed limit on this stretch of the I-5.

4.13 Overweight baggage, Part I. Suppose weights of the checked baggage of airline passengers follow a nearly normal distribution with mean 45 pounds and standard deviation 3.2 pounds. Most airlines charge a fee for baggage that weigh in excess of 50 pounds. Determine what percent of airline passengers incur this fee.

4.14 Find the SD. Find the standard deviation of the distribution in the following situations.

(a) MENSA is an organization whose members have IQs in the top 2% of the population. IQs are normally distributed with mean 100, and the minimum IQ score required for admission to MENSA is 132.

(b) Cholesterol levels for women aged 20 to 34 follow an approximately normal distribution with mean 185 milligrams per deciliter (mg/dl). Women with cholesterol levels above 220 mg/dl are considered to have high cholesterol and about 18.5% of women fall into this category.

4.15 Buying books on Ebay. The textbook you need to buy for your chemistry class is expensive at the college bookstore, so you consider buying it on Ebay instead. A look at past auctions suggest that the prices of that chemistry textbook have an approximately normal distribution with mean $89 and standard deviation $15.

(a) What is the probability that a randomly selected auction for this book closes at more than $100?

(b) Ebay allows you to set your maximum bid price so that if someone outbids you on an auction you can automatically outbid them, up to the maximum bid price you set. If you are only bidding on one auction, what are the advantages and disadvantages of setting a bid price too high or too low? What if you are bidding on multiple auctions?

(c) If you watched 10 auctions, roughly what percentile might you use for a maximum bid cutoff to be somewhat sure that you will win one of these ten auctions? Is it possible to find a cutoff point that will ensure that you win an auction?

(d) If you are willing to track up to ten auctions closely, about what price might you use as your maximum bid price if you want to be somewhat sure that you will buy one of these ten books?

---

[39]S. Johnson and D. Murray. "Empirical Analysis of Truck and Automobile Speeds on Rural Interstates: Impact of Posted Speed Limits". In: *Transportation Research Board 89th Annual Meeting*. 2010.

4.16 SAT scores. SAT scores (out of 2400) are distributed normally with a mean of 1500 and a standard deviation of 300. Suppose a school council awards a certificate of excellence to all students who score at least 1900 on the SAT, and suppose we pick one of the recognized students at random. What is the probability this student's score will be at least 2100? (The material covered in Section 3.2 would be useful for this question.)

4.17 Scores on stats final, Part I. Below are final exam scores of 20 Introductory Statistics students.

$$\overset{1}{57}, \overset{2}{66}, \overset{3}{69}, \overset{4}{71}, \overset{5}{72}, \overset{6}{73}, \overset{7}{74}, \overset{8}{77}, \overset{9}{78}, \overset{10}{78}, \overset{11}{79}, \overset{12}{79}, \overset{13}{81}, \overset{14}{81}, \overset{15}{82}, \overset{16}{83}, \overset{17}{83}, \overset{18}{88}, \overset{19}{89}, \overset{20}{94}$$

The mean score is 77.7 points. with a standard deviation of 8.44 points. Use this information to determine if the scores approximately follow the 68-95-99.7% Rule.

4.18 Heights of female college students, Part I. Below are heights of 25 female college students.

$$\overset{1}{54}, \overset{2}{55}, \overset{3}{56}, \overset{4}{56}, \overset{5}{57}, \overset{6}{58}, \overset{7}{58}, \overset{8}{59}, \overset{9}{60}, \overset{10}{60}, \overset{11}{60}, \overset{12}{61}, \overset{13}{61}, \overset{14}{62}, \overset{15}{62}, \overset{16}{63}, \overset{17}{63}, \overset{18}{63}, \overset{19}{64}, \overset{20}{65}, \overset{21}{65}, \overset{22}{67}, \overset{23}{67}, \overset{24}{69}, \overset{25}{73}$$

The mean height is 61.52 inches with a standard deviation of 4.58 inches. Use this information to determine if the heights approximately follow the 68-95-99.7% Rule.

4.19 Scores on stats final, Part II. Exercise 4.17 lists the final exam scores of 20 Introductory Statistics students. Do these data appear to follow a normal distribution? Explain your reasoning using the graphs provided below.

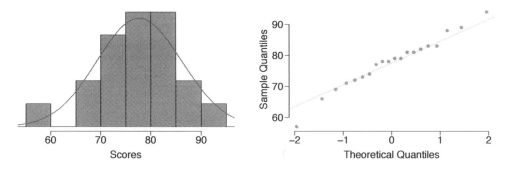

4.20 Heights of female college students, Part II. Exercise 4.18 lists the heights of 25 female college students. Do these data appear to follow a normal distribution? Explain your reasoning using the graphs provided below.

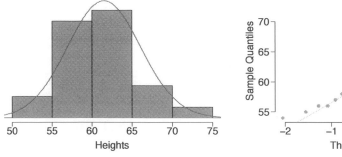
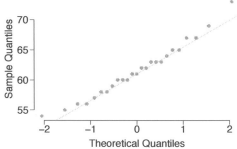

4.21 Lemonade at The Cafe. Drink pitchers at The Cafe are intended to hold about 64 ounces of lemonade and glasses hold about 12 ounces. However, when the pitchers are filled by a server, they do not always fill it with exactly 64 ounces. There is some variability. Similarly, when they pour out some of the lemonade, they do not pour exactly 12 ounces. The amount of lemonade in a pitcher is normally distributed with mean 64 ounces and standard deviation 1.732 ounces. The amount of lemonade in a glass is normally distributed with mean 12 ounces and standard deviation 1 ounce.

(a) How much lemonade would you expect to be left in a pitcher after pouring one glass of lemonade?

(b) What is the standard deviation of the amount left in a pitcher after pouring one glass of lemonade?

(c) What is the probability that more than 50 ounces of lemonade is left in a pitcher after pouring one glass of lemonade?

4.22 Spray paint, Part I. Suppose the area that can be painted using a single can of spray paint is slightly variable and follows a nearly normal distribution with a mean of 25 square feet and a standard deviation of 3 square feet. Suppose also that you buy three cans of spray paint.

(a) How much area would you expect to cover with these three cans of spray paint?

(b) What is the standard deviation of the area you expect to cover with these three cans of spray paint?

(c) The area you wanted to cover is 80 square feet. What is the probability that you will be able to cover this entire area with these three cans of spray paint?

4.23 GRE scores, Part III. In Exercise 4.3 we saw two distributions for GRE scores: $N(\mu = 151, \sigma = 7)$ for the verbal part of the exam and $N(\mu = 153, \sigma = 7.67)$ for the quantitative part. Suppose performance on these two sections is independent. Use this information to compute each of the following:

(a) The probability of a combined (verbal + quantitative) score above 320.

(b) The score of a student who scored better than 90% of the test takers overall.

4.24 Betting on dinner, Part I. Suppose a restaurant is running a promotion where prices of menu items are determined randomly following some underlying distribution. This means that if you're lucky you can get a basket of fries for $3, or if you're not so lucky you might end up having to pay $10 for the same menu item. The price of basket of fries is drawn from a normal distribution with mean 6 and standard deviation of 2. The price of a fountain drink is drawn from a normal distribution with mean 3 and standard deviation of 1. What is the probability that you pay more than $10 for a dinner consisting of a basket of fries and a fountain drink?

4.6.2 Sampling distribution of a sample mean

4.25 Ages of pennies, Part I. The histogram below shows the distribution of ages of pennies at a bank.

(a) Describe the distribution.

(b) Sampling distributions for means from simple random samples of 5, 30, and 100 pennies is shown in the histograms below. Describe the shapes of these distributions and comment on whether they look like what you would expect to see based on the Central Limit Theorem.

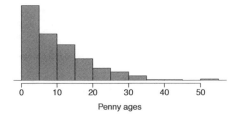

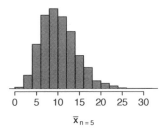

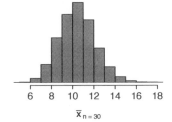

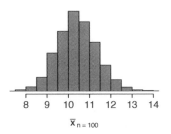

4.26 Ages of pennies, Part II. The mean age of the pennies from Exercise 4.25 is 10.44 years with a standard deviation of 9.2 years. Using the Central Limit Theorem, calculate the means and standard deviations of the distribution of the mean from random samples of size 5, 30, and 100. Comment on whether the sampling distributions shown in Exercise 4.25 agree with the values you compute.

4.27 Housing prices, Part I. A housing survey was conducted to determine the price of a typical home in Topanga, CA. The mean price of a house was roughly $1.3 million with a standard deviation of $300,000. There were no houses listed below $600,000 but a few houses above $3 million.

(a) Is the distribution of housing prices in Topanga symmetric, right skewed, or left skewed? *Hint:* Sketch the distribution.

(b) Would you expect most houses in Topanga to cost more or less than $1.3 million?

(c) Can we estimate the probability that a randomly chosen house in Topanga costs more than $1.4 million using the normal distribution?

(d) What is the probability that the mean of 60 randomly chosen houses in Topanga is more than $1.4 million?

(e) How would doubling the sample size affect the standard deviation of the mean?

4.28 Stats final scores. Each year about 1500 students take the introductory statistics course at a large university. This year scores on the final exam are distributed with a median of 74 points, a mean of 70 points, and a standard deviation of 10 points. There are no students who scored above 100 (the maximum score attainable on the final) but a few students scored below 20 points.

(a) Is the distribution of scores on this final exam symmetric, right skewed, or left skewed?

(b) Would you expect most students to have scored above or below 70 points?

(c) Can we calculate the probability that a randomly chosen student scored above 75 using the normal distribution?

(d) What is the probability that the average score for a random sample of 40 students is above 75?

(e) How would cutting the sample size in half affect the standard deviation of the mean?

4.29 Identify distributions, Part I. Four plots are presented below. The plot at the top is a distribution for a population. The mean is 10 and the standard deviation is 3. Also shown below is a distribution of (1) a single random sample of 100 values from this population, (2) a distribution of 100 sample means from random samples with size 5, and (3) a distribution of 100 sample means from random samples with size 25. Determine which plot (A, B, or C) is which and explain your reasoning.

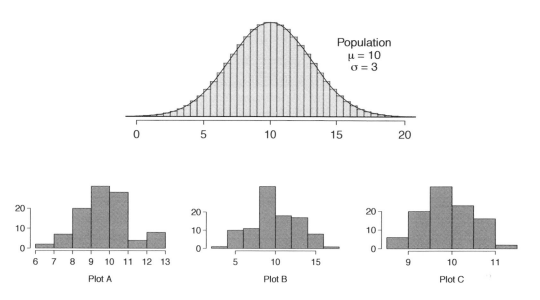

4.30 Identify distributions, Part II. Four plots are presented below. The plot at the top is a distribution for a population. The mean is 60 and the standard deviation is 18. Also shown below is a distribution of (1) a single random sample of 500 values from this population, (2) a distribution of 500 sample means from random samples of each size 18, and (3) a distribution of 500 sample means from random samples of each size 81. Determine which plot (A, B, or C) is which and explain your reasoning.

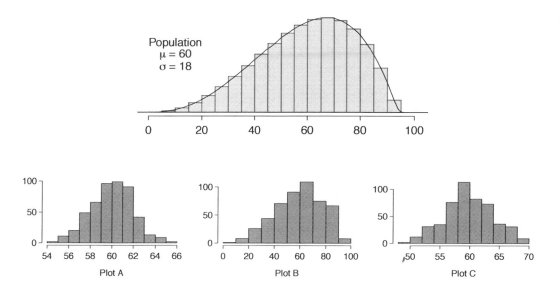

4.31   Weights of pennies. The distribution of weights of United States pennies is approximately normal with a mean of 2.5 grams and a standard deviation of 0.03 grams.

(a) What is the probability that a randomly chosen penny weighs less than 2.4 grams?

(b) Describe the sampling distribution of the mean weight of 10 randomly chosen pennies.

(c) What is the probability that the mean weight of 10 pennies is less than 2.4 grams?

(d) Sketch the two distributions (population and sampling) on the same scale.

(e) Could you estimate the probabilities from (a) and (c) if the weights of pennies had a skewed distribution?

4.32   CFLs. A manufacturer of compact fluorescent light bulbs advertises that the distribution of the lifespans of these light bulbs is nearly normal with a mean of 9,000 hours and a standard deviation of 1,000 hours.

(a) What is the probability that a randomly chosen light bulb lasts more than 10,500 hours?

(b) Describe the distribution of the mean lifespan of 15 light bulbs.

(c) What is the probability that the mean lifespan of 15 randomly chosen light bulbs is more than 10,500 hours?

(d) Sketch the two distributions (population and sampling) on the same scale.

(e) Could you estimate the probabilities from parts (a) and (c) if the lifespans of light bulbs had a skewed distribution?

4.33   Songs on an iPod. Suppose an iPod has 3,000 songs. The histogram below shows the distribution of the lengths of these songs. We also know that, for this iPod, the mean length is 3.45 minutes and the standard deviation is 1.63 minutes.

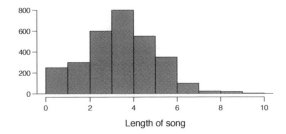

Length of song

(a) Calculate the probability that a randomly selected song lasts more than 5 minutes.

(b) You are about to go for an hour run and you make a random playlist of 15 songs. What is the probability that your playlist lasts for the entire duration of your run? *Hint:* If you want the playlist to last 60 minutes, what should be the minimum average length of a song?

(c) You are about to take a trip to visit your parents and the drive is 6 hours. You make a random playlist of 100 songs. What is the probability that your playlist lasts the entire drive?

4.34   Spray paint, Part II. As described in Exercise 4.22, the area that can be painted using a single can of spray paint is slightly variable and follows a nearly normal distribution with a mean of 25 square feet and a standard deviation of 3 square feet.

(a) What is the probability that the area covered by a can of spray paint is more than 27 square feet?

(b) Suppose you want to spray paint an area of 540 square feet using 20 cans of spray paint. On average, how many square feet must each can be able to cover to spray paint all 540 square feet?

(c) What is the probability that you can cover a 540 square feet area using 20 cans of spray paint?

(d) If the area covered by a can of spray paint had a slightly skewed distribution, could you still calculate the probabilities in parts (a) and (c) using the normal distribution?

4.35 Wireless routers. John is shopping for wireless routers and is overwhelmed by the number of available options. In order to get a feel for the average price, he takes a random sample of 75 routers and finds that the average price for this sample is $75 and the standard deviation is $25.

(a) Based on this information, how much variability should he expect to see in the mean prices of repeated samples, each containing 75 randomly selected wireless routers?

(b) A consumer website claims that the average price of routers is $80. Is a true average of $80 consistent with John's sample?

4.36 Chocolate chip cookies. Students are asked to count the number of chocolate chips in 22 cookies for a class activity. The packaging for these cookies claims that there are an average of 20 chocolate chips per cookie with a standard deviation of 4.37 chocolate chips.

(a) Based on this information, about how much variability should they expect to see in the mean number of chocolate chips in random samples of 22 chocolate chip cookies?

(b) What is the probability that a random sample of 22 cookies will have an average less than 14.77 chocolate chips if the companies claim on the packaging is true?

(c) Assume the students got 14.77 as the average in their sample of 22 cookies. Do you have confidence or not in the company's claim that the true average is 20? Explain your reasoning.

4.37 Overweight baggage, Part II. Suppose weights of the checked baggage of airline passengers follow a nearly normal distribution with mean 45 pounds and standard deviation 3.2 pounds. What is the probability that the *total* weight of 10 bags is greater than 460 lbs?

4.38 Betting on dinner, Part II. Exercise 4.24 introduces a promotion at a restaurant where prices of menu items are determined randomly following some underlying distribution. We are told that the price of basket of fries is drawn from a normal distribution with mean 6 and standard deviation of 2. You want to get 5 baskets of fries but you only have $28 in your pocket. What is the probability that you would have enough money to pay for all five baskets of fries?

4.6.3 Geometric distribution

4.39 Is it Bernoulli? Determine if each trial can be considered an independent Bernouilli trial for the following situations.

(a) Cards dealt in a hand of poker.

(b) Outcome of each roll of a die.

4.40 With and without replacement. In the following situations assume that half of the specified population is male and the other half is female.

(a) Suppose you're sampling from a room with 10 people. What is the probability of sampling two females in a row when sampling with replacement? What is the probability when sampling without replacement?

(b) Now suppose you're sampling from a stadium with 10,000 people. What is the probability of sampling two females in a row when sampling with replacement? What is the probability when sampling without replacement?

(c) We often treat individuals who are sampled from a large population as independent. Using your findings from parts (a) and (b), explain whether or not this assumption is reasonable.

4.41 Married women. The 2010 American Community Survey estimates that 47.1% of women ages 15 years and over are married.[40]

(a) We randomly select three women between these ages. What is the probability that the third woman selected is the only one who is married?

(b) What is the probability that all three randomly selected women are married?

(c) On average, how many women would you expect to sample before selecting a married woman? What is the standard deviation?

(d) If the proportion of married women was actually 30%, how many women would you expect to sample before selecting a married woman? What is the standard deviation?

(e) Based on your answers to parts (c) and (d), how does decreasing the probability of an event affect the mean and standard deviation of the wait time until success?

4.42 Defective rate. A machine that produces a special type of transistor (a component of computers) has a 2% defective rate. The production is considered a random process where each transistor is independent of the others.

(a) What is the probability that the 10th transistor produced is the first with a defect?

(b) What is the probability that the machine produces no defective transistors in a batch of 100?

(c) On average, how many transistors would you expect to be produced before the first with a defect? What is the standard deviation?

(d) Another machine that also produces transistors has a 5% defective rate where each transistor is produced independent of the others. On average how many transistors would you expect to be produced with this machine before the first with a defect? What is the standard deviation?

(e) Based on your answers to parts (c) and (d), how does increasing the probability of an event affect the mean and standard deviation of the wait time until success?

4.43 Eye color, Part I. A husband and wife both have brown eyes but carry genes that make it possible for their children to have brown eyes (probability 0.75), blue eyes (0.125), or green eyes (0.125).

(a) What is the probability the first blue-eyed child they have is their third child? Assume that the eye colors of the children are independent of each other.

(b) On average, how many children would such a pair of parents have before having a blue-eyed child? What is the standard deviation of the number of children they would expect to have until the first blue-eyed child?

4.44 Speeding on the I-5, Part II. Exercise 4.12 states that the distribution of speeds of cars traveling on the Interstate 5 Freeway (I-5) in California is nearly normal with a mean of 72.6 miles/hour and a standard deviation of 4.78 miles/hour. The speed limit on this stretch of the I-5 is 70 miles/hour.

(a) A highway patrol officer is hidden on the side of the freeway. What is the probability that 5 cars pass and none are speeding? Assume that the speeds of the cars are independent of each other.

(b) On average, how many cars would the highway patrol officer expect to watch until the first car that is speeding? What is the standard deviation of the number of cars he would expect to watch?

---

[40]U.S. Census Bureau, 2010 American Community Survey, Marital Status.

4.6.4 Binomial distribution

4.45 Underage drinking, Part II. We learned in Exercise 3.35 that about 70% of 18-20 year olds consumed alcoholic beverages in 2008. We now consider a random sample of fifty 18-20 year olds.

(a) How many people would you expect to have consumed alcoholic beverages? And with what standard deviation?

(b) Would you be surprised if there were 45 or more people who have consumed alcoholic beverages?

(c) What is the probability that 45 or more people in this sample have consumed alcoholic beverages? How does this probability relate to your answer to part (b)?

4.46 Chickenpox, Part II. We learned in Exercise 3.36 that about 90% of American adults had chickenpox before adulthood. We now consider a random sample of 120 American adults.

(a) How many people in this sample would you expect to have had chickenpox in their childhood? And with what standard deviation?

(b) Would you be surprised if there were 105 people who have had chickenpox in their childhood?

(c) What is the probability that 105 or fewer people in this sample have had chickenpox in their childhood? How does this probability relate to your answer to part (b)?

4.47 University admissions. Suppose a university announced that it admitted 2,500 students for the following year's freshman class. However, the university has dorm room spots for only 1,786 freshman students. If there is a 70% chance that an admitted student will decide to accept the offer and attend this university, what is the approximate probability that the university will not have enough dormitory room spots for the freshman class?

4.48 Survey response rate. Pew Research reported in 2012 that the typical response rate to their surveys is only 9%. If for a particular survey 15,000 households are contacted, what is the probability that at least 1,500 will agree to respond?[41]

4.49 Game of dreidel. A dreidel is a four-sided spinning top with the Hebrew letters *nun*, *gimel*, *hei*, and *shin*, one on each side. Each side is equally likely to come up in a single spin of the dreidel. Suppose you spin a dreidel three times. Calculate the probability of getting

(a) at least one *nun*?

(b) exactly 2 *nuns*?

(c) exactly 1 *hei*?

(d) at most 2 *gimels*?

Photo by Staccabees, cropped
(http://flic.kr/p/7gLZTf)
CC BY 2.0 license

4.50 Arachnophobia. A 2005 Gallup Poll found that 7% of teenagers (ages 13 to 17) suffer from arachnophobia and are extremely afraid of spiders. At a summer camp there are 10 teenagers sleeping in each tent. Assume that these 10 teenagers are independent of each other.[42]

(a) Calculate the probability that at least one of them suffers from arachnophobia.

(b) Calculate the probability that exactly 2 of them suffer from arachnophobia?

(c) Calculate the probability that at most 1 of them suffers from arachnophobia?

(d) If the camp counselor wants to make sure no more than 1 teenager in each tent is afraid of spiders, does it seem reasonable for him to randomly assign teenagers to tents?

---

[41]The Pew Research Center for the People and the Press, Assessing the Representativeness of Public Opinion Surveys, May 15, 2012.
[42]Gallup Poll, What Frightens America's Youth?, March 29, 2005.

4.51   Eye color, Part II. Exercise 4.43 introduces a husband and wife with brown eyes who have 0.75 probability of having children with brown eyes, 0.125 probability of having children with blue eyes, and 0.125 probability of having children with green eyes.

(a) What is the probability that their first child will have green eyes and the second will not?

(b) What is the probability that exactly one of their two children will have green eyes?

(c) If they have six children, what is the probability that exactly two will have green eyes?

(d) If they have six children, what is the probability that at least one will have green eyes?

(e) What is the probability that the first green eyed child will be the 4th child?

(f) Would it be considered unusual if only 2 out of their 6 children had brown eyes?

4.52   Sickle cell anemia. Sickle cell anemia is a genetic blood disorder where red blood cells lose their flexibility and assume an abnormal, rigid, "sickle" shape, which results in a risk of various complications. If both parents are carriers of the disease, then a child has a 25% chance of having the disease, 50% chance of being a carrier, and 25% chance of neither having the disease nor being a carrier. If two parents who are carriers of the disease have 3 children, what is the probability that

(a) two will have the disease?

(b) none will have the disease?

(c) at least one will neither have the disease nor be a carrier?

(d) the first child with the disease will the be 3rd child?

4.53   Roulette winnings. In the game of roulette, a wheel is spun and you place bets on where it will stop. One popular bet is that it will stop on a red slot; such a bet has an 18/38 chance of winning. If it stops on red, you double the money you bet. If not, you lose the money you bet. Suppose you play 3 times, each time with a $1 bet. Let Y represent the total amount won or lost. Write a probability model for Y.

4.54   Multiple choice quiz. In a multiple choice quiz there are 5 questions and 4 choices for each question (a, b, c, d). Robin has not studied for the quiz at all, and decides to randomly guess the answers. What is the probability that

(a) the first question she gets right is the 3rd question?

(b) she gets exactly 3 or exactly 4 questions right?

(c) she gets the majority of the questions right?

4.6.5   Sampling distribution of a sample proportion

4.55   Distribution of $\hat{p}$. Suppose the true population proportion were $p = 0.95$. The figure below shows what the distribution of a sample proportion looks like when the sample size is $n = 20$, $n = 100$, and $n = 500$. (a) What does each point (observation) in each of the samples represent? (b) Describe the distribution of the sample proportion, $\hat{p}$. How does the distribution of the sample proportion change as n becomes larger?

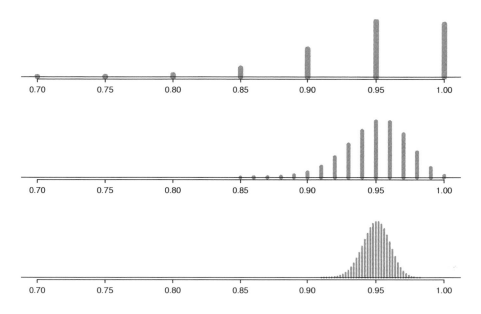

4.56   Distribution of $\hat{p}$. Suppose the true population proportion were $p = 0.5$. The figure below shows what the distribution of a sample proportion looks like when the sample size is $n = 20$, $n = 100$, and $n = 500$. What does each point (observation) in each of the samples represent? Describe how the distribution of the sample proportion, $\hat{p}$, changes as n becomes larger.

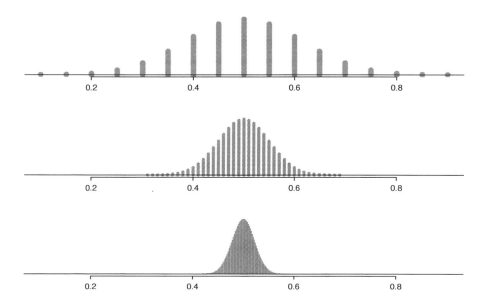

4.57 Distribution of $\hat{p}$. Suppose the true population proportion were $p = 0.5$ and a researcher takes a simple random sample of size $n = 50$.

(a) Find and interpret the standard deviation of the sample proportion $\hat{p}$.

(b) Calculate the probability that the sample proportion will be larger than 0.55 for a random sample of size 50.

4.58 Distribution of $\hat{p}$. Suppose the true population proportion were $p = 0.6$ and a researcher takes a simple random sample of size $n = 50$.

(a) Find and interpret the standard deviation of the sample proportion $\hat{p}$.

(b) Calculate the probability that the sample proportion will be larger than 0.65 for a random sample of size 50.

4.59 Nearsighted children. It is believed that nearsightedness affects about 8% of all children. We are interested in finding the probability that fewer than 12 out of 200 randomly sampled children will be nearsighted.

(a) Estimate this probability using the normal approximation to the binomial distribution.

(b) Estimate this probability using the distribution of the sample proportion.

(c) How do your answers from parts (a) and (b) compare?

4.60 Poverty in the US. The 2013 Current Population Survey (CPS) estimates that 22.5% of Mississippians live in poverty, which makes Mississippi the state with the highest poverty rate in the United States.[43] We are interested in finding out the probability that at least 250 people among a random sample of 1,000 Mississippians live in poverty.

(a) Estimate this probability using the normal approximation to the binomial distribution.

(b) Estimate this probability using the distribution of the sample proportion.

(c) How do your answers from parts (a) and (b) compare?

4.61 Young Hispanics in the US. The 2012 Current Population Survey (CPS) estimates that 38.9% of the people of Hispanic origin in the Unites States are under 21 years old.[44] Calculate the probability that at least 35 people among a random sample of 100 Hispanic people living in the United States are under 21 years old.

4.62 Social network use. The Pew Research Center estimates that as of January 2014, 89% of 18-29 year olds in the United States use social networking sites.[45] Calculate the probability that at least 95% of 500 randomly sampled 18-29 year olds use social networking sites.

---

[43]United States Census Bureau. 2013 Current Population Survey.Historical Poverty Tables - People. Web.

[44]United States Census Bureau.2012 Current Population Survey.The Hispanic Population in the United States: 2012. Web.

[45]Pew Research Center, Washington, D.C. Social Networking Fact Sheet, accessed on May 9, 2015.

Chapter 5

Foundation for inference

In the last chapter we encountered a probability problem in which we calculated the chance of getting less than 15% smokers in a sample, if we *knew* the true proportion of smokers in the population was 0.20. This chapter introduces the topic of inference, that is, the methods of drawing conclusions when the population value is *unknown*.

Probability versus inference

Probability Probability involves using a known population value (parameter) to make a prediction about the likelihood of a particular sample value (statistic).

Inference Inference involves using a calculated sample value (statistic) to estimate or better understand an unknown population value (parameter).

Statistical inference is concerned primarily with understanding the quality of parameter estimates. In this chapter, we will focus on the case of estimating a proportion from a random sample. While the equations and details change depending on the setting, the foundations for inference are the same throughout all of statistics. We introduce these common themes in this chapter, setting the stage for inference on other parameters. Understanding this chapter will make the rest of this book, and indeed the rest of statistics, seem much more familiar.

5.1   Estimating unknown parameters ▶

5.1.1   Point estimates

● **Example 5.1** We take a sample of size $n = 80$ from a particular county and find that 12 of the 80 people smoke. Estimate the **population proportion** based on the sample. Note that this example differs from Example 4.59 of the previous chapter in that we are not trying to predict what will happen in a sample. Instead, we have a sample, and we are trying to infer something about the true proportion.

The most intuitive way to go about doing this is to simply take the **sample proportion**. That is, $\hat{p} = \frac{12}{80} = 0.15$ is our best estimate for p, the population proportion.

The sample proportion $\hat{p} = 0.15$ is called a **point estimate** of the population proportion: if we can only choose one value to estimate the population proportion, this is our best guess. Suppose we take a new sample of 80 people and recompute the proportion of smokers in the sample; we will probably not get the exact same answer that we got the first time. Estimates generally vary from one sample to another, and this **sampling variation** tells us how close we expect our estimate to be to the true parameter.

● **Example 5.2**   In Chapter 2, we found the summary statistics for the number of characters in a set of 50 email data. These values are summarized below.

$\bar{x}$	11,160
median	6,890
s_x	13,130

Estimate the **population mean** based on the sample.

The best estimate for the population mean is the **sample mean**. That is, $\bar{x} = 11,160$ is our best estimate for μ.

⊙ **Guided Practice 5.3**   Using the email data, what quantity should we use as a point estimate for the population standard deviation σ?[1]

5.1.2   Introducing the standard error

Point estimates only approximate the population parameter, and they vary from one sample to another. It will be useful to quantify how variable an estimate is from one sample to another. For a random sample, when this variability is small we can have greater confidence that our estimate is close to the true value.

How can we quantify the expected variability in a point estimate $\hat{p}$? The discussion in Section 4.5 tells us how. The variability in the distribution of $\hat{p}$ is given by its standard deviation.

$$SD_{\hat{p}} = \sqrt{\frac{p(1-p)}{n}}$$

---

[1]Again, intuitively we would use the sample standard deviation $s = 13,130$ as our best estimate for σ.

● **Example 5.4** Calculate the standard deviation of $\hat{p}$ for smoking example, where $\hat{p}$ = 0.15 is the proportion in a sample of size 80 that smoke.

It may seem easy to calculate the SD at first glance, but there is a serious problem: p is *unknown*. In fact, when doing inference, p must be unknown, otherwise it is illogical to try to estimate it. We cannot calculate the SD, but we can estimate it using, you might have guessed, the sample proportion $\hat{p}$.

This estimate of the standard deviation is known as the **standard error**, or **SE** for short.

$$SE_{\hat{p}} = \sqrt{\frac{\hat{p}(1 - \hat{p})}{n}}$$

● **Example 5.5** Calculate and interpret the SE of $\hat{p}$ for the previous example.

$$SE_{\hat{p}} = \sqrt{\frac{\hat{p}(1 - \hat{p})}{n}} = \sqrt{\frac{0.15(1 - 0.15)}{80}} = 0.04$$

The average or expected error in our estimate is 4%.

● **Example 5.6** If we quadruple the sample size from 80 to 240, what will happen to the SE?

$$SE_{\hat{p}} = \sqrt{\frac{\hat{p}(1 - \hat{p})}{n}} = \sqrt{\frac{0.15(1 - 0.15)}{240}} = 0.02$$

The larger the sample size, the smaller our standard error. This is consistent with intuition: the more data we have, the more reliable an estimate will tend to be. However, quadrupling the sample size does not reduce the error by a factor of 4. Because of the square root, the effect is to reduce the error by a factor $\sqrt{4}$, or 2.

5.1.3 Basic properties of point estimates

We achieved three goals in this section. First, we determined that point estimates from a sample may be used to estimate population parameters. We also determined that these point estimates are not exact: they vary from one sample to another. Lastly, we quantified the uncertainty of the sample proportion using what we call the standard error. We will learn how to calculate the standard error for other point estimates such as a mean, a difference in means, or a difference in proportions in the chapters that follow.

5.2 Confidence intervals

A point estimate provides a single plausible value for a parameter. However, a point estimate is rarely perfect; usually there is some error in the estimate. In addition to supplying a point estimate of a parameter, a next logical step would be to provide a plausible *range of values* for the parameter.

5.2.1 Capturing the population parameter

A plausible range of values for the population parameter is called a **confidence interval**. Using only a point estimate is like fishing in a murky lake with a spear, and using a confidence interval is like fishing with a net. We can throw a spear where we saw a fish, but we will probably miss. On the other hand, if we toss a net in that area, we have a good chance of catching the fish.

   If we report a point estimate, we probably will not hit the exact population parameter. On the other hand, if we report a range of plausible values – a confidence interval – we have a good shot at capturing the parameter.

⊙ **Guided Practice 5.7**   If we want to be very confident we capture the population parameter, should we use a wider interval or a smaller interval?[2]

5.2.2 Constructing a 95% confidence interval

A point estimate is our best guess for the value of the parameter, so it makes sense to build the confidence interval around that value. The standard error, which is a measure of the uncertainty associated with the point estimate, provides a guide for how large we should make the confidence interval.

Constructing a 95% confidence interval

When the sampling distribution of a point estimate can reasonably be modeled as normal, the point estimate we observe will be within 1.96 standard errors of the true value of interest about 95% of the time. Thus, a **95% confidence interval** for such a point estimate can be constructed:

$$\text{point estimate } \pm \ 1.96 \times SE \tag{5.8}$$

We can be **95% confident** this interval captures the true value.

⊙ **Guided Practice 5.9**   Compute the area between -1.96 and 1.96 for a normal distribution with mean 0 and standard deviation 1.[3]

● **Example 5.10**   The point estimate from the smoking example was 15%. In the next chapters we will determine when we can apply a normal model to a point estimate. For now, assume that the normal model is reasonable. The standard error for this point estimate was calculated to be $SE = 0.04$. Construct a 95% confidence interval.

$$\text{point estimate } \pm \ 1.96 \times SE$$
$$0.15 \ \pm \ 1.96 \times 0.04$$
$$(0.0716,\ 0.2284)$$

We are 95% confident that the true percent of smokers in this population is between 7.16% and 22.84%.

---

[2]If we want to be more confident we will capture the fish, we might use a wider net. Likewise, we use a wider confidence interval if we want to be more confident that we capture the parameter.

[3]We will leave it to you to draw a picture. The Z-scores are $Z_{left} = -1.96$ and $Z_{right} = 1.96$. The area between these two Z-scores is 0.9500. This is where "1.96" comes from in the 95% confidence interval formula.

● **Example 5.11** Based on the confidence interval above, is there evidence that a smaller proportion smoke in this county than in the state as a whole? The proportion that smoke in the state is known to be 0.20.

While the point estimate of 0.15 is lower than 0.20, this deviation is likely due to random chance. Because the confidence interval *includes* the value 0.20, 0.20 is a reasonable value for the proportion of smokers in the county. Therefore, based on this confidence interval, we do not have evidence that a smaller proportion smoke in the county than in the state.

In Section 1.1 we encountered an experiment that examined whether implanting a stent in the brain of a patient at risk for a stroke helps reduce the risk of a stroke. The results from the first 30 days of this study, which included 451 patients, are summarized in Table 5.1. These results are surprising! The point estimate suggests that patients who received stents may have a *higher* risk of stroke: $p_{trmt} - p_{ctrl} = 0.090$.

	stroke	no event	Total
treatment	33	191	224
control	13	214	227
Total	46	405	451

Table 5.1: Descriptive statistics for 30-day results for the stent study.

● **Example 5.12** Consider the stent study and results. The conditions necessary to ensure the point estimate $p_{trmt} - p_{ctrl} = 0.090$ is nearly normal have been verified for you, and the estimate's standard error is $SE = 0.028$. Construct a 95% confidence interval for the change in 30-day stroke rates from usage of the stent.

The conditions for applying the normal model have already been verified, so we can proceed to the construction of the confidence interval:

$$\text{point estimate } \pm \ 1.96 \times SE$$
$$0.090 \ \pm \ 1.96 \times 0.028$$
$$(0.035 \ , \ 0.145)$$

We are 95% confident that implanting a stent in a stroke patient's brain. Since the entire interval is greater than 0, it means the data provide statistically significant evidence that the stent used in the study *increases* the risk of stroke, contrary to what researchers had expected before this study was published!

We can be 95% confident that a 95% confidence interval contains the true population parameter. However, confidence intervals are imperfect. About 1-in-20 (5%) properly constructed 95% confidence intervals will fail to capture the parameter of interest. Figure 5.2 shows 25 confidence intervals for a proportion that were constructed from simulations where the true proportion was $p = 0.3$. However, 1 of these 25 confidence intervals happened not to include the true value.

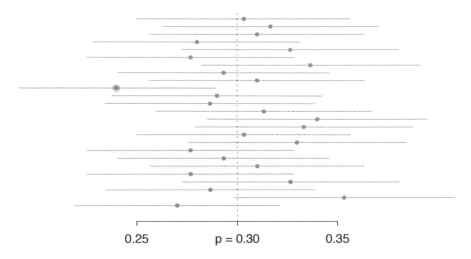

Figure 5.2: Twenty-five samples of size $n = 300$ were simulated when $p = 0.30$. For each sample, a confidence interval was created to try to capture the true proportion p. However, 1 of these 25 intervals did not capture $p = 0.30$.

⊙ **Guided Practice 5.13**   In Figure 5.2, one interval does not contain the true proportion, $p = 0.3$. Does this imply that there was a problem with the simulations?[4]

5.2.3   Changing the confidence level

Suppose we want to consider confidence intervals where the confidence level is somewhat higher than 95%: perhaps we would like a confidence level of 99%.

● **Example 5.14**   Would a 99% confidence interval be wider or narrower than a 95% confidence interval?

Using a previous analogy: if we want to be more confident that we will catch a fish, we should use a wider net, not a smaller one. To be 99% confidence of capturing the true value, we must use a wider interval. On the other hand, if we want an interval with lower confidence, such as 90%, we would use a narrower interval.

The 95% confidence interval structure provides guidance in how to make intervals with new confidence levels. Below is a general 95% confidence interval for a point estimate that comes from a nearly normal distribution:

$$\text{point estimate} \pm 1.96 \times SE \tag{5.15}$$

There are three components to this interval: the point estimate, "1.96", and the standard error. The choice of $1.96 \times SE$ was based on capturing 95% of the distribution since the estimate is within 1.96 standard deviations of the true value about 95% of the time. The choice of 1.96 corresponds to a 95% confidence level.

---

[4]No. Just as some observations occur more than 1.96 standard deviations from the mean, some point estimates will be more than 1.96 standard errors from the parameter. A confidence interval only provides a plausible range of values for a parameter. While we might say other values are implausible based on the data, this does not mean they are impossible.

⊙ **Guided Practice 5.16** If X is a normally distributed random variable, how often will X be within 2.58 standard deviations of the mean?[5]

To create a 99% confidence interval, change 1.96 in the 95% confidence interval formula to be 2.58. Guided Practice 5.16 highlights that 99% of the time a normal random variable will be within 2.58 standard deviations of its mean. This approach – using the Z-scores in the normal model to compute confidence levels – is appropriate when the point estimate is associated with a normal distribution and we can properly compute the standard error. Thus, the formula for a 99% confidence interval is

$$\text{point estimate } \pm\ 2.58 \times SE \qquad (5.17)$$

Figure 5.3 provides a picture of how to identify $z^\star$ based on a confidence level.

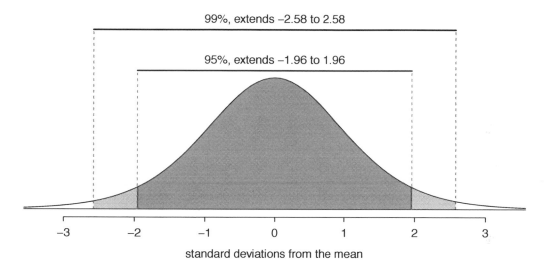

Figure 5.3: The area between $-z^\star$ and $z^\star$ increases as $|z^\star|$ becomes larger. If the confidence level is 99%, we choose $z^\star$ such that 99% of the normal curve is between $-z^\star$ and $z^\star$, which corresponds to 0.5% in the lower tail and 0.5% in the upper tail: $z^\star = 2.58$.

⊙ **Guided Practice 5.18** Create a 99% confidence interval for the impact of the stent on the risk of stroke using the data from Example 5.12. The point estimate is 0.090, and the standard error is $SE = 0.028$. It has been verified for you that the point estimate can reasonably be modeled by a normal distribution.[6]

---

[5]This is equivalent to asking how often the Z-score will be larger than -2.58 but less than 2.58. (For a picture, see Figure 5.3.) There is ≈ 0.99 probability that the unobserved random variable X will be within 2.58 standard deviations of the mean.

[6]Since the necessary conditions for applying the normal model have already been checked for us, we can go straight to the construction of the confidence interval: point estimate $\pm\ 2.58 \times SE \rightarrow (0.018, 0.162)$. We are 99% confident that implanting a stent in the brain of a patient who is at risk of stroke increases the risk of stroke within 30 days by a rate of 0.018 to 0.162 (assuming the patients are representative of the population).

Confidence interval for any confidence level

If the point estimate follows the normal model with standard error SE, then a confidence interval for the population parameter is

$$\text{point estimate } \pm \ z^{\star} \times SE$$

where $z^{\star}$ depends on the confidence level selected.

Finding the value of $z^{\star}$ that corresponds to a particular confidence level is most easily accomplished by using a new table, called the t-table. For now, what is noteworthy about this table is that the bottom row corresponds to confidence levels. The numbers inside the table are the critical values, but which row should we use? Later in this book, we will see that a t curve with infinite degrees of freedom corresponds to the normal curve. For this reason, when finding using the t-table to find the appropriate $z^{\star}$, always use row ∞.

	one tail	0.100	0.050	0.025	0.010	0.005
df	1	3.078	6.314	12.71	31.82	63.66
	2	1.886	2.920	4.303	6.965	9.925
	3	1.638	2.353	3.182	4.541	5.841
	$\vdots$	$\vdots$	$\vdots$	$\vdots$	$\vdots$	$\vdots$
	1000	1.282	1.646	1.962	2.330	2.581
	∞	1.282	1.645	1.960	2.326	2.576
Confidence level C		80%	90%	95%	98%	99%

Table 5.4: An abbreviated look at the t-table. The columns correspond to confidence levels. Row ∞ corresponds to the normal curve.

TIP: Finding $z^{\star}$ for a particular confidence level
We select $z^{\star}$ so that the area between $-z^{\star}$ and $z^{\star}$ in the normal model corresponds to the confidence level. Use the t-table at row ∞ to find the critical value $z^{\star}$.

⊙ **Guided Practice 5.19**   In Example 5.12 we found that implanting a stent in the brain of a patient at risk for a stroke *increased* the risk of a stroke. The study estimated a 9% increase in the number of patients who had a stroke, and the standard error of this estimate was about $SE = 2.8\%$ or 0.028. Compute a 90% confidence interval for the effect. Note: the conditions for normality had earlier been confirmed for us.[7]

---

[7]We must find $z^{\star}$ such that 90% of the distribution falls between $-z^{\star}$ and $z^{\star}$ in the standard normal model. Using the t-table with a confidence level of 90% at row ∞ gives 1.645. Thus $z^{\star} = 1.645$. The 90% confidence interval can then be computed as

$$\text{point estimate } \pm \ z^{\star} \times SE$$
$$0.09 \ \pm \ 1.645 \times 0.028$$
$$(0.044 \ , \ 0.136)$$

That is, we are 90% confident that implanting a stent in a stroke patient's brain increased the risk of stroke within 30 days by 4.4% to 13.6%.

The normal approximation is crucial to the precision of these confidence intervals. The next two chapters provides detailed discussions about when the normal model can safely be applied to a variety of situations. When the normal model is not a good fit, we will use alternate distributions that better characterize the sampling distribution.

5.2.4 Margin of error

The confidence intervals we have encountered thus far have taken the form

$$\text{point estimate } \pm \ z^* \times SE$$

Confidence intervals are also often reported as

$$\text{point estimate } \pm \ \text{margin of error}$$

For example, instead of reporting an interval as $0.09 \ \pm \ 1.645 \times 0.028$ or $(0.044, 0.136)$, it could be reported as $0.09 \ \pm \ 0.046$.

The **margin of error** is the distance between the point estimate and the lower or upper bound of a confidence interval.

Margin of error

A confidence interval can be written as point estimate $\pm$ margin of error.
For a confidence interval for a proportion, the margin of error is $z^\star \times SE$.

⊙ **Guided Practice 5.20** To have a smaller margin or error, should one use a larger sample or a smaller sample?[8]

⊙ **Guided Practice 5.21** What is the margin of error for the confidence interval: $(0.035, 0.145)$?[9]

5.2.5 Interpreting confidence intervals

A careful eye might have observed the somewhat awkward language used to describe confidence intervals. Correct interpretation:

We are [XX]% confident that the population parameter is between...

Incorrect language might try to describe the confidence interval as capturing the population parameter with a certain probability.[10] This is one of the most common errors: while it might be useful to think of it as a probability, the confidence level only quantifies how plausible it is that the parameter is in the interval.

As we saw in Figure 5.2, the 95% confidence interval *method* has a 95% probability of producing an interval that will contain the population parameter. A correct interpretation

---

[8]Intuitively, a larger sample should tend to yield less error. We can also note that n, the sample size is in the denominator of the SE formula, so a n goes up, the SE and thus the margin of error go down.

[9]Because we both add and subtract the margin of error to get the confidence interval, the margin of error is *half* of the width of the interval. $(0.145 - 0.035)/2 = 0.055$.

[10]To see that this interpretation is incorrect, imagine taking two random samples and constructing two 95% confidence intervals for an unknown proportion. If these intervals are disjoint, can we say that there is a 95%+95%=190% chance that the first or the second interval captures the true value?

of the confidence *level* is that such intervals will contain the population parameter that percent of the time. However, each individual *interval* either does or does not contain the population parameter. A correct interpretation of an individual confidence interval cannot involve the vocabulary of probability.

Another especially important consideration of confidence intervals is that they *only try to capture the population parameter*. Our intervals say nothing about the confidence of capturing individual observations, a proportion of the observations, or about capturing point estimates. Confidence intervals only attempt to capture population parameters.

5.2.6 Using confidence intervals: a stepwise approach

Follow these six steps when carrying out any confidence interval problem.

Steps for using confidence intervals (AP exam tip) The AP exam is scored in a standardized way, so to ensure full points for a problem, make sure to complete each of the following steps.

1. State the name of the CI being used.

2. Verify conditions to ensure the standard error estimate is reasonable and the point estimate is unbiased and follows the expected distribution, often a normal distribution.

3. Plug in the numbers and write the interval in the form

$$\text{point estimate} \pm \text{critical value} \times \text{SE of estimate}$$

   So far, the **critical value** has taken the form $z^\star$.

4. Evaluate the CI and write in the form (___, ___).

5. Interpret the interval: "We are [XX]% confident that the true [describe the parameter in context] falls between [identify the upper and lower endpoints of the calculated interval].

6. State your conclusion to the original question. (Sometimes, as in the case of the examples in this section, no conclusion is necessary.)

5.3 Introducing hypothesis testing ▶

● **Example 5.22** Suppose your professor splits the students in class into two groups: students on the left and students on the right. If $\hat{p}_L$ and $\hat{p}_R$ represent the proportion of students who own an Apple product on the left and right, respectively, would you be surprised if $\hat{p}_L$ did not exactly equal $\hat{p}_R$?

While the proportions would probably be close to each other, they are probably not exactly the same. We would probably observe a small difference due to chance.

Studying randomness of this form is a key focus of statistics. How large would the observed difference in these two proportions need to be for us to believe that there is a real difference in Apple ownership? In this section, we'll explore this type of randomness in the context of an unknown proportion, and we'll learn new tools and ideas that will be applied throughout the rest of the book.

5.3.1 Case study: medical consultant

People providing an organ for donation sometimes seek the help of a special medical consultant. These consultants assist the patient in all aspects of the surgery, with the goal of reducing the possibility of complications during the medical procedure and recovery. Patients might choose a consultant based in part on the historical complication rate of the consultant's clients.

One consultant tried to attract patients by noting the overall complication rate for liver donor surgeries in the US is about 10%, but her clients have had only 9 complications in the 142 liver donor surgeries she has facilitated. She claims this is strong evidence that her work meaningfully contributes to reducing complications (and therefore she should be hired!).

● **Example 5.23** We will let p represent the true complication rate for liver donors working with this consultant. Estimate p using the data, and label this value $\hat{p}$.

The sample proportion for the complication rate is 9 complications divided by the 142 surgeries the consultant has worked on: $\hat{p} = 9/142 = 0.063$.

● **Example 5.24** Is it possible to prove that the consultant's work reduces complications?

No. The claim implies that there is a causal connection, but the data are observational. For example, maybe patients who can afford a medical consultant can afford better medical care, which can also lead to a lower complication rate.

● **Example 5.25** While it is not possible to assess the causal claim, it is still possible to ask whether the low complication rate of $\hat{p} = 0.063$ provides evidence that the consultant's true complication rate is different than the US complication rate. Why might we be tempted to immediately conclude that the consultant's true complication rate is different than the US complication rate? Can we draw this conclusion?

Her sample complication rate is $\hat{p} = 0.063$, which is 0.037 lower than the US complication rate of 10%. However, we cannot yet be sure if the observed difference represents a real difference or is just the result of random variation. We wouldn't expect the sample proportion to be *exactly* 0.10, even if the truth was that her real complication rate was 0.10.

5.3.2   Setting up the null and alternate hypothesis

We can set up two competing hypotheses about the consultant's true complication rate. The first is call the **null hypothesis** and represents either a skeptical perspective or a perspective of no difference. The second is called the **alternative hypothesis** (or alternate hypothesis) and represents a new perspective such as the possibility that there has been a change or that there is a treatment effect in an experiment.

Null and alternative hypotheses

The **null hypothesis** is abbreviated H_0. It states that nothing has changed and that any deviation from what was expected is due to chance error.

The **alternative hypothesis** is abbreviated H_A. It asserts that there has been a change and that the observed deviation is too large to be explained by chance alone.

● **Example 5.26** Identify the null and alternative claim regarding the consultant's complication rate.

H_0: The true complication rate for the consultant's clients is the *same as* the US complication rate of 10%.

H_A: The true complication rate for the consultant's clients is different than 10%.

Often it is convenient to write the null and alternative hypothesis in mathematical or numerical terms. To do so, we must first identify the quantity of interest. This quantity of interest is known as the parameter for a hypothesis test.

Parameters and point estimates

A **parameter** for a hypothesis test is the "true" value of the population of interest. When the parameter is a proportion, we call it p.

A **point estimate** is calculated from a sample. When the point estimate is a proportion, we call it $\hat{p}$.

The observed or sample proportion 0f 0.063 is a point estimate for the true proportion. The parameter in this problem is the true proportion of complications for this consultant's clients. The parameter is unknown, but the null hypothesis is that it equals the overall proportion of complications: $p = 0.10$. This hypothesized value is called the null value.

Null value of a hypothesis test

The **null value** is the value hypothesized for the parameter in H_0, and it is sometimes represented with a subscript 0, e.g. p_0 (just like H_0).

In the medical consultant case study, the parameter is p and the null value is $p_0 = 0.10$. We can write the null and alternative hypothesis as numerical statements as follows.

- H_0: $p = 0.10$ (The complication rate for the consultant's clients is equal to the US complication rate of 10%.)

- H_A: $p \neq 0.10$ (The complication rate for the consultant's clients is not equal to the US complication rate of 10%.)

Hypothesis testing

These hypotheses are part of what is called a **hypothesis test**. A hypothesis test is a statistical technique used to evaluate competing claims using data. Often times, the null hypothesis takes a stance of *no difference* or *no effect*. If the null hypothesis and the data notably disagree, then we will reject the null hypothesis in favor of the alternative hypothesis.

Don't worry if you aren't a master of hypothesis testing at the end of this section. We'll discuss these ideas and details many times in this chapter and the two chapters that follow.

The null claim is always framed as an equality: it tells us what quantity we should use for the parameter when carrying out calculations for the hypothesis test. There are three choices for the alternative hypothesis, depending upon whether the researcher is trying to prove that the value of the parameter is greater than, less than, or not equal to the null value.

TIP: Always write the null hypothesis as an equality
We will find it most useful if we always list the null hypothesis as an equality (e.g. $p = 7$) while the alternative always uses an inequality (e.g. $p \neq 0.7$, $p > 0.7$, or $p < 0.7$).

⊙ **Guided Practice 5.27** According to US census data, in 2013 the percent of male residents in the state of Alaska was 52.4%.[11] A researcher plans to take a random sample of residents from Alaska to test whether or not this is still the case. Write out the hypotheses that the researcher should test in both plain and statistical language. [12]

When the alternative claim uses a $\neq$, we call the test a **two-sided** test, because either extreme provides evidence against H_0. When the alternative claim uses a $<$ or a $>$, we call it a **one-sided** test.

---

[11]quickfacts.census.gov/qfd/states/02000.html
[12]H_0: $p = 0.524$; The proportion of male residents in Alaska is *unchanged* from 2012. H_A: $p \neq 0.524$; The proportion of male residents in Alaska has changed from 2012. Note that it could have increased or decreased.

TIP: One-sided and two-sided tests

If the researchers are only interested in showing an increase or a decrease, but not both, use a one-sided test. If the researchers would be interested in any difference from the null value – an increase or decrease – then the test should be two-sided.

● **Example 5.28**   For the example of the consultant's complication rate, we knew that her sample complication rate was 0.063, which was lower than the US complication rate of 0.10. Why did we conduct a two-sided hypothesis test for this setting?

The setting was framed in the context of the consultant being helpful, but what if the consultant actually performed worse than the US complication rate? Would we care? More than ever! Since we care about a finding in either direction, we should run a two-sided test.

Caution: One-sided hypotheses are allowed only *before* seeing data

After observing data, it is tempting to turn a two-sided test into a one-sided test. Avoid this temptation. Hypotheses must be set up *before* observing the data. If they are not, the test must be two-sided.

5.3.3   Evaluating the hypotheses with a p-value

● **Example 5.29**   There were 142 patients in the consultant's sample. If the null claim is true, how many would we expect to have had a complication?

If the null claim is true, we would expect about 10% of the patients, or about 14.2 to have a complication.

The consultant's complication rate for her 142 clients was 0.063 ($0.063 \times 142 \approx 9$). What is the probability that a sample would produce a number of complications this far from the expected value of 14.2, *if her true complication rate were* 0.10, that is, if H_0 were true? The probability, which is estimated in Section 5.5 on page 231, is about 0.1754. We call this quantity the **p-value**.

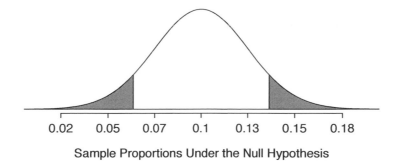

Sample Proportions Under the Null Hypothesis

Figure 5.5: The shaded area represents the p-value. We observed $\hat{p} = 0.063$, so any observations smaller than this are at least as extreme relative to the null value, $p_0 = 0.1$, and so the lower tail is shaded. However, since this is a two-sided test, values above 0.137 are also at least as extreme as 0.063 (relative to 0.1), and so they also contribute to the p-value. The tail areas together total of about 0.1754 when calculated using a simulation technique in Section 5.3.4.

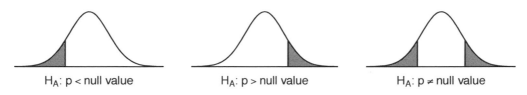

H_A: p < null value H_A: p > null value H_A: p ≠ null value

Figure 5.6: When the alternative hypothesis takes the form $p <$ null value, the p-value is represented by the lower tail. When it takes the form $p >$ null value, the p-value is represented by the upper tail. When using $p \neq$ null value, then the p-value is represented by both tails.

Finding and interpreting the p-value

When examining a proportion, we find and interpret the **p-value** according to the nature of the alternative hypothesis.

$H_A : p > p_0$. The p-value is the probability of observing a sample proportion as large as we saw in our sample, if the null hypothesis were true. The p-value corresponds to the area in the *upper* tail.

$H_A : p < p_0$. The p-value is the probability of observing a sample proportion as small as we saw in our sample, if the null hypothesis were true. The p-value corresponds to the area in the *lower* tail.

$H_A : p \neq p_0$. The p-value is the probability of observing a sample proportion as far from the null value as what was observed in the current data set, if the null hypothesis were true. The p-value corresponds to the area in *both* tails.

When the p-value is small, i.e. less than a previously set threshold, we say the results are **statistically significant**. This means the data provide such strong evidence against

α
significance
level of a
hypothesis test

H_0 that we reject the null hypothesis in favor of the alternative hypothesis. The threshold, called the **significance level** and often represented by α (the Greek letter *alpha*), is typically set to $\alpha = 0.05$, but can vary depending on the field or the application.

Statistical significance

If the p-value is less than the significance level α (usually 0.05), we say that the result is **statistically significant**. We reject H_0, and we have strong evidence favoring H_A.

If the p-value is greater than the significance level α, we say that the result is not statistically significant. We do not reject H_0, and we do not have strong evidence for H_A.

Recall that the null claim is the claim of no difference. If we reject H_0, we are asserting that there is a real difference. If we do not reject H_0, we are saying that the null claim is reasonable, but *it has not been proven*.

⊙ **Guided Practice 5.30**   Because the p-value is 0.1754, which is larger than the significance level 0.05, we do not reject the null hypothesis. Explain what this means in the context of the problem using plain language.[13]

● **Example 5.31**   In the previous exercise, we did not reject H_0. This means that we did not disprove the null claim. Is this equivalent to proving the null claim is true?

No. We did not prove that the consultant's complication rate is *exactly* equal to 10%. Recall that the test of hypothesis starts by *assuming the null claim is true*. That is, the test proceeds as an argument by contradiction. *If the null claim is true*, there is a 0.1754 chance of seeing sample data as divergent from 10% as we saw in our sample. Because 0.1754 is large, it is within the realm of chance error and we cannot say the null hypothesis is unreasonable.[14]

TIP: Double negatives can sometimes be used in statistics
In many statistical explanations, we use double negatives. For instance, we might say that the null hypothesis is *not implausible* or we *failed to reject* the null hypothesis. Double negatives are used to communicate that while we are not rejecting a position, we are also not saying that we know it to be true.

● **Example 5.32**   Does the conclusion in Guided Practice 5.30 ensure that there is no real association between the surgical consultant's work and the risk of complications? Explain.

No. It is possible that the consultant's work is associated with a lower or higher risk of complications. However, the data did not provide enough information to reject the null hypothesis (the sample was too small).

---

[13]The data do not provide evidence that the consultant's complication rate is significantly lower or higher that the US complication rate of 10%.

[14]The p-value is actually a conditional probability. It is P(getting data at least as divergent from the null value as we observed | H_0 is true). It is NOT P(H_0 is true | we got data this divergent from the null value.

● **Example 5.33** An experiment was conducted where study participants were randomly divided into two groups. Both were given the opportunity to purchase a DVD, but the one half was reminded that the money, if not spent on the DVD, could be used for other purchases in the future while the other half was not. The half that were reminded that the money could be used on other purchases were 20% less likely to continue with a DVD purchase. We determined that such a large difference would only occur about 1-in-150 times if the reminder actually had no influence on student decision-making. What is the p-value in this study? Was the result statistically significant?

The p-value was 0.006 (about 1/150). Since the p-value is less than 0.05, the data provide statistically significant evidence that US college students were actually influenced by the reminder.

What's so special about 0.05?

We often use a threshold of 0.05 to determine whether a result is statistically significant. But why 0.05? Maybe we should use a bigger number, or maybe a smaller number. If you're a little puzzled, that probably means you're reading with a critical eye – good job! We've made a video to help clarify *why 0.05*:

www.openintro.org/why05

Sometimes it's also a good idea to deviate from the standard. We'll discuss when to choose a threshold different than 0.05 in Section 5.3.7.

Statistical inference is the practice of making decisions and conclusions from data in the context of uncertainty. Just as a confidence interval may occasionally fail to capture the true parameter, a test of hypothesis may occasionally lead us to an incorrect conclusion. While a given data set may not always lead us to a correct conclusion, statistical inference gives us tools to control and evaluate how often these errors occur.

5.3.4   Calculating the p-value by simulation (special topic)

When conditions for the applying the normal model are met, we use the normal model to find the p-value of a test of hypothesis. In the complication rate example, the distribution is not normal. It is, however, *binomial*, because we are interested in how many out of 142 patients will have complications.

We could calculate the p-value of this test using binomial probabilities. A more general approach, though, for calculating p-values when the normal model does not apply is to use what is known as **simulation**. While performing this procedure is outside of the scope of the course, we provide an example here in order to better understand the concept of a p-value.

We simulate 142 new patients to see what result might happen if the complication rate really is 0.10. To do this, we could use a deck of cards. Take one red card, nine black cards, and mix them up. If the cards are well-shuffled, drawing the top card is one way of simulating the chance a patient has a complication if the true rate is 0.10: if the card is red, we say the patient had a complication, and if it is black then we say they did not have a complication. If we repeat this process 142 times and compute the proportion of simulated patients with complications, $\hat{p}_{sim}$, then this simulated proportion is exactly a draw from the null distribution.

There were 12 simulated cases with a complication and 130 simulated cases without a complication: $\hat{p}_{sim} = 12/142 = 0.085$.

One simulation isn't enough to get a sense of the null distribution, so we repeated the simulation 10,000 times using a computer. Figure 5.7 shows the null distribution from these 10,000 simulations. The simulated proportions that are less than or equal to $\hat{p} = 0.063$ are shaded. There were 0.0877 simulated sample proportions with $\hat{p}_{sim} \leq 0.063$, which represents a fraction 0.0877 of our simulations:

$$\text{left tail} = \frac{\text{Number of observed simulations with } \hat{p}_{sim} \leq 0.063}{10000} = \frac{877}{10000} = 0.0877$$

However, this is not our p-value! Remember that we are conducting a two-sided test, so we should double the one-tail area to get the p-value:[15]

$$\text{p-value} = 2 \times \text{left tail} = 2 \times 0.0877 = 0.1754$$

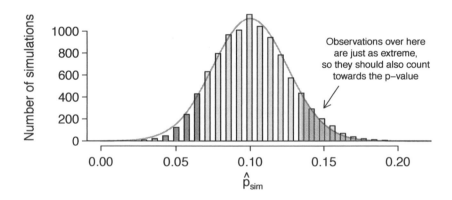

Figure 5.7: The null distribution for $\hat{p}$, created from 10,000 simulated studies. The left tail contains 8.77% of the simulations. For a two-sided test, we double the tail area to get the p-value. This doubling accounts for the observations we might have observed in the upper tail, which are also at least as extreme (relative to 0.10) as what we observed, $\hat{p} = 0.063$.

---

[15]This doubling approach is preferred even when the distribution isn't symmetric, as in this case.

5.3.5 Formal hypothesis testing: a stepwise approach

Carrying out a formal test of hypothesis (AP exam tip)

Follow these seven steps when carrying out a hypothesis test.

1. State the name of the test being used.

2. Verify conditions to ensure the standard error estimate is reasonable and the point estimate follows the appropriate distribution and is unbiased.

3. Write the hypotheses in plain language, then set them up in mathematical notation.

4. Identify the significance level α.

5. Calculate the test statistic, often Z, using an appropriate point estimate of the parameter of interest and its standard error.

$$\text{test statistic} = \frac{\text{point estimate} - \text{null value}}{\text{SE of estimate}}$$

6. Find the p-value, compare it to α, and state whether to reject or not reject the null hypothesis.

7. Write your conclusion in context.

5.3.6 Decision errors

The hypothesis testing framework is a very general tool, and we often use it without a second thought. If a person makes a somewhat unbelievable claim, we are initially skeptical. However, if there is sufficient evidence that supports the claim, we set aside our skepticism. The hallmarks of hypothesis testing are also found in the US court system.

● **Example 5.34** A US court considers two possible claims about a defendant: she is either innocent or guilty. If we set these claims up in a hypothesis framework, which would be the null hypothesis and which the alternative?

The jury considers whether the evidence is so convincing (strong) that there is no reasonable doubt of the person's guilt. That is, the skeptical perspective (null hypothesis) is that the person is innocent until evidence is presented that convinces the jury that the person is guilty (alternative hypothesis). In statistics, our evidence comes in the form of data, and we use the significance level to decide what is beyond a reasonable doubt.

Jurors examine the evidence to see whether it convincingly shows a defendant is guilty. Notice that a jury finds a defendent either guilty or not guilty. They either reject the null claim or they do not reject the null claim. They never prove the null claim, that is, they never find the defendant innocent. If a jury finds a defendant *not guilty*, this does not necessarily mean the jury is confident in the person's innocence. They are simply not convinced of the alternative that the person is guilty.

This is also the case with hypothesis testing: *even if we fail to reject the null hypothesis, we typically do not accept the null hypothesis as truth.* Failing to find strong evidence for the alternative hypothesis is not equivalent to providing evidence that the null hypothesis is true.

Hypothesis tests are not flawless. Just think of the court system: innocent people are sometimes wrongly convicted and the guilty sometimes walk free. Similarly, data can point to the wrong conclusion. However, what distinguishes statistical hypothesis tests from a court system is that our framework allows us to quantify and control how often the data lead us to the incorrect conclusion.

There are two competing hypotheses: the null and the alternative. In a hypothesis test, we make a statement about which one might be true, but we might choose incorrectly. There are four possible scenarios in a hypothesis test, which are summarized in Table 5.8.

		Test conclusion	
		do not reject H_0	reject H_0 in favor of H_A
Truth	H_0 true	okay	Type 1 Error
	H_A true	Type 2 Error	okay

Table 5.8: Four different scenarios for hypothesis tests.

Type 1 and Type 2 Errors

A **Type 1 Error** is rejecting the null hypothesis when H_0 is actually true. When we reject the null hypothesis, it is possible that we make a Type 1 Error.

A **Type 2 Error** is failing to reject the null hypothesis when H_A is actually true.

● **Example 5.35**   In a US court, the defendant is either innocent (H_0) or guilty (H_A). What does a Type 1 Error represent in this context? What does a Type 2 Error represent? Table 5.8 may be useful.

If the court makes a Type 1 Error, this means the defendant is innocent (H_0 true) but wrongly convicted. A Type 2 Error means the court failed to reject H_0 (i.e. failed to convict the person) when she was in fact guilty (H_A true).

● **Example 5.36**   How could we reduce the Type 1 Error rate in US courts? What influence would this have on the Type 2 Error rate?

To lower the Type 1 Error rate, we might raise our standard for conviction from "beyond a reasonable doubt" to "beyond a conceivable doubt" so fewer people would be wrongly convicted. However, this would also make it more difficult to convict the people who are actually guilty, so we would make more Type 2 Errors.

⊙ **Guided Practice 5.37**   How could we reduce the Type 2 Error rate in US courts? What influence would this have on the Type 1 Error rate?[16]

---

[16]To lower the Type 2 Error rate, we want to convict more guilty people. We could lower the standards for conviction from "beyond a reasonable doubt" to "beyond a little doubt". Lowering the bar for guilt will also result in more wrongful convictions, raising the Type 1 Error rate.

⊙ **Guided Practice 5.38** A group of women bring a class action lawsuit that claims discrimination in promotion rates. What would a Type 1 Error represent in this context?[17]

The example and Exercise above provide an important lesson: if we reduce how often we make one type of error, we generally make more of the other type.

5.3.7 Choosing a significance level

If H_0 is true, what is the probability that we will incorrectly reject it? In hypothesis testing, we perform calculations under the premise that H_0 is true, and we reject H_0 if the p-value is smaller than the significance level α. That is, α *is* the probability of making a Type 1 Error. The choice of what to make α is not arbitrary. It depends on the gravity of the consequences of a Type 1 Error.

Relationship between Type 1 and Type 2 Errors

The probability of a Type 1 Error is called α and corresponds to the significance level of a test. The probability of a Type 2 Error is called β. As we make α smaller, β typically gets larger, and vice versa.

● **Example 5.39** If making a Type 1 Error is especially dangerous or especially costly, should we choose a smaller significance level or a higher significance level?

Under this scenario, we want to be very cautious about rejecting the null hypothesis, so we demand very strong evidence before we are willing to reject the null hypothesis. Therefore, we want a smaller significance level, maybe $\alpha = 0.01$.

● **Example 5.40** If making a Type 2 Error is especially dangerous or especially costly, should we choose a smaller significance level or a higher significance level?

We should choose a higher significance level (e.g. 0.10). Here we want to be cautious about failing to reject H_0 when the null is actually false.

TIP: Significance levels should reflect consequences of errors
The significance level selected for a test should reflect the real-world consequences associated with making a Type 1 or Type 2 Error. If a Type 1 Error is very dangerous, make α smaller.

---

[17]We must first identify which is the null hypothesis and which is the alternative. The alternative hypothesis is the one that bears the burden of proof, so the null hypothesis is that there was no discrimination and the alternative hypothesis is that there was discrimination. Making a Type 1 Error in this context would mean that in fact there was no discrimination, even though we concluded that women were discriminated against. Notice that this does *not* necessarily mean something was wrong with the data or that we made a computational mistake. Sometimes data simply point us to the wrong conclusion, which is why scientific studies are often repeated to check initial findings.

5.3.8 Statistical power of a hypothesis test

When the alternative hypothesis is true, the probability of <u>not</u> making a Type 2 Error is called **power**. It is common for researchers to perform a power analysis to ensure their study collects enough data to detect the effects they anticipate finding. As you might imagine, if the effect they care about is small or subtle, then if the effect is real, the researchers will need to collect a large sample size in order to have a good chance of detecting the effect. However, if they are interested in large effect, they need not collect as much data.

The Type 2 Error rate β and the magnitude of the error for a point estimate are controlled by the sample size. As the sample size n goes up, the Type 2 Error rate goes down, and power goes up. Real differences from the null value, even large ones, may be difficult to detect with small samples. However, if we take a very large sample, we might find a statistically significant difference but the magnitude might be so small that it is of no practical value.

5.4   Does it make sense?

5.4.1   When to retreat

Statistical tools rely on conditions. When the conditions are not met, these tools are unreliable and drawing conclusions from them is treacherous. The conditions for these tools typically come in two forms.

- **The individual observations must be independent.** A random sample from less than 10% of the population ensures the observations are independent. In experiments, we generally require that subjects are randomized into groups. If independence fails, then advanced techniques must be used, and in some such cases, inference may not be possible.

- **Other conditions focus on sample size and skew.** For example, if the sample size is too small, the skew too strong, or extreme outliers are present, then the normal model for the sample mean will fail.

Verification of conditions for statistical tools is always necessary. Whenever conditions are not satisfied for a statistical technique, there are three options. The first is to learn new methods that are appropriate for the data. The second route is to consult a statistician.[18] The third route is to ignore the failure of conditions. This last option effectively invalidates any analysis and may discredit novel and interesting findings.

Finally, we caution that there may be no inference tools helpful when considering data that include unknown biases, such as convenience samples. For this reason, there are books, courses, and researchers devoted to the techniques of sampling and experimental design. See Sections 1.3-1.5 for basic principles of data collection.

---

[18]If you work at a university, then there may be campus consulting services to assist you. Alternatively, there are many private consulting firms that are also available for hire.

5.4.2 Statistical significance versus practical significance

When the sample size becomes larger, point estimates become more precise and any real differences in the mean and null value become easier to detect and recognize. Even a very small difference would likely be detected if we took a large enough sample. Sometimes researchers will take such large samples that even the slightest difference is detected. While we still say that difference is **statistically significant**, it might not be **practically significant**.

Statistically significant differences are sometimes so minor that they are not practically relevant. This is especially important to research: if we conduct a study, we want to focus on finding a meaningful result. We don't want to spend lots of money finding results that hold no practical value.

The role of a statistician in conducting a study often includes planning the size of the study. The statistician might first consult experts or scientific literature to learn what would be the smallest meaningful difference from the null value. She also would obtain some reasonable estimate for the standard deviation. With these important pieces of information, she would choose a sufficiently large sample size so that the power for the meaningful difference is perhaps 80% or 90%. While larger sample sizes may still be used, she might advise against using them in some cases, especially in sensitive areas of research.

5.5   Exercises

5.5.1   Estimating unknown parameters

5.1   Identify the parameter, Part I. For each of the following situations, state whether the parameter of interest is a mean or a proportion. It may be helpful to examine whether individual responses are numerical or categorical.

(a) In a survey, one hundred college students are asked how many hours per week they spend on the Internet.

(b) In a survey, one hundred college students are asked: "What percentage of the time you spend on the Internet is part of your course work?"

(c) In a survey, one hundred college students are asked whether or not they cited information from Wikipedia in their papers.

(d) In a survey, one hundred college students are asked what percentage of their total weekly spending is on alcoholic beverages.

(e) In a sample of one hundred recent college graduates, it is found that 85 percent expect to get a job within one year of their graduation date.

5.2   Identify the parameter, Part II. For each of the following situations, state whether the parameter of interest is a mean or a proportion.

(a) A poll shows that 64% of Americans personally worry a great deal about federal spending and the budget deficit.

(b) A survey reports that local TV news has shown a 17% increase in revenue between 2009 and 2011 while newspaper revenues decreased by 6.4% during this time period.

(c) In a survey, high school and college students are asked whether or not they use geolocation services on their smart phones.

(d) In a survey, internet users are asked whether or not they purchased any Groupon coupons.

(e) In a survey, internet users are asked how many Groupon coupons they purchased over the last year.

5.3   College credits. A college counselor is interested in estimating how many credits a student typically enrolls in each semester. The counselor decides to randomly sample 100 students by using the registrar's database of students. The histogram below shows the distribution of the number of credits taken by these students. Sample statistics for this distribution are also provided.

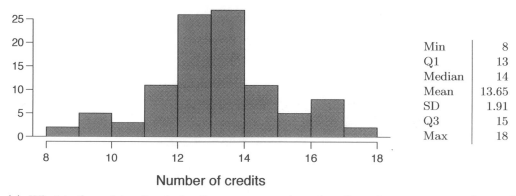

Min	8
Q1	13
Median	14
Mean	13.65
SD	1.91
Q3	15
Max	18

(a) What is the point estimate for the average number of credits taken per semester by students at this college? What about the median?

(b) What is the point estimate for the standard deviation of the number of credits taken per semester by students at this college? What about the IQR? **(See the next page for parts (c)-(e).)**

(c) Is a load of 16 credits unusually high for this college? What about 18 credits? Explain your reasoning. *Hint:* Observations farther than two standard deviations from the mean are usually considered to be unusual.

(d) The college counselor takes another random sample of 100 students and this time finds a sample mean of 14.02 units. Should she be surprised that this sample statistic is slightly different than the one from the original sample? Explain your reasoning.

(e) The sample means given above are point estimates for the mean number of credits taken by all students at that college. What measures do we use to quantify the variability of this estimate (Hint: recall that $SD_{\bar{x}} = \frac{\sigma}{\sqrt{n}}$)? Compute this quantity using the data from the original sample.

5.4 Heights of adults. Researchers studying anthropometry collected body girth measurements and skeletal diameter measurements, as well as age, weight, height and gender, for 507 physically active individuals. The histogram below shows the sample distribution of heights in centimeters.[19]

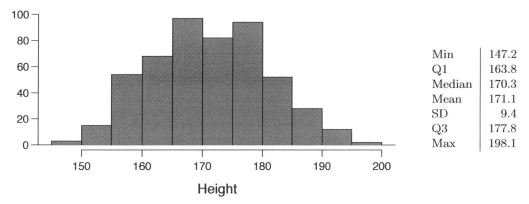

Min	147.2
Q1	163.8
Median	170.3
Mean	171.1
SD	9.4
Q3	177.8
Max	198.1

(a) What is the point estimate for the average height of active individuals? What about the median?

(b) What is the point estimate for the standard deviation of the heights of active individuals? What about the IQR?

(c) Is a person who is 1m 80cm (180 cm) tall considered unusually tall? And is a person who is 1m 55cm (155cm) considered unusually short? Explain your reasoning.

(d) The researchers take another random sample of physically active individuals. Would you expect the mean and the standard deviation of this new sample to be the ones given above? Explain your reasoning.

(e) The sample means obtained are point estimates for the mean height of all active individuals, if the sample of individuals is equivalent to a simple random sample. What measure do we use to quantify the variability of such an estimate (Hint: recall that $SD_{\bar{x}} = \frac{\sigma}{\sqrt{n}}$)? Compute this quantity using the data from the original sample under the condition that the data are a simple random sample.

[19]G. Heinz et al. "Exploring relationships in body dimensions". In: *Journal of Statistics Education* 11.2 (2003).

5.5 Hen eggs. The distribution of the number of eggs laid by a certain species of hen during their breeding period is 35 eggs with a standard deviation of 18.2. Suppose a group of researchers randomly samples 45 hens of this species, counts the number of eggs laid during their breeding period, and records the sample mean. They repeat this 1,000 times, and build a distribution of sample means.

(a) What is this distribution called?

(b) Would you expect the shape of this distribution to be symmetric, right skewed, or left skewed? Explain your reasoning.

(c) Calculate the variability of this distribution and state the appropriate term used to refer to this value.

(d) Suppose the researchers' budget is reduced and they are only able to collect random samples of 10 hens. The sample mean of the number of eggs is recorded, and we repeat this 1,000 times, and build a new distribution of sample means. How will the variability of this new distribution compare to the variability of the original distribution?

5.6 Art after school. Elijah and Tyler, two high school juniors, conducted a survey on 15 students at their school, asking the students whether they would like the school to offer an after-school art program, counted the number of "yes" answers, and recorded the sample proportion. 14 out of the 15 students responded "yes". They repeated this 100 times and built a distribution of sample means.

(a) What is this distribution called?

(b) Would you expect the shape of this distribution to be symmetric, right skewed, or left skewed? Explain your reasoning.

(c) Calculate the variability of this distribution and state the appropriate term used to refer to this value.

(d) Suppose that the students were able to recruit a few more friends to help them with sampling, and are now able to collect data from random samples of 25 students. Once again, they record the number of "yes" answers, and record the sample proportion, and repeat this 100 times to build a new distribution of sample proportions. How will the variability of this new distribution compare to the variability of the original distribution?

5.5.2 Confidence intervals

5.7 Chronic illness, Part I. In 2013, the Pew Research Foundation reported that "45% of U.S. adults report that they live with one or more chronic conditions".[20] However, this value was based on a sample, so it may not be a perfect estimate for the population parameter of interest on its own. The study reported a standard error of about 1.2%, and a normal model may reasonably be used in this setting. Create a 95% confidence interval for the proportion of U.S. adults who live with one or more chronic conditions. Also interpret the confidence interval in the context of the study.

5.8 Twitter users and news, Part I. A poll conducted in 2013 found that 52% of U.S. adult Twitter users get at least some news on Twitter.[21]. The standard error for this estimate was 2.4%, and a normal distribution may be used to model the sample proportion. Construct a 99% confidence interval for the fraction of U.S. adult Twitter users who get some news on Twitter, and interpret the confidence interval in context.

[20]Pew Research Center, Washington, D.C. The Diagnosis Difference, November 26, 2013.
[21]Pew Research Center, Washington, D.C. Twitter News Consumers: Young, Mobile and Educated, November 4, 2013.

5.9 Chronic illness, Part II. In 2013, the Pew Research Foundation reported that "45% of U.S. adults report that they live with one or more chronic conditions", and the standard error for this estimate is 1.2%. Identify each of the following statements as true or false. Provide an explanation to justify each of your answers.

(a) We can say with certainty that the confidence interval from Exercise 5.7 contains the true percentage of U.S. adults who suffer from a chronic illness.

(b) If we repeated this study 1,000 times and constructed a 95% confidence interval for each study, then approximately 950 of those confidence intervals would contain the true fraction of U.S. adults who suffer from chronic illnesses.

(c) The poll provides statistically significant evidence (at the $\alpha = 0.05$ level) that the percentage of U.S. adults who suffer from chronic illnesses is below 50%.

(d) Since the standard error is 1.2%, only 1.2% of people in the study communicated uncertainty about their answer.

5.10 Twitter users and news, Part II. A poll conducted in 2013 found that 52% of U.S. adult Twitter users get at least some news on Twitter, and the standard error for this estimate was 2.4%. Identify each of the following statements as true or false. Provide an explanation to justify each of your answers.

(a) The data provide statistically significant evidence that more than half of U.S. adult Twitter users get some news through Twitter. Use a significance level of $\alpha = 0.01$.

(b) Since the standard error is 2.4%, we can conclude that 97.6% of all U.S. adult Twitter users were included in the study.

(c) If we want to reduce the standard error of the estimate, we should collect less data.

(d) If we construct a 90% confidence interval for the percentage of U.S. adults Twitter users who get some news through Twitter, this confidence interval will be wider than a corresponding 99% confidence interval.

5.11 Relaxing after work. The 2010 General Social Survey asked the question: "After an average work day, about how many hours do you have to relax or pursue activities that you enjoy?" to a random sample of 1,155 Americans.[22] A 95% confidence interval for the mean number of hours spent relaxing or pursuing activities they enjoy was (1.38, 1.92).

(a) Interpret this interval in context of the data.

(b) Suppose another set of researchers reported a confidence interval with a larger margin of error based on the same sample of 1,155 Americans. How does their confidence level compare to the confidence level of the interval stated above?

(c) Suppose next year a new survey asking the same question is conducted, and this time the sample size is 2,500. Assuming that the population characteristics, with respect to how much time people spend relaxing after work, have not changed much within a year. How will the margin of error of the 95% confidence interval constructed based on data from the new survey compare to the margin of error of the interval stated above?

---

[22]National Opinion Research Center, General Social Survey, 2010.

5.12 Take a walk. The Centers for Disease Control monitors the physical activity level of Americans. A recent survey on a random sample of 23,129 Americans yielded a 95% confidence interval of 61.1% to 62.9% for the proportion of Americans who walk for at least 10 minutes per day.

(a) Interpret this interval in context of the data.

(b) Suppose another set of researchers reported a confidence interval with a larger margin of error based on the same sample of 23,129 Americans. How does their confidence level compare to the confidence level of the interval stated above?

(c) Suppose next year a new survey asking the same question is conducted, and this time the sample size is 10,000. Assuming that the population characteristics, with respect to walking habits, have not changed much within a year, how will the width of the confidence interval constructed based on data from the new survey compare to the width of the interval stated above?

5.13 Women leaders, Part I. A November 2014 Pew Research poll on women and leadership asked respondents what they believed is holding women back from top jobs. 43% of the respondents said that women are held to higher standards than men when being considered for top executive business positions. This result is based on 1,835 randomly sampled national adults.[23]

(a) Construct a 95% confidence interval for the proportion of Americans who believe women are held to higher standards than men when being considered for top executive business positions.

(b) How would you expect the width of a 90% confidence interval to compare to the interval you calculated in part (a)? Explain your reasoning.

(c) Now construct the 90% confidence interval, and comment on whether your answer to part (b) is confirmed.

5.14 Women leaders, Part II. The poll introduced in Exercise 5.13 also asked whether respondents expected to see a female president in their lifetime. 78% of the 1,835 respondents said "yes".

(a) Construct a 90% confidence interval for the proportion of Americans who expect to see a female president in their lifetime, and interpret this interval in context of the data.

(b) How would you expect the width of a 98% confidence interval to compare to the interval you calculated in part (a)? Explain your reasoning.

(c) Now construct the 98% confidence interval, and comment on whether your answer to part (b) is confirmed.

---

[23]Pew Research Center, Washington, D.C. Women and Leadership: Public Says Women are Equally Qualified, but Barriers Persist, January 14, 2015.

5.5.3 Introducing hypothesis testing

5.15 Social experiment, Part I. A "social experiment" conducted by a TV program questioned what people do when they see a very obviously bruised woman getting picked on by her boyfriend. On two different occasions at the same restaurant, the same couple was depicted. In one scenario the woman was dressed "provocatively" and in the other scenario the woman was dressed "conservatively". The table below shows how many restaurant diners were present under each scenario, and whether or not they intervened.

| | | Scenario | | |
		Provocative	Conservative	Total
Intervene	Yes	5	15	20
	No	15	10	25
	Total	20	25	45

A simulation was conducted to test if people react differently under the two scenarios. 10,000 simulated differences were generated to construct the null distribution shown. The value $\hat{p}_{pr,sim}$ represents the proportion of diners who intervened in the simulation for the provocatively dressed woman, and $\hat{p}_{con,sim}$ is the proportion for the conservatively dressed woman.

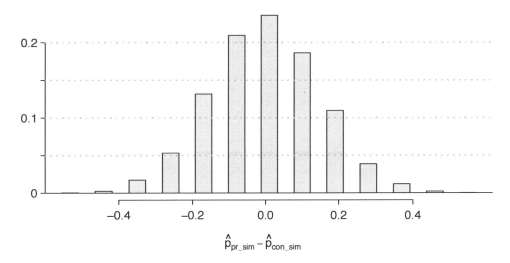

(a) What are the hypotheses? For the purposes of this exercise, you may assume that each observed person at the restaurant behaved independently, though we would want to evaluate this assumption more rigorously if we were reporting these results.

(b) Calculate the observed difference between the rates of intervention under the provocative and conservative scenarios: $\hat{p}_{pr} - \hat{p}_{con}$.

(c) Estimate the p-value using the figure above and determine the conclusion of the hypothesis test.

5.16   Is yawning contagious, Part I. An experiment conducted by the *MythBusters*, a science entertainment TV program on the Discovery Channel, tested if a person can be subconsciously influenced into yawning if another person near them yawns. 50 people were randomly assigned to two groups: 34 to a group where a person near them yawned (treatment) and 16 to a group where there wasn't a person yawning near them (control). The following table shows the results of this experiment.[24]

		Group		
		Treatment	Control	Total
Result	Yawn	10	4	14
	Not Yawn	24	12	36
	Total	34	16	50

A simulation was conducted to understand the distribution of the test statistic under the assumption of independence: having someone yawn near another person has no influence on if the other person will yawn. In order to conduct the simulation, a researcher wrote yawn on 14 index cards and not yawn on 36 index cards to indicate whether or not a person yawned. Then he shuffled the cards and dealt them into two groups of size 34 and 16 for treatment and control, respectively. He counted how many participants in each simulated group yawned in an apparent response to a nearby yawning person, and calculated the difference between the simulated proportions of yawning as $\hat{p}_{trtmt,sim} - \hat{p}_{ctrl,sim}$. This simulation was repeated 10,000 times using software to obtain 10,000 differences that are due to chance alone. The histogram shows the distribution of the simulated differences.

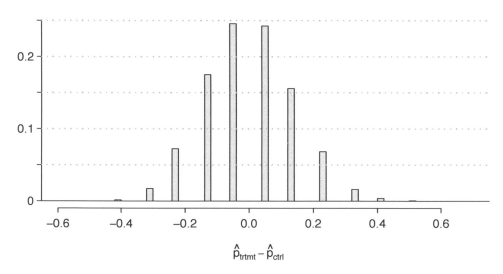

(a) What are the hypotheses?

(b) Calculate the observed difference between the yawning rates under the two scenarios.

(c) Estimate the p-value using the figure above and determine the conclusion of the hypothesis test.

5.17   Social experiment, Part II. In Exercise 5.15, we encountered a scenario where researchers were evaluating the impact of the way someone is dressed against the actions of people around them. In that exercise, researchers may have believed that dressing provocatively may reduce the chance of bystander intervention. One might be tempted to use a one-sided hypothesis test for this study. Discuss the drawbacks of doing so in 1-3 sentences.

---

[24]MythBusters, Season 3, Episode 28.

5.18 Is yawning contagious, Part II. Exercise 5.16 describes an experiment by Myth Busters, where they examined whether a person yawning would affect whether others to yawn. The traditional belief is that yawning is contagious – one yawn can lead to another yawn, which might lead to another, and so on. In that exercise, there was the option of selecting a one-sided or two-sided test. Which would you recommend (or which did you choose)? Justify your answer in 1-3 sentences.

5.19 The Egyptian Revolution. A popular uprising that started on January 25, 2011 in Egypt led to the 2011 Egyptian Revolution. Polls show that about 69% of American adults followed the news about the political crisis and demonstrations in Egypt closely during the first couple weeks following the start of the uprising. Among a random sample of 30 high school students, it was found that only 17 of them followed the news about Egypt closely during this time.[25]

(a) Write the hypotheses for testing if the proportion of high school students who followed the news about Egypt is different than the proportion of American adults who did.

(b) Calculate the proportion of high schoolers in this sample who followed the news about Egypt closely during this time.

(c) Describe how to perform a simulation and, once you had results, how to estimate the p-value.

(d) Below is a histogram showing the distribution of $\hat{p}_{sim}$ in 10,000 simulations under the null hypothesis. Estimate the p-value using the plot and determine the conclusion of the hypothesis test.

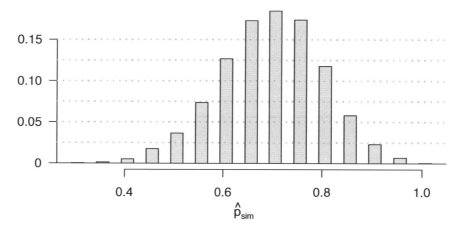

---

[25]Gallup Politics, Americans' Views of Egypt Sharply More Negative, data collected February 2-5, 2011.

5.20   Assisted Reproduction. Assisted Reproductive Technology (ART) is a collection of techniques that help facilitate pregnancy (e.g. in vitro fertilization). A 2008 report by the Centers for Disease Control and Prevention estimated that ART has been successful in leading to a live birth in 31% of cases[26]. A new fertility clinic claims that their success rate is higher than average. A random sample of 30 of their patients yielded a success rate of 40%. A consumer watchdog group would like to determine if this provides strong evidence to support the company's claim.

(a) Write the hypotheses to test if the success rate for ART at this clinic is significantly higher than the success rate reported by the CDC.

(b) Describe a setup for a simulation that would be appropriate in this situation and how the p-value can be calculated using the simulation results.

(c) Below is a histogram showing the distribution of $\hat{p}_{sim}$ in 10,000 simulations under the null hypothesis. Estimate the p-value using the plot and use it to evaluate the hypotheses.

(d) After performing this analysis, the consumer group releases the following news headline: "In-fertility clinic falsely advertises better success rates". Comment on the appropriateness of this statement.

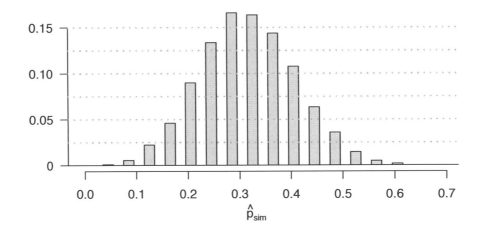

5.21   Spam mail, Part I. The 2004 National Technology Readiness Survey sponsored by the Smith School of Business at the University of Maryland surveyed 418 randomly sampled Americans, asking them how many spam emails they receive per day. The survey was repeated on a new random sample of 499 Americans in 2009.[27]

(a) What are the hypotheses for evaluating if the average spam emails per day has changed from 2004 to 2009.

(b) In 2004 the mean was 18.5 spam emails per day, and in 2009 this value was 14.9 emails per day. What is the point estimate for the difference between the two population means?

(c) A report on the survey states that the observed difference between the sample means is not statistically significant. Explain what this means in context of the hypothesis test and data.

(d) Would you expect a confidence interval for the difference between the two population means to contain 0? Explain your reasoning.

---

[26]CDC. *2008 Assisted Reproductive Technology Report.*
[27]Rockbridge, 2009 National Technology Readiness Survey SPAM Report.

5.22   Nearsightedness. It is believed that nearsightedness affects about 8% of all children. In a random sample of 194 children, 21 are nearsighted.

(a) Construct hypotheses appropriate for the following question: do these data provide evidence that the 8% value is inaccurate?

(b) What proportion of children in this sample are nearsighted?

(c) Given that the standard error of the sample proportion is 0.0195 and the point estimate follows a nearly normal distribution, calculate the test statistic (the Z-statistic).

(d) What is the p-value for this hypothesis test?

(e) What is the conclusion of the hypothesis test?

5.23   Spam mail, Part II. The National Technology Readiness Survey from Exercise 5.21 also asked Americans how often they delete spam emails. 23% of the respondents in 2004 said they delete their spam mail once a month or less, and in 2009 this value was 16%.

(a) What are the hypotheses for evaluating if the proportion of those who delete their email once a month or less has changed from 2004 to 2009?

(b) What is the point estimate for the difference between the two population proportions?

(c) A report on the survey states that the observed decrease from 2004 to 2009 is statistically significant. Explain what this means in context of the hypothesis test and the data.

(d) Would you expect a confidence interval for the difference between the two population proportions to contain 0? Explain your reasoning.

5.24   Unemployment and relationship problems. A USA Today/Gallup poll conducted between 2010 and 2011 asked a group of unemployed and underemployed Americans if they have had major problems in their relationships with their spouse or another close family member as a result of not having a job (if unemployed) or not having a full-time job (if underemployed). 27% of the 1,145 unemployed respondents and 25% of the 675 underemployed respondents said they had major problems in relationships as a result of their employment status.

(a) What are the hypotheses for evaluating if the proportions of unemployed and underemployed people who had relationship problems were different?

(b) The p-value for this hypothesis test is approximately 0.35. Explain what this means in context of the hypothesis test and the data.

5.25   Testing for Fibromyalgia. A patient named Diana was diagnosed with Fibromyalgia, a long-term syndrome of body pain, and was prescribed anti-depressants. Being the skeptic that she is, Diana didn't initially believe that anti-depressants would help her symptoms. However after a couple months of being on the medication she decides that the anti-depressants are working, because she feels like her symptoms are in fact getting better.

(a) Write the hypotheses in words for Diana's skeptical position when she started taking the anti-depressants.

(b) What is a Type 1 Error in this context?

(c) What is a Type 2 Error in this context?

5.26 Testing for food safety. A food safety inspector is called upon to investigate a restaurant with a few customer reports of poor sanitation practices. The food safety inspector uses a hypothesis testing framework to evaluate whether regulations are not being met. If he decides the restaurant is in gross violation, its license to serve food will be revoked.

(a) Write the hypotheses in words.

(b) What is a Type 1 Error in this context?

(c) What is a Type 2 Error in this context?

(d) Which error is more problematic for the restaurant owner? Why?

(e) Which error is more problematic for the diners? Why?

(f) As a diner, would you prefer that the food safety inspector requires strong evidence or very strong evidence of health concerns before revoking a restaurant's license? Explain your reasoning.

5.27 True / False. Determine whether the following statement is true or false, and explain your reasoning: "A cutoff of $\alpha = 0.05$ is the ideal value for all hypothesis tests."

5.28 True / False. Determine whether the following statement is true or false, and explain your reasoning: "Power of a test and the probability of making a Type 1 Error are complements."

5.29 Which is higher? In each part below, there is a value of interest and two scenarios (I and II). For each part, report if the value of interest is larger under scenario I, scenario II, or whether the value is equal under the scenarios.

(a) The standard error of $\bar{x}$ when $s = 120$ and (I) n $= 25$ or (II) n $= 125$.

(b) The margin of error of a confidence interval when the confidence level is (I) 90% or (II) 80%.

(c) The p-value for a Z-statistic of 2.5 when (I) n $= 500$ or (II) n $= 1000$.

(d) The probability of making a Type 2 Error when the alternative hypothesis is true and the significance level is (I) 0.05 or (II) 0.10.

5.30 True or false. Determine if the following statements are true or false, and explain your reasoning. If false, state how it could be corrected.

(a) If a given value (for example, the null hypothesized value of a parameter) is within a 95% confidence interval, it will also be within a 99% confidence interval.

(b) Decreasing the significance level (α) will increase the probability of making a Type 1 Error.

(c) Suppose the null hypothesis is $\mu = 5$ and we fail to reject H_0. Under this scenario, the true population mean is 5.

(d) If the alternative hypothesis is true, then the probability of making a Type 2 Error and the power of a test add up to 1.

(e) With large sample sizes, even small differences between the null value and the true value of the parameter, a difference often called the effect size, will be identified as statistically significant.

5.5.4 Does it make sense?

5.31 True / False. Determine whether the following statement is true or false, and explain your reasoning: "With large sample sizes, even small differences between the null value and the point estimate can be statistically significant."

5.32 Same observation, different sample size. Suppose you conduct a hypothesis test based on a sample where the sample size is $n = 50$, and arrive at a p-value of 0.08. You then refer back to your notes and discover that you made a careless mistake, the sample size should have been $n = 500$. How, if at all, will your p-value change (increase or decrease)? Explain.

Chapter 6

Inference for categorical data

Chapter 5 introduced the logic and the steps for constructing confidence intervals and carrying out tests of hypothesis. We use these methods to answer questions like the following:

- What proportion of the American public approves of the job the Supreme Court is doing?

- The Pew Research Center conducted a poll about support for the 2010 health care law, and they used two forms of the survey question. Each respondent was randomly given one of the two questions. What is the difference in the support for respondents under the two question orderings?

The methods we learned in Chapter 5 are very useful in these settings. In this chapter we will consider the sampling distribution for a proportion and for the difference of two proportions, and we will examine the conditions under which a normal model is appropriate. We will also encounter a new distribution for hypothesis tests on frequency and contingency tables.

6.1 Inference for a single proportion

The distribution of a sample proportion, such as an estimate of the fraction of people who share a particular opinion in a poll, was introduced in Section 4.5.

6.1.1 Confidence intervals for a proportion

Suppose we want to construct a confidence interval for the proportion of Americans who approve of the job the Supreme Court is doing. In a simple random sample of $n = 976$, 44% of respondents approved.[1] In the examples below, we will construct a *1-proportion z-interval*.

Before constructing the confidence interval, we should determine whether we can model the sample proportion, $\hat{p} = 0.44$, using a normal model, which requires two conditions to be satisfied.

---

[1] nytimes.com/2012/06/08/us/politics/44-percent-of-americans-approve-of-supreme-court-in-new-poll.html

Conditions for the sampling distribution of $\hat{p}$ being nearly normal

The sampling distribution for $\hat{p}$, taken from a sample of size n from a population with a true proportion p, is nearly normal when

1. the sample observations are independent and

2. we expected to see at least 10 successes and 10 failures in our sample, i.e. $np \geq 10$ and $n(1 - p) \geq 10$. This is called the **success-failure condition**.

If these conditions are met, then the sampling distribution of $\hat{p}$ is nearly normal with mean $\mu_{\hat{p}} = p$ and standard deviation $\sigma_{\hat{p}} = \sqrt{\frac{p(1-p)}{n}}$.

● **Example 6.1** Verify that we can use a normal distribution to model $\hat{p} = 0.44$ for the Supreme Court poll of $n = 976$ US adults.

The data are from a simple random sample, so the independence condition is satisfied. To check the success-failure condition we want to check that np and $n(1 - p)$ are at least 10. However, p is unknown. Therefore, we will use the sample proportion $\hat{p}$ to check this condition.

$$n\hat{p} = 976 \times 0.44 = 429 \text{ ("successes")}$$
$$n(1 - \hat{p}) = 976 \times (1 - 0.44) = 547 \text{ ("failures")}$$

The second condition is satisfied since 429 and 547 are both larger than 10. With the two conditions satisfied, we can model the sample proportion $\hat{p} = 0.44$ using a normal model.

TIP: Reminder on checking independence of observations
If data come from a simple random sample, then the independence assumption is generally reasonable. Alternatively, if the data come from a random process, we must evaluate the independence condition more carefully.

● **Example 6.2** The general form of a confidence interval is:

$$\text{point estimate } \pm \text{ critical value} \times SE$$

What should we use as the point estimate for the confidence interval?

The best estimate for the unknown parameter p (the proportion of Americans who approve of the job the Supreme Court is doing) is the sample proportion. When constructing a confidence interval for a single proportion, we use $\hat{p} = 0.44$ as the point estimate for p.

● **Example 6.3** Calculate the standard error for the confidence interval.

In Section 4.5, we learned that the formula for the standard deviation of $\hat{p}$ is

$$\sigma_{\hat{p}} = \sqrt{\frac{p(1-p)}{n}}$$

The proportion p is unknown, but we can use sample proportion $\hat{p}$ instead when finding the SE in a confidence interval:

$$SE = \sqrt{\frac{\hat{p}(1-\hat{p})}{n}} = \sqrt{\frac{0.44(1-0.44)}{976}} = 0.016$$

When the conditions for a normal model are met, we use $z^\star$ for the critical value. An appropriate value for $z^\star$ can most easily be found in the t-table on page 452 in the last row (∞), where the column corresponds to the desired confidence level.

● **Example 6.4** Construct a 90% confidence interval for p, the proportion of Americans who approve of the job the Supreme Court is doing.

Using the point estimate $\hat{p} = 0.44$ and standard error $SE = 0.16$ computed earlier, we can construct the confidence interval:

$$\text{point estimate} \pm \text{critical value} \times SE$$
$$0.44 \pm 1.65 \times 0.016$$
$$(0.414, \ 0.466)$$

The critical value $z^\star$ was found by looking in the 90% column in the t-table on page 452.

We are 90% confident that the true proportion of Americans who approve of the job the Supreme Court is doing is between 41.4% and 46.6%. Because the entire interval is below 0.5, we have evidence that the true percent that approve is less than 50%.

Constructing a confidence interval for a proportion

A complete solution to a confidence interval question for a single proportion includes the following steps:

1. State the name of the confidence interval being used.

   - 1-proportion z-interval

2. Verify conditions.

   - A simple random sample.
   - $n\hat{p} \geq 10$ and $n(1 - \hat{p}) \geq 10$.

3. Plug in the numbers and write the interval in the form

$$\text{point estimate } \pm z^{\star} \times \text{SE of estimate}$$

   - The point estimate is $\hat{p}$.
   - Critical value $z^{\star} = 1.96$ a 95% CI,
     otherwise find $z^{\star}$ using the t-table at row ∞.
   - Use $SE = \sqrt{\frac{\hat{p}(1-\hat{p})}{n}}$.

4. Evaluate the CI and write in the form (___, ___).

5. Interpret the interval: "We are [XX]% confident that the true proportion of [...] is between [...] and [...]."

6. State the conclusion to the original question.

⊙ **Guided Practice 6.5** Identify each of the six steps for constructing a confidence interval in the Supreme Court description and examples.[2]

● **Example 6.6** A poll randomly selected 1,042 adults residing in the state of New York and asked the question, "Regardless of whether the person shows symptoms or not, do you support or oppose a 21 day quarantine for anyone who has come in contact with someone with the Ebola virus?"[3] Among the sample, 82% said "support", 15% said "oppose", and 3% said "unsure". Carry out the appropriate 95% confidence interval procedure to estimate the true proportion of adults in New York who supported a quarantine. Is there evidence that the true percent is greater than 75%?

We will construct a 1-proportion z-interval. The poll can be considered a simple random sample of adults form New York. Also, $1042 \times 0.82 \geq 10$ and $1042 \times (1 - 0.82) \geq 10$, so the conditions for the confidence interval are satisfied. The standard

---

[2]The following are the required components for constructing a confidence interval for a single proportion. 1. The last sentence in the first paragraph of Section 6.1.1. 2. in the solution to Example 6.1, the first sentence to cover independence and the two calculations verifying $n\hat{p}$ and $n(1 - \hat{p})$ were at least 10. 3. The last formula in the solution to Example 6.3 and the calculations and identification of $z^{\star}$ in Example 6.3. 4. At the end of the calculations in the solution to Example 6.3. Items 5 and 6 were contained in the last paragraph of Example 6.2's solution.

[3]NBC 4 New York / The Wall Street Journal / Marist Poll. October 31, 2014.

error and confidence interval can be calculated as

$$SE = \sqrt{\frac{0.82(1 - 0.82)}{1042}} = 0.0119$$
$$0.82 \pm 1.96 \times 0.0119$$
$$(0.796, \ 0.843)$$

We are 95% confident that the true proportion of adults in New York who supported a 21 day quarantine for anyone who has come in contact with someone with the Ebola virus lies between 0.796 and 0.843. Because the entire interval is above 0.75, the interval provides evidence that the true percent is greater than 75%.

6.1.2 Hypothesis testing for a proportion

While a confidence interval provides a reasonable range of values for an unknown parameter, a hypothesis test evaluates a specific claim. In a hypothesis test, we declare what test we will use, check that the test is reasonable for the context, and construct appropriate null and alternative hypotheses. We then construct a p-value for the test and use it to assess the hypotheses, which allows us to form a conclusion based on the data.

● **Example 6.7** Deborah Toohey is running for Congress, and her campaign manager claims she has more than 50% from the district's electorate. A newspaper collects a simple random sample of 500 likely voters in the district and estimates Toohey's support to be 52%.

(a) What is the name of the test that is appropriate for this context?

(b) State the alternate hypothesis. What value we should use as the null value, p_0?

(c) Can we model $\hat{p} = 0.52$ using a normal model? Check the conditions.

(a) The name of the test we will use is the *1-proportion z-test*.

(b) The alternate hypothesis, the one that bears the burden of proof, argues that Toohey has more than 50% support. Therefore, H_A will be one-sided and the null value will be $p_0 = 50\% = 0.5$. H_A: $p > 0.5$.

(c) The calculations in a hypothesis test for a proportion assume the value p_0 for the unknown p, so we use p_0 rather than $\hat{p}$ when verifying the hypothesis test conditions:

$$np_0 \geq 10 \quad \rightarrow \quad 500 \times 0.5 \geq 10$$
$$n(1 - p_0) \geq 10 \quad \rightarrow \quad 500 \times (1 - 0.5) \geq 10$$

The conditions for a normal model are met.

In Chapter 5, we saw that the general form of the test statistic for a hypothesis test took the following form:

$$\text{test statistic} = \frac{\text{point estimate} - \text{null value}}{\text{SE of estimate}}$$

When the conditions for a normal model are met,

- we use Z as the test statistic,

- the point estimate is $\hat{p}$ (just like for a confidence interval), and

- since we compute the test statistic under the null hypothesized value of $p = p_0$, we compute the standard error as

$$SE = \sqrt{\frac{p_0(1 - p_0)}{n}}$$

● **Example 6.8**   Deborah Toohey is running for Congress, and her campaign manager claimed she has more than 50% support from the district's electorate. A newspaper poll finds that 52% of 500 likely voters who were sampled support Toohey. Does this provide convincing evidence for the claim by Toohey's manager at the 5% significance level?

We will use a one-sided test with the following hypotheses:

H_0: $p = 0.5$. Toohey's support is 50%.

H_A: $p > 0.5$. Toohey's manager is correct, and her support is higher than 50%.

We will use a significance level of $\alpha = 0.05$ for the test. We can compute the standard error as

$$SE = \sqrt{\frac{p_0(1 - p_0)}{n}} = \sqrt{\frac{0.5(1 - 0.5)}{500}} = 0.022$$

The test statistic can be computed as:

$$Z = \frac{\hat{p} - p_0}{SE} = \frac{0.52 - 0.50}{0.022} = 0.89$$

A picture featuring the p-value is shown in Figure 6.1 as the shaded region. Using a table or a calculator, we can get the p-value as about 0.19, which is larger than $\alpha = 0.05$, so we do not reject H_0. That is, we do not have strong evidence to support Toohey's campaign manager's claims that she has more than 50% support within the district.

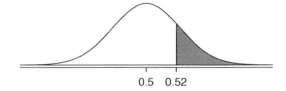

0.5 0.52

Figure 6.1: Sampling distribution of the sample proportion if the null hypothesis is true for Example 6.8. The p-value for the test is shaded.

Hypothesis test for a proportion

A complete solution to a test of hypothesis problem for a single proportion should include the following steps:

1. State the name of the test being used.

   - 1-proportion z-test

2. Verify conditions to ensure the standard error estimate is reasonable and the point estimate is nearly normal and unbiased.

   - A simple random sample.
   - $np_0 \geq 10$ and $n(1 - p_0) \geq 10$ (use hypothesized p, not sample $\hat{p}$).

3. Write the hypotheses in plain language and mathematical notation.

   - $H_0 : p = p_0$, where p_0 is the hypothesized value of p
   - $H_A : p \neq$ or $<$ or $> p_0$

4. Identify the significance level α.

5. Calculate the test statistic: $Z = \frac{\text{point estimate} - \text{null value}}{\text{SE of estimate}}$

   - The point estimate is $\hat{p}$.
   - Use $SE = \sqrt{\frac{p_0(1 - p_0)}{n}}$ (plug in hypothesized p, not sample $\hat{p}$).

6. Find the p-value and compare it to α to determine whether to reject or not reject H_0.

7. Write the conclusion in the context of the question.

⊙ **Guided Practice 6.9**   Identify each of the seven steps for conducting a hypothesis test in the example for Toohey's support.[4]

⊙ **Guided Practice 6.10**   In Example 6.8, the data did not show strong evidence that Toohey's campaign manager was correct. Does this mean the manager was wrong?[5]

---

[4]The following are the required components for running a hypothesis test for a single proportion. Items 1 and 2 are contained in Example 6.7. Items 3-7 are covered in Example 6.8.

[5]Not necessarily. While we did not reject the null hypothesis, that does not mean it is true. It is possible that Toohey does have support above 50%, but that the sample did not provide enough evidence to convincingly show this.

⊙ **Guided Practice 6.11**   A Gallup poll conducted in March of 2015 found that
51% of respondents support nuclear energy.[6] The survey was based on telephone
interviews from a random sample of 1,025 adults in the United States. Before the
poll was conducted, a nuclear energy advocacy group claimed a majority of US adults
support nuclear energy. Does the poll provide strong evidence that supports their
claim? Carry out an appropriate test at the 0.10 significance level.[7]

6.1.3   Calculator: the 1-proportion z-test and z-interval

We can use a calculator to compute a confidence interval or to evaluate the test statistic
and the p-value. Remember to show work and first substitute in all numbers before using
the calculator.

📹 TI-83/84: 1-proportion z-interval

Use STAT, TESTS, 1-PropZInt.

1. Choose STAT.

2. Right arrow to TESTS.

3. Down arrow and choose A:1-PropZInt.

4. Let x be the *number* of yes's (must be an integer).

5. Let n be the sample size.

6. Let C-Level be the desired confidence level.

7. Choose Calculate and hit ENTER, which returns

    (__,__)   the confidence interval
    p̂   the sample proportion
    n   the sample size

---

[6]www.gallup.com/poll/182180/support-nuclear-energy.aspx

[7]We will perform a 1-proportion z-test. We will assume that the sample can be treated as a simple
random sample of adults from the United States. Our null value will be $p_0 = 0.5$, and $1025 \times 0.5 =
1025 \times (1 - 0.5) \geq 10$ so the conditions for the test are satisfied. We will use a one-sided test with the
following hypotheses:

H_0: $p = 0.5$. Support for nuclear energy is 50%.

H_A: $p > 0.5$. Support for nuclear energy is higher than 50%.

We will use a significance level of $\alpha = 0.10$ for the test. We can compute the standard error as $SE =
\sqrt{\frac{p_0(1-p_0)}{n}} = \sqrt{\frac{0.5(1-0.5)}{1025}} = 0.0156$. The test statistic can be computed as: $Z = \frac{\hat{p}-p_0}{SE} = \frac{0.51-0.50}{0.0156} = 0.656$
and the p-value for this one-sided test is 0.256. $0.256 > 0.10$, so we do not reject H_0. We do not have
strong evidence that the true percent of adults in the United States that support nuclear energy is greater
than 50%. That is, the poll does not provide evidence supporting the nuclear energy advocacy group's
claim.

▶ Casio fx-9750GII: 1-proportion z-interval

1. Navigate to STAT (MENU button, then hit the 2 button or select STAT).

2. Choose the INTR option (F4 button).

3. Choose the Z option (F1 button).

4. Choose the 1-P option (F3 button).

5. Specify the interval details:

   - Confidence level of interest for C-Level.
   - Enter the number of successes, x.
   - Enter the sample size, n.

6. Hit the EXE button, which returns

Left, Right	ends of the confidence interval
p̂	sample proportion
n	sample size

⊙ **Guided Practice 6.12**   Using a calculator, confirm the earlier result from Example 6.4: a 90% confidence interval for the percent of Americans who approve of the job the Supreme Court is doing is between 41.4% and 47.1%. The sample percent was 44% and $n = 976$.

▶ TI-83/84: 1-proportion z-test

Use STAT, TESTS, 1-PropZTest.

1. Choose STAT.

2. Right arrow to TESTS.

3. Down arrow and choose 5:1-PropZTest.

4. Let p_0 be the null or hypothesized value of p.

5. Let x be the *number* of yes's (must be an integer).

6. Let n be the sample size.

7. Choose $\neq$, $<$, or $>$ to correspond to H_A.

8. Choose Calculate and hit ENTER, which returns

z	Z-statistic
p	p-value
p̂	the sample proportion
n	the sample size

▶ Casio fx-9750GII: 1-proportion z-test

The steps closely match those of the 1-proportion confidence interval.

1. Navigate to STAT (MENU button, then hit the 2 button or select STAT).
2. Choose the TEST option (F3 button).
3. Choose the Z option (F1 button).
4. Choose the 1-P option (F3 button).
5. Specify the test details:

   - Specify the sidedness of the test using the F1, F2, and F3 keys.
   - Enter the null value, p0.
   - Enter the number of successes, x.
   - Enter the sample size, n.

6. Hit the EXE button, which returns

   z   Z-statistic
   p   p-value
   p̂   the sample proportion
   n   the sample size

⊙ **Guided Practice 6.13**   Using a calculator, confirm the earlier result from Example 6.8, that we do not have strong evidence that Toohey's voter support is above 50% because the p-value is 0.19. The sample percent was 52% and $n = 500$.

6.1.4   Choosing a sample size when estimating a proportion

Planning a sample size before collecting data is important. If we collect too little data, the standard error of our point estimate may be so large that the data are not very useful. On the other hand, collecting data in some contexts is time-consuming and expensive, so we don't want to waste resources on collecting more data than we need.

When considering the sample size, we want to put an upper bound on the margin of error. The **margin of error** is defined as quantity that follows the +/- in the confidence interval. It is half the total width of the confidence interval.

Margin of error

The margin of error of a confidence interval is given by:

$$ME = \text{critical value} \times SE$$

The margin of error is affected by both the sample size and the confidence level.

● **Example 6.14** All other things being equal, will the margin of error be bigger for a 90% confidence interval or a 95% confidence interval?

A 95% confidence interval is wider than a 90% confidence interval, so the 95% confidence interval will have a larger margin of error.

● **Example 6.15** All other things being equal, what happens to the margin of error as the sample size increases?

As the sample size n increases, the SE will decrease, so the margin of error will decrease as n increases. This makes sense as we expect less error with a larger sample.

● **Example 6.16** Suppose we are conducting a university survey to determine whether students support a \$200 per year increase in fees to pay for a new football stadium, how big of a sample is needed to be sure the margin of error is less than 0.04 using a 95% confidence level? Find the smallest sample size n so that the margin of error of the point estimate $\hat{p}$ will be no larger than $m = 0.04$ when using a 95% confidence interval.

For a 95% confidence level, the value $z^\star$ corresponds to 1.96. We want:

$$ME \leq 0.04$$
$$1.96 \times SE \leq 0.04$$
$$1.96 \times \sqrt{\frac{p(1-p)}{n}} \leq 0.04$$

There are two unknowns in the equation: p and n. If we have an estimate of p, perhaps from a similar survey, we could use that value. If we have no such estimate, we must use some other value for p. It turns out that the margin of error is largest when p is 0.5, so we typically use this *worst case estimate* if no other estimate is available:

$$1.96 \times \sqrt{\frac{0.5(1-0.5)}{n}} \leq 0.04$$
$$1.96^2 \times \frac{0.5(1-0.5)}{n} \leq 0.04^2$$
$$1.96^2 \times \frac{0.5(1-0.5)}{0.04^2} \leq n$$
$$600.25 \leq n$$
$$n = 601$$

The sample size must be an integer and we round up because n must be greater than or equal to 600.25. We need at least 601 participants to ensure the sample proportion is within 0.04 of the true proportion with 95% confidence.

No estimate of the true proportion is required in sample size computations for a proportion. However, if we have an estimate of the proportion, we should use it in place of the worst case estimate of the proportion, 0.5.

⊙ **Guided Practice 6.17**   A manager is about to oversee the mass production of a new tire model in her factory, and she would like to estimate what proportion of these tires will be rejected through quality control. The quality control team has monitored the last three tire models produced by the factory, failing 1.7% of tires in the first model, 6.2% of the second model, and 1.3% of the third model. The manager would like to examine enough tires to estimate the failure rate of the new tire model to within about 2% with a 90% confidence level.[8]

(a) There are three different failure rates to choose from. Perform the sample size computation for each separately, and identify three sample sizes to consider.

(b) The sample sizes vary widely. Which of the three would you suggest using? What would influence your choice?

⊙ **Guided Practice 6.18**   A recent estimate of Congress' approval rating was 17%.[9] What sample size does this estimate suggest we should use for a margin of error of 0.04 with 95% confidence?[10]

---

[8](a) For the 1.7% estimate of p, we estimate the appropriate sample size as follows:

$$1.645 \times \sqrt{\frac{0.017(1 - 0.017)}{n}} \leq 0.02$$

$$n \geq 113.7$$

$$n = 114$$

Using the estimate from the first model, we would suggest examining 114 tires (round up!). A similar computation can be accomplished using 0.062 and 0.013 for p: 396 and 88.

(b) We could examine which of the old models is most like the new model, then choose the corresponding sample size. Or if two of the previous estimates are based on small samples while the other is based on a larger sample, we should consider the value corresponding to the larger sample. (Answers will vary.)

[9]www.gallup.com/poll/155144/Congress-Approval-June.aspx

[10]We complete the same computations as before, except now we use 0.17 instead of 0.5 for p:

$$1.96 \times \sqrt{\frac{0.17(1 - 0.17)}{n}} \leq 0.04 \quad \rightarrow \quad n \geq 338.8 \quad \rightarrow \quad n = 339$$

A sample size of 339 or more would be reasonable.

6.2 Difference of two proportions

We would like to make conclusions about the difference in two population proportions: $p_1 - p_2$. We consider three examples. In the first, we compare the approval of the 2010 healthcare law under two different question phrasings. In the second application, a company weighs whether they should switch to a higher quality parts manufacturer. In the last example, we examine the cancer risk to dogs from the use of yard herbicides.

In our investigations, we first identify a reasonable point estimate of $p_1 - p_2$ based on the sample. You may have already guessed its form: $\hat{p}_1 - \hat{p}_2$. Next, we develop a formula for the standard deviation of $\hat{p}_1 - \hat{p}_2$.

6.2.1 Sampling distribution of the difference of two proportions

The mean or expected value of $\hat{p}_1 - \hat{p}_2$ is $p_1 - p_2$. The standard deviation can be computed as:

$$SD_{\hat{p}_1 - \hat{p}_2} = \sqrt{SD_{\hat{p}_1}^2 + SD_{\hat{p}_2}^2} = \sqrt{\frac{p_1(1 - p_1)}{n_1} + \frac{p_2(1 - p_2)}{n_2}}$$

In addition to the mean and the standard deviation of $\hat{p}_1 - \hat{p}_2$, we would like to the know the shape of its distribution. First, the sampling distribution for each sample proportion must be nearly normal, and secondly, the samples must be independent. Under these two conditions, the sampling distribution of $\hat{p}_1 - \hat{p}_2$ may be well approximated using the normal model.

Conditions for the sampling distribution of $\hat{p}_1 - \hat{p}_2$ to be normal

The difference $\hat{p}_1 - \hat{p}_2$ tends to follow a normal model when

- each proportion separately follows a normal model (check $n_1 p_1 \geq 10$, $n_1(1 - p_1) \geq 10$, $n_2 p_2 \geq 10$, and $n_2(1 - p_2) \geq 10$) and
- the two samples are independent of each other.

The standard deviation of the difference in sample proportions is

$$SD_{\hat{p}_1 - \hat{p}_2} = \sqrt{\frac{p_1(1 - p_1)}{n_1} + \frac{p_2(1 - p_2)}{n_2}} \tag{6.19}$$

where p_1 and p_2 represent the population proportions, and n_1 and n_2 represent the sample sizes.

	Sample size (n_i)	Approve law (%)	Disapprove law (%)	Other
"people who cannot afford it will receive financial help from the government" is given second	771	47	49	3
"people who do not buy it will pay a penalty" is given second	732	34	63	3

Table 6.2: Results for a Pew Research Center poll where the ordering of two statements in a question regarding healthcare were randomized.

6.2.2 Confidence Interval for $p_1 - p_2$

In the setting of confidence intervals, the sample proportions are used in place of the population proportions to verify the success-failure condition and also compute standard error, just as was the case with a single proportion.

● **Example 6.20** The way a question is phrased can influence a person's response. For example, Pew Research Center conducted a survey with the following question:[11]

> As you may know, by 2014 nearly all Americans will be required to have health insurance. [People who do not buy insurance will pay a penalty] while [People who cannot afford it will receive financial help from the government]. Do you approve or disapprove of this policy?

For each randomly sampled respondent, the statements in brackets were randomized: either they were kept in the order given above, or the two statements were reversed. Table 6.2 shows the results of this experiment. Create and interpret a 90% confidence interval of the difference in approval.

First the conditions must be verified. Because each group is a simple random sample, the observations are independent, both within the samples and between the samples. The success-failure condition should also be verified:

$$771 \times 0.47 \geq 10 \qquad 771 \times 0.53 \geq 10 \qquad 732 \times 0.34 \geq 10 \qquad 732 \times 0.66 \geq 10$$

Because all conditions are met, the normal model can be used for the point estimate of the difference in support, where p_1 corresponds to the original ordering and p_2 to the reversed ordering:

$$\hat{p}_1 - \hat{p}_2 = 0.47 - 0.34 = 0.13$$

The standard error may be computed from Equation (6.19) using the sample proportions in place of the population proportions:

$$SE = \sqrt{\frac{0.47(1 - 0.47)}{771} + \frac{0.34(1 - 0.34)}{732}} = 0.025$$

For a 90% confidence interval, we use $z^\star = 1.645$:

$$\text{point estimate} \ \pm \ z^\star SE \quad \rightarrow \quad 0.13 \ \pm \ 1.645 \times 0.025 \quad \rightarrow \quad (0.09, 0.17)$$

---

[11]www.people-press.org/2012/03/26/public-remains-split-on-health-care-bill-opposed-to-mandate. Sample sizes for each polling group are approximate.

We are 90% confident that the approval rating for the 2010 healthcare law changes between 9% and 17% due to the ordering of the two statements in the survey question. Because the entire interval is positive, we have evidence that the approval rating *increased*. The Pew Research Center reported that this modestly large difference suggests that the opinions of much of the public are still fluid on the health insurance mandate.

Constructing a confidence interval for the difference of two proportion

1. State the name of the CI being used.

   - 2-proportion z-interval

2. Verify conditions.

   - 2 independent random samples OR 2 randomly allocated treatments.
   - $n_1 \hat{p}_1 \geq 10$, $n_1(1 - \hat{p}_1) \geq 10$
     $n_2 \hat{p}_2 \geq 10$, $n_2(1 - \hat{p}_2) \geq 10$

3. Plug in the numbers and write the interval in the form

   $$\text{point estimate} \pm z^{\star} \times \text{SE of estimate}$$

   - The point estimate is $\hat{p}_1 - \hat{p}_2$.
   - Use critical value $z^* = 1.96$ for a 95% CI, otherwise find z^* using the t-table at row ∞.
   - Use $\text{SE} = \sqrt{\frac{\hat{p}_1(1-\hat{p}_1)}{n_1} + \frac{\hat{p}_2(1-\hat{p}_2)}{n_2}}$.

4. Evaluate the CI and write in the form (____, ____).

5. Interpret the interval: "We are [XX]% confident that the true difference in the proportion of [...] is between [...] and [...].

6. State the conclusion to the original question.

● **Example 6.21** A remote control car company is considering a new manufacturer for wheel gears. The new manufacturer would be more expensive but their higher quality gears are more reliable, resulting in happier customers and fewer warranty claims. However, management must be convinced that the more expensive gears are worth the conversion before they approve the switch. The quality control engineer collects a sample of gears, examining 1000 gears from each company and finds that 899 gears pass inspection from the current supplier and 958 pass inspection from the prospective supplier. Using these data, construct a 95% confidence interval for the difference in the proportion that pass inspection.

We will calculate a 2-proportion z-interval.

The samples are independent, but not necessarily random, so to proceed we must assume the gears are all independent. For this sample we will suppose this assumption is reasonable, but the engineer would be more knowledgeable as to whether this assumption is appropriate. We also must verify the minimum sample size conditions:

$$1000 \times \frac{899}{1000} \geq 10 \quad 1000 \times \frac{101}{1000} \geq 10 \quad 1000 \times \frac{958}{1000} \geq 10 \quad 1000 \times \frac{42}{1000} \geq 10$$

To construct a confidence interval, we first identify the point estimate and standard error, then we can construct the confidence interval:

$$\text{point estimate} = 0.958 - 0.899 = 0.059$$

$$SE = \sqrt{\frac{0.899(1 - 0.899)}{1000} + \frac{0.958(1 - 0.958)}{1000}} = 0.0114$$

$$0.059 \pm 1.96 \times 0.0114$$

$$(0.037, 0.081)$$

We are 95% confident that the true difference in proportion of current and prospective gears that pass inspection is between 0.037 and 0.081, favoring the prospective gears. Because the entire interval is above zero, the data provide strong evidence that the prospective gears pass inspection more often than the current gears. The remote control car company should go with the new manufacturer.

6.2.3   Hypothesis testing when $H_0 : p_1 = p_2$

Here we use a new example to examine a special estimate of the standard error when $H_0 : p_1 = p_2$. We investigate whether there is an increased risk of cancer in dogs that are exposed to the herbicide 2,4-dichlorophenoxyacetic acid (2,4-D). A study in 1994 examined 491 dogs that had developed cancer and 945 dogs as a control group.[12] Of these two groups, researchers identified which dogs had been exposed to 2,4-D in their owner's yard. The results are shown in Table 6.3.

	cancer	no cancer
2,4-D	191	304
no 2,4-D	300	641

Table 6.3: Summary results for cancer in dogs and the use of 2,4-D by the dog's owner.

⊙ **Guided Practice 6.22**   Is this study an experiment or an observational study?[13]

⊙ **Guided Practice 6.23**   Set up hypotheses to test whether 2,4-D and the occurrence of cancer in dogs are related. Use a one-sided test and compare across the cancer and no cancer groups.[14]

● **Example 6.24**   Are the conditions for using the normal model and make inference on the results?

(1) It is unclear whether this is a random sample. However, if we believe the dogs in both the cancer and no cancer groups are representative of each respective population and that the dogs in the study do not interact in any way, then we may find it reasonable to assume independence between observations. (2) The success-failure condition (minimums of 10) easily holds for each sample.

Under the assumption of independence, we can use the normal model and make statements regarding the canine population based on the data.

In the hypotheses for Guided Practice 6.23, the null is that the proportion of dogs with exposure to 2,4-D is the same in each group. The point estimate of the difference in sample proportions is $\hat{p}_c - \hat{p}_n = 0.067$. To identify the p-value for this test, we first check conditions (Example 6.24) and compute the standard error of the difference.

---

[12]Hayes HM, Tarone RE, Cantor KP, Jessen CR, McCurnin DM, and Richardson RC. 1991. Case-Control Study of Canine Malignant Lymphoma: Positive Association With Dog Owner's Use of 2, 4-Dichlorophenoxyacetic Acid Herbicides. Journal of the National Cancer Institute 83(17):1226-1231.

[13]The owners were not instructed to apply or not apply the herbicide, so this is an observational study. This question was especially tricky because one group was called the *control group*, which is a term usually seen in experiments.

[14]Using the proportions within the cancer and no cancer groups may seem odd. We intuitively may desire to compare the fraction of dogs with cancer in the 2,4-D and no 2,4-D groups, since the herbicide is an explanatory variable. However, the cancer rates in each group do not necessarily reflect the real cancer rates due to the way the data were collected. For this reason, computing cancer rates may greatly alarm dog owners.

H_0: the proportion of dogs with exposure to 2,4-D is the same in "cancer" and "no cancer" dogs, $p_c - p_n = 0$.

H_A: dogs with cancer are more likely to have been exposed to 2,4-D than dogs without cancer, $p_c - p_n > 0$.

The standard deviation is given by

$$SD = \sqrt{\frac{p_c(1-p_c)}{n_c} + \frac{p_n(1-p_n)}{n_n}}$$

In a hypothesis test, the distribution of the test statistic is always examined as though the null hypothesis is true, i.e. in this case, $p_c = p_n$. The standard deviation formula should reflect this equality in the null hypothesis. We will use p to represent the common rate of dogs that are exposed to 2,4-D in the two groups:

$$SD = \sqrt{\frac{p(1-p)}{n_c} + \frac{p(1-p)}{n_n}}$$

$$= \sqrt{p(1-p)}\sqrt{\frac{1}{n_c} + \frac{1}{n_n}}$$

We don't know the exposure rate, p, but we can obtain a good estimate of it by *pooling* the results of both samples to find $\hat{p}$:

$$\hat{p} = \frac{\# \text{ of "successes"}}{\# \text{ of cases}} = \frac{191 + 300}{191 + 300 + 304 + 641} = 0.342$$

This is called the **pooled estimate** of the sample proportion, and we use it to compute the standard error when the null hypothesis is that $p_1 = p_2$ (e.g. $p_c = p_n$ or $p_c - p_n = 0$). We also typically use it to verify the success-failure condition.

Pooled estimate of a proportion

When the null hypothesis is $p_1 = p_2$, it is useful to find the pooled estimate of the shared proportion:

$$\hat{p} = \frac{\text{number of "successes"}}{\text{number of cases}} = \frac{x_1 + x_2}{n_1 + n_2} = \frac{\hat{p}_1 n_1 + \hat{p}_2 n_2}{n_1 + n_2}$$

Here x_1 represents the number of successes in sample 1. x_1 can be computed as $\hat{p}_1 n_1$ if it is unknown. Similarly, x_2 represents the number of successes in sample 2. It also can be computed as $\hat{p}_2 n_2$.

TIP: Use the pooled proportion estimate when $H_0 : p_1 = p_2$

When the null hypothesis suggests the proportions are equal, we use the pooled proportion estimate ($\hat{p}$) to verify the success-failure condition and also to estimate the standard error:

$$SE = \sqrt{\hat{p}(1-\hat{p})}\sqrt{\frac{1}{n_1} + \frac{1}{n_2}} \tag{6.25}$$

⊙ **Guided Practice 6.26** Using Equation (6.25), $\hat{p} = 0.342$, $n_1 = 491$, and $n_2 = 945$, verify the standard error estimate in the context of a hypothesis test is $SE = 0.026$.

● **Example 6.27** Complete the hypothesis test using a significance level of 0.01.

We will complete a 2-proportion z-test. The conditions are met - we will assume that there two independent random samples. Using the pooled proportion:

$$n_1\hat{p} = 491 \times 0.342 = 167.9 \qquad n_1(1-\hat{p}) = 491 \times 0.658 = 323.1$$
$$n_2\hat{p} = 945 \times 0.342 = 323.2 \qquad n_2(1-\hat{p}) = 945 \times 0.658 = 621.8$$

are all at least 10. Now we set up hypotheses, which were identified in Guided Practice 6.23:

H_0: The proportion of dogs with exposure to 2,4-D is the same in "cancer" and "no cancer" dogs, $p_c - p_n = 0$.

H_A: Dogs with cancer are more likely to have been exposed to 2,4-D than dogs without cancer, $p_c - p_n > 0$.

We will use a significance level of $\alpha = 0.01$. All values are much larger than 10. Under the assumption that there were two independent random samples, we can proceed.

Next, we compute the test statistic using the standard error using the result of Guided Practice 6.26:

$$Z = \frac{\text{point estimate} - \text{null value}}{SE} = \frac{0.067 - 0}{0.026} = 2.58$$

Looking up $Z = 2.58$ in the normal probability table: 0.9951. However this is the lower tail, and the upper tail represents the p-value: $1 - 0.9951 = 0.0049$. Because the p-value is smaller than $\alpha = 0.01$, we reject the null hypothesis and conclude that there is an association between dogs getting cancer and owners using 2,4-D.

Hypothesis test for the difference of two proportions

1. State the name of the test being used.

   - 2-proportion z-test

2. Verify conditions to ensure the standard error estimate is reasonable and the point estimate is nearly normal and unbiased.

   - 2 independent random samples OR 2 randomly allocated treatments.
   - Calculate the pooled sample proportion $\hat{p}$ and verify $n_1\hat{p}$, $n_2\hat{p}$, $n_1(1-\hat{p})$, and $n_2(1-\hat{p})$ are greater than or equal to 10.

3. Write the hypotheses in plain language and using mathematical notation.

   - $H_0 : p_1 = p_2$ (or $p_1 - p_2 = 0$)
   - $H_A : p_1 \neq$ or $<$ or $> p_2$

4. Identify the significance level α.

5. Calculate the test statistic: $Z = \frac{\text{point estimate} - \text{null value}}{\text{SE of estimate}}$

   - The point estimate is $\hat{p}_1 - \hat{p}_2$.
   - Use $SE = \sqrt{\hat{p}(1-\hat{p})}\sqrt{\frac{1}{n_1} + \frac{1}{n_2}}$.

6. Find the p-value and compare it to α to determine whether to reject or not reject H_0.

7. Write the conclusion in the context of the question.

6.2.4 Calculator: the 2-proportion z-test and z-interval

TI-83/84: 2-proportion z-interval

Use STAT, TESTS, 2-PropZInt.

1. Choose STAT.

2. Right arrow to TESTS.

3. Down arrow and choose B:2-PropZInt.

4. Let x1 be the *number* of yes's (must be an integer) in sample 1 and let n1 be the size of sample 1.

5. Let x2 be the *number* of yes's (must be an integer) in sample 2 and let n2 be the size of sample 2.

6. Let C-Level be the desired confidence level.

7. Choose Calculate and hit ENTER, which returns:

   (__,__)   the confidence interval
   $\hat{p}_1$ sample 1 proportion n_1 size of sample 1
   $\hat{p}_2$ sample 2 proportion n_2 size of sample 2

Casio fx-9750GII: 2-proportion z-interval

1. Navigate to STAT (MENU button, then hit the 2 button or select STAT).

2. Choose the INTR option (F4 button).

3. Choose the Z option (F1 button).

4. Choose the 2-P option (F4 button).

5. Specify the interval details:

   - Confidence level of interest for C-Level.
   - Enter the number of successes for each group, x1 and x2.
   - Enter the sample size for each group, n1 and n2.

6. Hit the EXE button, which returns

   Left, Right   the ends of the confidence interval
   $\hat{p}1$, $\hat{p}2$ the sample proportions
   n1, n2   sample sizes

⊙ **Guided Practice 6.28** Use the data in Table 6.4 and a calculator to find a 95% confidence interval for the difference in proportion of dogs with cancer that have been exposed to 2,4-D versus not exposed to 2,4-D.[15]

---

[15] Correctly going through the calculator steps should lead to an interval of (0.01484, 0.11926). There is no value given for the pooled proportion since we do not pool for confidence intervals.

	cancer	no cancer
2,4-D	191	304
no 2,4-D	300	641

Table 6.4: Summary results for cancer in dogs and the use of 2,4-D by the dog's owner.

TI-83/84: 2-proportion z-test

Use STAT, TESTS, 2-PropZTest.

1. Choose STAT.

2. Right arrow to TESTS.

3. Down arrow and choose 6:2-PropZTest.

4. Let x1 be the *number* of yes's (must be an integer) in sample 1 and let n1 be the size of sample 1.

5. Let x2 be the *number* of yes's (must be an integer) in sample 2 and let n2 be the size of sample 2.

6. Choose $\neq$, $<$, or $>$ to correspond to H_A.

7. Choose Calculate and hit ENTER, which returns:
   z   Z-statistic   p   p-value
   $\hat{p}_1$ sample 1 proportion $\hat{p}$ pooled sample proportion
   $\hat{p}_2$ sample 2 proportion

Casio fx-9750GII: 2-proportion z-test

1. Navigate to STAT (MENU button, then hit the 2 button or select STAT).

2. Choose the TEST option (F3 button).

3. Choose the Z option (F1 button).

4. Choose the 2-P option (F4 button).

5. Specify the test details:

   - Specify the sidedness of the test using the F1, F2, and F3 keys.
   - Enter the number of successes for each group, x1 and x2.
   - Enter the sample size for each group, n1 and n2.

6. Hit the EXE button, which returns
   z   Z-statistic $\hat{p}1$, $\hat{p}2$ sample proportions
   p   p-value $\hat{p}$ pooled proportion
                           n1, n2   sample sizes

⊙ **Guided Practice 6.29** Use the data in Table 6.4 and a calculator to find the Z-score and p-value for one-sided test with H_A: dogs with cancer are more likely to have been exposed to 2,4-D than dogs without cancer, $p_c - p_n > 0$.[16]

6.3   Testing for goodness of fit using chi-square

In this section, we develop a method for assessing a null model when the data are binned. This technique is commonly used in two circumstances:

- Given a sample of cases that can be classified into several groups, determine if the sample is representative of the general population.

- Evaluate whether data resemble a particular distribution, such as a normal distribution or a geometric distribution.

Each of these scenarios can be addressed using the same statistical test: a chi-square test.

In the first case, we consider data from a random sample of 275 jurors in a small county. Jurors identified their racial group, as shown in Table 6.5, and we would like to determine if these jurors are racially representative of the population. If the jury is representative of the population, then the proportions in the sample should roughly reflect the population of eligible jurors, i.e. registered voters.

Race	White	Black	Hispanic	Other	Total
Representation in juries	205	26	25	19	275
Registered voters	0.72	0.07	0.12	0.09	1.00

Table 6.5: Representation by race in a city's juries and population.

While the proportions in the juries do not precisely represent the population proportions, it is unclear whether these data provide convincing evidence that the sample is not representative. If the jurors really were randomly sampled from the registered voters, we might expect small differences due to chance. However, unusually large differences may provide convincing evidence that the juries were not representative.

A second application, assessing the fit of a distribution, is presented at the end of this section. Daily stock returns from the S&P500 for the years 1990-2011 are used to assess whether stock activity each day is independent of the stock's behavior on previous days.

In these problems, we would like to examine all bins simultaneously, not simply compare one or two bins at a time, which will require us to develop a new test statistic.

6.3.1   Creating a test statistic for one-way tables

● **Example 6.30** Of the people in the city, 275 served on a jury. If the individuals are randomly selected to serve on a jury, about how many of the 275 people would we expect to be white? How many would we expect to be black?

About 72% of the population is white, so we would expect about 72% of the jurors to be white: $0.72 \times 275 = 198$.

Similarly, we would expect about 7% of the jurors to be black, which would correspond to about $0.07 \times 275 = 19.25$ black jurors.

---

[16]Correctly going through the calculator steps should lead to a solution with $Z = 2.55$ and p-value = 0.0055. The pooled proportion is $\hat{p} = 0.342$.

⊙ **Guided Practice 6.31**   Twelve percent of the population is Hispanic and 9% represent other races. How many of the 275 jurors would we expect to be Hispanic or from another race? Answers can be found in Table 6.6.

Race	White	Black	Hispanic	Other	Total
Observed data	205	26	25	19	275
Expected counts	198	19.25	33	24.75	275

Table 6.6: Actual and expected make-up of the jurors.

The sample proportion represented from each race among the 275 jurors was not a precise match for any ethnic group. While some sampling variation is expected, we would expect the sample proportions to be fairly similar to the population proportions if there is no bias on juries. We need to test whether the differences are strong enough to provide convincing evidence that the jurors are not a random sample. These ideas can be organized into hypotheses:

H_0: The jurors are a random sample, i.e. there is no racial bias in who serves on a jury, and the observed counts reflect natural sampling fluctuation.

H_A: The jurors are not randomly sampled, i.e. there is racial bias in juror selection.

To evaluate these hypotheses, we quantify how different the observed counts are from the expected counts. Strong evidence for the alternative hypothesis would come in the form of unusually large deviations in the groups from what would be expected based on sampling variation alone.

6.3.2   The chi-square test statistic

In previous hypothesis tests, we constructed a test statistic of the following form:

$$Z = \frac{\text{point estimate} - \text{null value}}{\text{SE of point estimate}}$$

This construction was based on (1) identifying the difference between a point estimate and an expected value if the null hypothesis was true, and (2) standardizing that difference using the standard error of the point estimate. These two ideas will help in the construction of an appropriate test statistic for count data.

In this example we have four categories: white, black, hispanic, and other. Because we have four values rather than just one or two, we need a new tool to analyze the data. Our strategy will be to find a test statistic that measures the overall deviation between the observed and the expected counts. We first find the difference between the observed and expected counts for the four groups:

	White	Black	Hispanic	Other
observed - expected	$205 - 198$	$26 - 19.25$	$25 - 33$	$19 - 24.75$

Next, we square the differences:

	White	Black	Hispanic	Other
(observed - expected)^2	$(205 - 198)^2$	$(26 - 19.25)^2$	$(25 - 33)^2$	$(19 - 24.75)^2$

We must standardize each term. To know whether the squared difference is large, we compare it to what was expected. If the expected count was 5, a squared difference of 25 is very large. However, if the expected count was 1,000, a squared difference of 25 is very small. We will divide each of the squared differences by the corresponding expected count.

	White	Black	Hispanic	Other
$\dfrac{(\text{observed - expected})^2}{\text{expected}}$	$\dfrac{(205-198)^2}{198}$	$\dfrac{(26-19.25)^2}{19.25}$	$\dfrac{(25-33)^2}{33}$	$\dfrac{(19-24.75)^2}{24.75}$

Finally, to arrive at the overall measure of deviation between the observed counts and the expected counts, we add up the terms.

$$X^2 = \sum \frac{(\text{observed - expected})^2}{\text{expected}}$$
$$= \frac{(205-198)^2}{198} + \frac{(26-19.25)^2}{19.25} + \frac{(25-33)^2}{33} + \frac{(19-24.75)^2}{24.75}$$

We can write an equation for X^2 using the observed counts and expected counts:

$$X^2 = \frac{(\text{observed count}_1 - \text{expected count}_1)^2}{\text{expected count}_1} + \cdots + \frac{(\text{observed count}_4 - \text{expected count}_4)^2}{\text{expected count}_4}$$

X^2
chi-square
test statistic

The final number X^2 summarizes how strongly the observed counts tend to deviate from the null counts.

In Section 6.3.4, we will see that if the null hypothesis is true, then X^2 follows a new distribution called a *chi-square distribution*. Using this distribution, we will be able to obtain a p-value to evaluate whether there appears to be racial bias in the juries for the city we are considering.

6.3.3 The chi-square distribution and finding areas

The **chi-square distribution** is sometimes used to characterize data sets and statistics that are always positive and typically right skewed. Recall the normal distribution had two parameters – mean and standard deviation – that could be used to describe its exact characteristics. The chi-square distribution has just one parameter called **degrees of freedom (df)**, which influences the shape, center, and spread of the distribution.

⊙ **Guided Practice 6.32** Figure 6.7 shows three chi-square distributions. (a) How does the center of the distribution change when the degrees of freedom is larger? (b) What about the variability (spread)? (c) How does the shape change?[17]

Figure 6.7 and Guided Practice 6.32 demonstrate three general properties of chi-square distributions as the degrees of freedom increases: the distribution becomes more symmetric, the center moves to the right, and the variability inflates.

Our principal interest in the chi-square distribution is the calculation of p-values, which (as we have seen before) is related to finding the relevant area in the tail of a distribution.

---

[17](a) The center becomes larger. If we look carefully, we can see that the center of each distribution is equal to the distribution's degrees of freedom. (b) The variability increases as the degrees of freedom increases. (c) The distribution is very strongly skewed for $df = 2$, and then the distributions become more symmetric for the larger degrees of freedom $df = 4$ and $df = 9$. In fact, as the degrees of freedom increase, the X^2 distribution approaches a normal distribution.

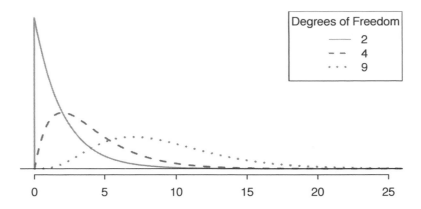

Figure 6.7: Three chi-square distributions with varying degrees of freedom.

To do so, a new table is needed: the **chi-square table**, partially shown in Table 6.8. A more complete table is presented in Appendix B.4 on page 454. This table is very similar to the t-table from Sections 7.1 and 7.3: we identify a range for the area, and we examine a particular row for distributions with different degrees of freedom. One important difference from the t-table is that the chi-square table only provides upper tail values.

Upper tail		0.3	0.2	0.1	0.05	0.02	0.01	0.005	0.001
df	1	1.07	1.64	2.71	3.84	5.41	6.63	7.88	10.83
df	2	2.41	**3.22**	**4.61**	5.99	7.82	9.21	10.60	13.82
	3	*3.66*	*4.64*	*6.25*	*7.81*	*9.84*	*11.34*	*12.84*	*16.27*
	4	4.88	5.99	7.78	9.49	11.67	13.28	14.86	18.47
	5	6.06	7.29	9.24	11.07	13.39	15.09	16.75	20.52
	6	7.23	8.56	10.64	12.59	15.03	16.81	18.55	22.46
	7	8.38	9.80	12.02	14.07	16.62	18.48	20.28	24.32

Table 6.8: A section of the chi-square table. A complete table is in Appendix B.4.

● **Example 6.33** Figure 6.9(b) shows a chi-square distribution with 3 degrees of freedom and an upper shaded tail starting at 6.25. Use Table 6.8 to estimate the shaded area.

This distribution has three degrees of freedom, so only the row with 3 degrees of freedom (df) is relevant. This row has been italicized in the table. Next, we see that the value – 6.25 – falls in the column with upper tail area 0.1. That is, the shaded upper tail of Figure 6.9(b) has area 0.1.

● **Example 6.34** We rarely observe the *exact* value in the table. For instance, Figure 6.9(a) shows the upper tail of a chi-square distribution with 2 degrees of freedom. The lower bound for this upper tail is at 4.3, which does not fall in Table 6.8. Find the approximate tail area.

The cutoff 4.3 falls between the second and third columns in the 2 degrees of freedom row. Because these columns correspond to tail areas of 0.2 and 0.1, we can be certain that the area shaded in Figure 6.9(a) is between 0.1 and 0.2.

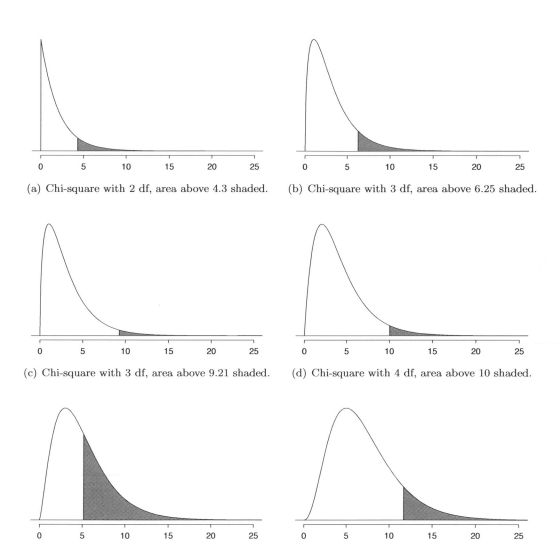

(a) Chi-square with 2 df, area above 4.3 shaded.

(b) Chi-square with 3 df, area above 6.25 shaded.

(c) Chi-square with 3 df, area above 9.21 shaded.

(d) Chi-square with 4 df, area above 10 shaded.

(e) Chi-square with 5 df, area above 5.1 shaded.

(f) Chi-square with 7 df, area above 11.7 shaded.

Figure 6.9: **(b)** Six chi-square distributions with different right tail areas shaded.

Using a calculator or statistical software allows us to get more precise areas under the chi-square curve than we can get from the table alone.

TI-84: Finding an upper tail area under the chi-square curve

Use the X^2cdf command to find areas under the chi-square curve.

1. Hit 2ND VARS (i.e. DISTR).
2. Choose 8:X^2cdf.
3. Enter the lower bound, which is generally the chi-square value.
4. Enter the upper bound. Use a large number, such as 1000.
5. Enter the degrees of freedom.
6. Choose Paste and hit ENTER.

TI-83: Do steps 1-2, then type the lower bound, upper bound, and degrees of freedom separated by commas. e.g. X^2cdf(5, 1000, 3), and hit ENTER.

Casio fx-9750GII: Finding an upper tail area under the chi-sq. curve

1. Navigate to STAT (MENU button, then hit the 2 button or select STAT).
2. Choose the DIST option (F5 button).
3. Choose the CHI option (F3 button).
4. Choose the Ccd option (F2 button).
5. If necessary, select the Var option (F2 button).
6. Enter the Lower bound (generally the chi-square value).
7. Enter the Upper bound (use a large number, such as 1000).
8. Enter the degrees of freedom, df.
9. Hit the EXE button.

⊙ **Guided Practice 6.35**   Figure 6.9(e) shows an upper tail for a chi-square distribution with 5 degrees of freedom and a cutoff of 5.1. Find the tail area using a calculator.[18]

⊙ **Guided Practice 6.36**   Figure 6.9(f) shows a cutoff of 11.7 on a chi-square distribution with 7 degrees of freedom. Find the area of the upper tail.[19]

⊙ **Guided Practice 6.37**   Figure 6.9(d) shows a cutoff of 10 on a chi-square distribution with 4 degrees of freedom. Find the area of the upper tail.[20]

---

[18]Using $df = 5$ and a *lower* bound of 5.1 for the tail, the upper tail area is 0.4038.
[19]The area is 0.1109.
[20]The area is 0.4043.

⊙ **Guided Practice 6.38** Figure 6.9(c) shows a cutoff of 9.21 with a chi-square distribution with 3 df. Find the area of the upper tail.[21]

6.3.4 Finding a p-value for a chi-square distribution

In Section 6.3.2, we identified a new test statistic (X^2) within the context of assessing whether there was evidence of racial bias in how jurors were sampled. The null hypothesis represented the claim that jurors were randomly sampled and there was no racial bias. The alternative hypothesis was that there was racial bias in how the jurors were sampled.

We determined that a large X^2 value would suggest strong evidence favoring the alternative hypothesis: that there was racial bias. However, we could not quantify what the chance was of observing such a large test statistic ($X^2 = 5.89$) if the null hypothesis actually was true. This is where the chi-square distribution becomes useful. If the null hypothesis was true and there was no racial bias, then X^2 would follow a chi-square distribution, with three degrees of freedom in this case. Under certain conditions, the statistic X^2 follows a chi-square distribution with $k - 1$ degrees of freedom, where k is the number of bins or categories of the variable.

● **Example 6.39** How many categories were there in the juror example? How many degrees of freedom should be associated with the chi-square distribution used for X^2?

In the jurors example, there were $k = 4$ categories: white, black, Hispanic, and other. According to the rule above, the test statistic X^2 should then follow a chi-square distribution with $k - 1 = 3$ degrees of freedom if H_0 is true.

Just like we checked sample size conditions to use the normal model in earlier sections, we must also check a sample size condition to safely apply the chi-square distribution for X^2. Each expected count must be at least 5. In the juror example, the expected counts were 198, 19.25, 33, and 24.75, all easily above 5, so we can apply the chi-square model to the test statistic, $X^2 = 5.89$.

● **Example 6.40** If the null hypothesis is true, the test statistic $X^2 = 5.89$ would be closely associated with a chi-square distribution with three degrees of freedom. Using this distribution and test statistic, identify the p-value and state whether or not there is evidence of racial bias in the juror selection.

The chi-square distribution and p-value are shown in Figure 6.10. Because larger chi-square values correspond to stronger evidence against the null hypothesis, we shade the upper tail to represent the p-value. Using the chi-square table in Appendix B.4 or the short table on page 276, we can determine that the area is between 0.1 and 0.2. That is, the p-value is larger than 0.1 but smaller than 0.2. Generally we do not reject the null hypothesis with such a large p-value. In other words, the data do not provide convincing evidence of racial bias in the juror selection.

The test that we just carried out regarding jury selection is known as the X^2 **goodness of fit test**. It is called "goodness of fit" because we test whether or not the proposed or expected distribution is a good fit for the observed data.

---

[21]The area is 0.0266.

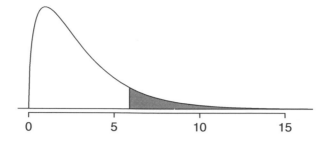

Figure 6.10: The p-value for the juror hypothesis test is shaded in the chi-square distribution with $df = 3$.

Chi-square goodness of fit test for one-way table

Suppose we are to evaluate whether there is convincing evidence that a set of observed counts O_1, O_2, ..., O_k in k categories are unusually different from what might be expected under a null hypothesis. Call the *expected counts* that are based on the null hypothesis E_1, E_2, ..., E_k. If each expected count is at least 5 and the null hypothesis is true, then the test statistic below follows a chi-square distribution with $k - 1$ degrees of freedom:

$$X^2 = \frac{(O_1 - E_1)^2}{E_1} + \frac{(O_2 - E_2)^2}{E_2} + \cdots + \frac{(O_k - E_k)^2}{E_k}$$

The p-value for this test statistic is found by looking at the upper tail of this chi-square distribution. We consider the upper tail because larger values of X^2 would provide greater evidence against the null hypothesis.

TIP: Conditions for the chi-square goodness of fit test
There are two conditions that must be checked before performing a chi-square goodness of fit test. If these conditions are not met, this test should not be used.

Simple random sample. The data must be arrived at by taking a simple random sample from the population of interest. The observed counts can then be organized into a list or one-way table.

All Expected Counts at least 5 Each particular scenario (i.e. cell count) must have at least 5 expected cases.

6.3.5 Evaluating goodness of fit for a distribution

Goodness of fit test for a one-way table

1. State the name of the test being used.
   - X^2 goodness of fit test.

2. Verify conditions.
   - A simple random sample.
   - All expected counts ≥ 5 (calculate and record expected counts).

3. Write the hypotheses in plain language. No mathematical notation is needed for this test.
   - H_0: The distribution of [...] matches [the expected distribution].
   - H_A: The distribution of [....] does not match [the expected distribution]

4. Identify the significance level α.

5. Calculate the test statistic and degrees of freedom.

$$X^2 = \sum \frac{(\text{observed counts - expected counts})^2}{\text{expected counts}}$$
$$df = (\# \text{ of categories} - 1)$$

6. Find the p-value and compare it to α to determine whether to reject or not reject H_0.

7. Write the conclusion in the context of the question.

Section 4.3 would be useful background reading for this example, but it is not a prerequisite.

We can apply our new chi-square testing framework to the second problem in this section: evaluating whether a certain statistical model fits a data set. Daily stock returns from the S&P500 for 1990-2011 can be used to assess whether stock activity each day is independent of the stock's behavior on previous days. This sounds like a very complex question, and it is, but a chi-square test can be used to study the problem. We will label each day as Up or Down (D) depending on whether the market was up or down that day. For example, consider the following changes in price, their new labels of up and down, and then the number of days that must be observed before each Up day:

Change in price	2.52	-1.46	0.51	-4.07	3.36	1.10	-5.46	-1.03	-2.99	1.71
Outcome	Up	D	Up	D	Up	Up	D	D	D	Up
Days to Up	1	-	2	-	2	1	-	-	-	4

If the days really are independent, then the number of days until a positive trading day should follow a geometric distribution. The geometric distribution describes the probability of waiting for the kth trial to observe the first success. Here each up day (Up) represents a success, and down (D) days represent failures. In the data above, it took only one day until the market was up, so the first wait time was 1 day. It took two more days before we observed our next Up trading day, and two more for the third Up day. We would like to determine if these counts (1, 2, 2, 1, 4, and so on) follow the geometric distribution. Table 6.11 shows the number of waiting days for a positive trading day during 1990-2011 for the S&P500.

Days	1	2	3	4	5	6	7+	Total
Observed	1532	760	338	194	74	33	17	2948

Table 6.11: Observed distribution of the waiting time until a positive trading day for the S&P500, 1990-2011.

We consider how many days one must wait until observing an Up day on the S&P500 stock index. If the stock activity was independent from one day to the next and the probability of a positive trading day was constant, then we would expect this waiting time to follow a *geometric distribution*. We can organize this into a hypothesis framework:

H_0: The stock market being up or down on a given day is independent from all other days. We will consider the number of days that pass until an Up day is observed. Under this hypothesis, the number of days until an Up day should follow a geometric distribution.

H_A: The stock market being up or down on a given day is not independent from all other days. Since we know the number of days until an Up day would follow a geometric distribution under the null, we look for deviations from the geometric distribution, which would support the alternative hypothesis.

There are important implications in our result for stock traders: if information from past trading days is useful in telling what will happen today, that information may provide an advantage over other traders.

We consider data for the S&P500 from 1990 to 2011 and summarize the waiting times in Table 6.12 and Figure 6.13. The S&P500 was positive on 53.2% of those days.

Because applying the chi-square framework requires expected counts to be at least 5, we have *binned* together all the cases where the waiting time was at least 7 days to ensure each expected count is well above this minimum. The actual data, shown in the *Observed* row in Table 6.12, can be compared to the expected counts from the *Geometric Model* row. The method for computing expected counts is discussed in Table 6.12. In general, the expected counts are determined by (1) identifying the null proportion associated with each bin, then (2) multiplying each null proportion by the total count to obtain the expected counts. That is, this strategy identifies what proportion of the total count we would expect to be in each bin.

Days	1	2	3	4	5	6	7+	Total
Observed	1532	760	338	194	74	33	17	2948
Geometric Model	1569	734	343	161	75	35	31	2948

Table 6.12: Distribution of the waiting time until a positive trading day. The expected counts based on the geometric model are shown in the last row. To find each expected count, we identify the probability of waiting D days based on the geometric model ($P(D) = (1 - 0.532)^{D-1}(0.532)$) and multiply by the total number of streaks, 2948. For example, waiting for three days occurs under the geometric model about $0.468^2 \times 0.532 = 11.65\%$ of the time, which corresponds to $0.1165 \times 2948 = 343$ streaks.

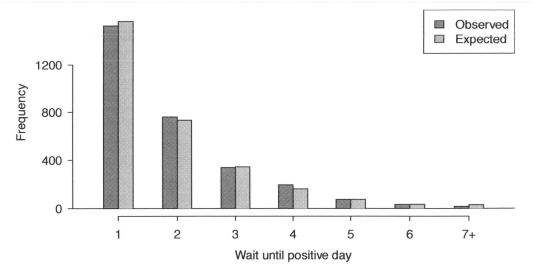

Figure 6.13: Side-by-side bar plot of the observed and expected counts for each waiting time.

● **Example 6.41**   Do you notice any unusually large deviations in the graph? Can you tell if these deviations are due to chance just by looking?

It is not obvious whether differences in the observed counts and the expected counts from the geometric distribution are significantly different. That is, it is not clear whether these deviations might be due to chance or whether they are so strong that the data provide convincing evidence against the null hypothesis. However, we can perform a chi-square test using the counts in Table 6.12.

⊙ **Guided Practice 6.42**   Table 6.12 provides a set of count data for waiting times ($O_1 = 1532$, $O_2 = 760$, ...) and expected counts under the geometric distribution ($E_1 = 1569$, $E_2 = 734$, ...). Compute the chi-square test statistic, X^2.[22]

⊙ **Guided Practice 6.43**   Because the expected counts are all at least 5, we can safely apply the chi-square distribution to X^2. However, how many degrees of freedom should we use?[23]

● **Example 6.44**   If the observed counts follow the geometric model, then the chi-square test statistic $X^2 = 15.08$ would closely follow a chi-square distribution with $df = 6$. Using this information, compute a p-value.

Figure 6.14 shows the chi-square distribution, cutoff, and the shaded p-value. If we look up the statistic $X^2 = 15.08$ in Appendix B.4, we find that the p-value is between 0.01 and 0.02. In other words, we have sufficient evidence to reject the notion that the wait times follow a geometric distribution, i.e. trading days are not independent and past days may help predict what the stock market will do today.

---

[22]$X^2 = \frac{(1532-1569)^2}{1569} + \frac{(760-734)^2}{734} + \cdots + \frac{(17-31)^2}{31} = 15.08$
[23]There are $k = 7$ groups, so we use $df = k - 1 = 6$.

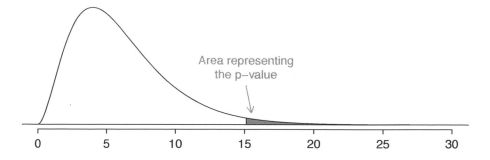

Figure 6.14: Chi-square distribution with 6 degrees of freedom. The p-value for the stock analysis is shaded.

● **Example 6.45** In Example 6.44, we rejected the null hypothesis that the trading days are independent. Why is this so important?

Because the data provided strong evidence that the geometric distribution is not appropriate, we reject the claim that trading days are independent. While it is not obvious how to exploit this information, it suggests there are some hidden patterns in the data that could be interesting and possibly useful to a stock trader.

6.3.6   Calculator: chi-square goodness of fit test

▶ TI-84: Chi-square goodness of fit test

Use STAT, TESTS, X^2GOF-Test.

1. Enter the observed counts into list L1 and the expected counts into list L2.
2. Choose STAT.
3. Right arrow to TESTS.
4. Down arrow and choose D:X^2GOF-Test.
5. Leave Observed: L1 and Expected: L2.
6. Enter the degrees of freedom after df:
7. Choose Calculate and hit ENTER, which returns:
   X^2 chi-square test statistic
   p   p-value
   df   degrees of freedom

TI-83: Unfortunately the TI-83 does not have this test built in. To carry out the test manually, make list L3 = (L1 - L2)^2 / L2 and do 1-Var-Stats on L3. The sum of L3 will correspond to the value of X^2 for this test.

▶ Casio fx-9750GII: Chi-square goodness of fit test

1. Navigate to STAT (MENU button, then hit the 2 button or select STAT).

2. Enter the observed counts into a list (e.g. List 1) and the expected counts into list (e.g. List 2).

3. Choose the TEST option (F3 button).

4. Choose the CHI option (F3 button).

5. Choose the GOF option (F1 button).

6. Adjust the Observed and Expected lists to the corresponding list numbers from Step 2.

7. Enter the degrees of freedom, df.

8. Specify a list where the contributions to the test statistic will be reported using CNTRB. This list number should be different from the others.

9. Hit the EXE button, which returns

x^2	chi-square test statistic
p	p-value
df	degrees of freedom
CNTRB	list showing the test statistic contributions

Days	1	2	3	4	5	6	7+	Total
Observed	1532	760	338	194	74	33	17	2948
Geometric Model	1569	734	343	161	75	35	31	2948

Table 6.15: Distribution of the waiting time until a positive trading day. The expected counts based on the geometric model are shown in the last row.

⊙ **Guided Practice 6.46**   Use the data above and a calculator to find the X^2 statistic, df, and p-value for chi-square goodness of fit test.[24]

6.4   Homogeneity and independence in two-way tables

Google is constantly running experiments to test new search algorithms. For example, Google might test three algorithms using a sample of 10,000 google.com search queries. Table 6.16 shows an example of 10,000 queries split into three algorithm groups.[25] The group sizes were specified before the start of the experiment to be 5000 for the current algorithm and 2500 for each test algorithm.

---

[24] You should find that $X^2 = 15.08$, $df = 6$, and p-value = 0.0196.

[25] Google regularly runs experiments in this manner to help improve their search engine. It is entirely possible that if you perform a search and so does your friend, that you will have different search results. While the data presented in this section resemble what might be encountered in a real experiment, these data are simulated.

Search algorithm	current	test 1	test 2	Total
Counts	5000	2500	2500	10000

Table 6.16: Google experiment breakdown of test subjects into three search groups.

● **Example 6.47** What is the ultimate goal of the Google experiment? What are the null and alternative hypotheses, in regular words?

The ultimate goal is to see whether there is a difference in the performance of the algorithms. The hypotheses can be described as the following:

H_0: The algorithms each perform equally well.

H_A: The algorithms do not perform equally well.

In this experiment, the explanatory variable is the search algorithm. However, an outcome variable is also needed. This outcome variable should somehow reflect whether the search results align with the user's interests. One possible way to quantify this is to determine whether (1) there was no new, related search, and the user clicked one of the links provided, or (2) there was a new, related search performed by the user. Under scenario (1), we might think that the user was satisfied with the search results. Under scenario (2), the search results probably were not relevant, so the user tried a second search.

Table 6.17 provides the results from the experiment. These data are very similar to the count data in Section 6.3. However, now the different combinations of two variables are binned in a *two-way* table. In examining these data, we want to evaluate whether there is strong evidence that at least one algorithm is performing better than the others. To do so, we apply a chi-square test to this two-way table. The ideas of this test are similar to those ideas in the one-way table case. However, degrees of freedom and expected counts are computed a little differently than before.

	Search algorithm			
	current	test 1	test 2	Total
No new search	3511	1749	1818	7078
New search	1489	751	682	2922
Total	5000	2500	2500	10000

Table 6.17: Results of the Google search algorithm experiment.

TIP: What is so different about one-way tables and two-way tables?
A one-way table describes counts for each outcome in a single variable. A two-way table describes counts for *combinations* of outcomes for two variables. When we consider a two-way table, we often would like to know, are these variables related in any way?

The hypothesis test for this Google experiment is really about assessing whether there is statistically significant evidence that the choice of the algorithm affects whether a user performs a second search. In other words, the goal is to check whether the three search algorithms perform differently.

6.4.1 Expected counts in two-way tables

● **Example 6.48** From the experiment, we estimate the proportion of users who were satisfied with their initial search (no new search) as 7078/10000 = 0.7078. If there really is no difference among the algorithms and 70.78% of people are satisfied with the search results, how many of the 5000 people in the "current algorithm" group would be expected to not perform a new search?

About 70.78% of the 5000 would be satisfied with the initial search:

$$0.7078 \times 5000 = 3539 \text{ users}$$

That is, if there was no difference between the three groups, then we would expect 3539 of the current algorithm users not to perform a new search.

⊙ **Guided Practice 6.49** Using the same rationale described in Example 6.48, about how many users in each test group would not perform a new search if the algorithms were equally helpful?[26]

We can compute the expected number of users who would perform a new search for each group using the same strategy employed in Example 6.48 and Guided Practice 6.49. These expected counts were used to construct Table 6.18, which is the same as Table 6.17, except now the expected counts have been added in parentheses.

Search algorithm	current		test 1		test 2		Total
No new search	3511	**(3539)**	1749	**(1769.5)**	1818	**(1769.5)**	7078
New search	1489	**(1461)**	751	**(730.5)**	682	**(730.5)**	2922
Total	5000		2500		2500		10000

Table 6.18: The observed counts and the **(expected counts)**.

The examples and exercises above provided some help in computing expected counts. In general, expected counts for a two-way table may be computed using the row totals, column totals, and the table total. For instance, if there was no difference between the groups, then about 70.78% of each column should be in the first row:

$$0.7078 \times (\text{column 1 total}) = 3539$$
$$0.7078 \times (\text{column 2 total}) = 1769.5$$
$$0.7078 \times (\text{column 3 total}) = 1769.5$$

Looking back to how the fraction 0.7078 was computed – as the fraction of users who did not perform a new search (7078/10000) – these three expected counts could have been computed as

$$\left(\frac{\text{row 1 total}}{\text{table total}}\right)(\text{column 1 total}) = 3539$$
$$\left(\frac{\text{row 1 total}}{\text{table total}}\right)(\text{column 2 total}) = 1769.5$$
$$\left(\frac{\text{row 1 total}}{\text{table total}}\right)(\text{column 3 total}) = 1769.5$$

---

[26]We would expect $0.7078 * 2500 = 1769.5$. It is okay that this is a fraction.

This leads us to a general formula for computing expected counts in a two-way table when we would like to test whether there is strong evidence of an association between the column variable and row variable.

Computing expected counts in a two-way table

To identify the expected count for the ith row and j^{th} column, compute

$$\text{Expected Count}_{\text{row } i, \text{ col } j} = \frac{(\text{row } i \text{ total}) \times (\text{column } j \text{ total})}{\text{table total}}$$

6.4.2   The chi-square test of homogeneity for two-way tables

The chi-square test statistic for a two-way table is found the same way it is found for a one-way table. For each table count, compute

General formula	$\dfrac{(\text{observed count} - \text{expected count})^2}{\text{expected count}}$
Row 1, Col 1	$\dfrac{(3511 - 3539)^2}{3539} = 0.222$
Row 1, Col 2	$\dfrac{(1749 - 1769.5)^2}{1769.5} = 0.237$
$\vdots$	$\vdots$
Row 2, Col 3	$\dfrac{(682 - 730.5)^2}{730.5} = 3.220$

Adding the computed value for each cell gives the chi-square test statistic X^2:

$$X^2 = 0.222 + 0.237 + \cdots + 3.220 = 6.120$$

Just like before, this test statistic follows a chi-square distribution. However, the degrees of freedom are computed a little differently for a two-way table.[27] For two way tables, the degrees of freedom is equal to

$$df = (\text{number of rows - 1}) \times (\text{number of columns - 1})$$

In our example, the degrees of freedom parameter is

$$df = (2 - 1) \times (3 - 1) = 2$$

If the null hypothesis is true (i.e. the algorithms are equally useful), then the test statistic $X^2 = 6.12$ closely follows a chi-square distribution with 2 degrees of freedom. Using this information, we can compute the p-value for the test, which is depicted in Figure 6.19.

---

[27]Recall: in the one-way table, the degrees of freedom was the number of groups minus 1.

Computing degrees of freedom for a two-way table

When applying the chi-square test to a two-way table, we use

$$df = (R - 1) \times (C - 1)$$

where R is the number of rows in the table and C is the number of columns.

TIP: Use two-proportion methods for 2-by-2 contingency tables

When analyzing 2-by-2 contingency tables, use the two-proportion methods introduced in Section 6.2.

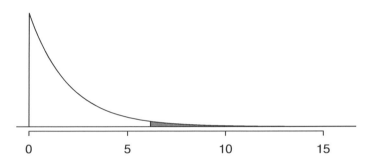

Figure 6.19: Computing the p-value for the Google hypothesis test.

TIP: Conditions for the chi-square test of homeneity

There are two conditions that must be checked before performing a chi-square test of homogeneity. If these conditions are not met, this test should not be used.

Mutliple random samples or randomly allocated treatments. Data collected by multiple independent random samples or multiple randomlly allocated treatments. Data can then be organized into a two-way table.

All Expected Counts at least 5. All of the expected counts must be at least 5.

● **Example 6.50** Compute the p-value and draw a conclusion about whether the search algorithms have different performances.

Looking in Appendix B.4 on page 454, we examine the row corresponding to 2 degrees of freedom. The test statistic, $X^2 = 6.120$, falls between the fourth and fifth columns, which means the p-value is between 0.02 and 0.05. Because we typically test at a significance level of $\alpha = 0.05$ and the p-value is less than 0.05, the null hypothesis is rejected. That is, the data provide convincing evidence that there is some difference in performance among the algorithms.

		Congress		
	Obama	Democrats	Republicans	Total
Approve	842	736	541	2119
Disapprove	616	646	842	2104
Total	1458	1382	1383	4223

Table 6.20: Pew Research poll results of a March 2012 poll.

6.4.3   The chi-square test of independence for two-way tables

The chi-square test of Independence proceeds exactly like the chi-square test of homogeneity, except that it applies when there is only one random sample (versus multiple random samples or an experiment with multiple randomly allocated treatments). The null claim is always that two variables are independent, while the alternate claim is that the variables are dependent.

● **Example 6.51** Table 6.20 summarizes the results of a Pew Research poll.[28] We would like to determine if three groups and approval ratings are associated. What are appropriate hypotheses for such a test?

H_0: The ratings are independent of the group. (There is no difference in approval ratings between the three groups.)

H_A: The ratings are dependent on the group. (There is some difference in approval ratings between the three groups, e.g. perhaps Obama's approval differs from Democrats in Congress.)

⊙ **Guided Practice 6.52**   A chi-square test for a two-way table may be used to test the hypotheses in Example 6.51. As a first step, compute the expected values for each of the six table cells.[29]

⊙ **Guided Practice 6.53**   Compute the chi-square test statistic.[30]

⊙ **Guided Practice 6.54**   Because there are 2 rows and 3 columns, the degrees of freedom for the test is $df = (2-1) \times (3-1) = 2$. Use $X^2 = 106.4$, $df = 2$, and the chi-square table on page 454 to evaluate whether to reject the null hypothesis.[31]

---

[28]See the Pew Research website: www.people-press.org/2012/03/14/romney-leads-gop-contest-trails-in-matchup-with-obama. The counts in Table 6.20 are approximate.

[29]The expected count for row one / column one is found by multiplying the row one total (2119) and column one total (1458), then dividing by the table total (4223): $\frac{2119 \times 1458}{4223} = 731.6$. Similarly for the first column and the second row: $\frac{2104 \times 1458}{4223} = 726.4$. Column 2: 693.5 and 688.5. Column 3: 694.0 and 689.0

[30]For each cell, compute $\frac{(obs-exp)^2}{exp}$. For instance, the first row and first column: $\frac{(842-731.6)^2}{731.6} = 16.7$. Adding the results of each cell gives the chi-square test statistic: $X^2 = 16.7 + \cdots + 34.0 = 106.4$.

[31]The test statistic is larger than the right-most column of the $df = 2$ row of the chi-square table, meaning the p-value is less than 0.001. That is, we reject the null hypothesis because the p-value is less than 0.05, and we conclude that Americans' approval has differences among Democrats in Congress, Republicans in Congress, and the president.

TIP: Conditions for the chi-square test of independence
There are two conditions that must be checked before performing a chi-square test of independence. If these conditions are not met, this test should not be used.

One simple random sample with two variables/questions. The data must be arrived at by taking a simple random sample. After the data is collected, it is separated and categorized according to two variables and can be organized into a two-way table.

All Expected Counts at least 5 All of the expected counts must be at least 5.

6.4.4 Summarizing the chi-square tests for two-way tables

X^2 test of homogeneity

1. State the name of the test being used.

   - X^2 test of homogeneity

2. Verify conditions.

   - Multiple random samples or treatments.
   - All expected counts ≥ 5 (calculate and record expected counts).

3. Write the hypotheses in plain language. No mathematical notation is needed for this test.

   - H_0: distribution of [variable 1] matches the distribution of [variable 2].
   - H_A: distribution of [variable 1] does not match the distribution of [variable 2].

4. Identify the significance level α.

5. Calculate the test statistic and degrees of freedom.

$$X^2 = \sum \frac{(\text{observed counts - expected counts})^2}{\text{expected counts}}$$
$$df = (\# \text{ of rows} - 1) \times (\# \text{ of columns} - 1)$$

6. Find the p-value and compare it to α to determine whether to reject or not reject H_0.

7. Write the conclusion in the context of the question.

X^2 test of independence

1. State the name of the test being used.
   - X^2 test of independence

2. Verify conditions.
   - A simple random sample.
   - All expected counts ≥ 5 (calculate and record expected counts).

3. Write the hypotheses in plain language. No mathematical notation is needed for this test.
   - H_0: [variable 1] and [variable 2] are independent.
   - H_A: [variable 1] and [variable 2] are dependent.

4. Identify the significance level α.

5. Calculate the test statistic and degrees of freedom.
$$X^2 = \sum \frac{(\text{observed counts - expected counts})^2}{\text{expected counts}}$$
$$df = (\# \text{ of rows} - 1) \times (\# \text{ of columns} - 1)$$

6. Find the p-value and compare it to α to determine whether to reject or not reject H_0.

7. Write the conclusion in the context of the question.

Example 6.55 A 2011 survey asked 806 randomly sampled adult Facebook users about their Facebook privacy settings. One of the questions on the survey was, "Do you know how to adjust your Facebook privacy settings to control what people can and cannot see?" The responses are cross-tabulated based on gender.[32]

| | | Gender | | |
		Male	Female	Total
	Yes	288	378	666
Response	No	61	62	123
	Not sure	10	7	17
	Total	359	447	806

Carry out an appropriate test at the 0.10 significance level to see if there is an association between gender and knowing how to adjust Facebook privacy settings to control what people can and cannot see.

According to the problem, there was one random sample taken. Two variables were recorded on the respondents: gender and response to the question regarding privacy settings. Because there was one random sample rather than two independent random samples, we carry out a X^2 test of independence.

H_0: Gender and knowing how to adjust Facebook privacy settings are independent.

---

[32]Survey USA, News Poll #17960, data collected February 16-17, 2011.

H_A: Gender and knowing how to adjust Facebook privacy settings are dependent.
$\alpha = 0.1$

Table of expected counts:

		Gender	
		Male	Female
	Yes	296.64	369.36
Response	No	54.785	68.215
	Not sure	7.572	9.428

All expected counts are ≥ 5. $X^2 = 3.13$; $df = 2$ p-value= $0.209 > \alpha$ We do not reject H_0. We do not have evidence that gender and knowing how to adjust Facebook privacy settings are dependent.

6.4.5 Calculator: chi-square test for two-way tables

TI-83/84: Entering data into a two-way table

1. Hit 2ND x^{-1} (i.e. MATRIX).

2. Right arrow to EDIT.

3. Hit 1 or ENTER to select matrix A.

4. Enter the dimensions by typing #rows, ENTER, #columns, ENTER.

5. Enter the data from the two-way table.

TI-83/84: Chi-square test of homogeneity and independence

Use STAT, TESTS, X^2-Test.

1. First enter two-way table data as described in the previous box.

2. Choose STAT.

3. Right arrow to TESTS.

4. Down arrow and choose C:X^2-Test.

5. Down arrow, choose Calculate, and hit ENTER, which returns

    X^2 chi-square test statistic
    p   p-value
    df   degrees of freedom

TI-83/84: Finding the expected counts

1. First enter two-way table data as described previously.

2. Carry out the chi-square test of homogeneity or independence as described in previous box.

3. Hit 2ND x^{-1} (i.e. MATRIX).

4. Right arrow to EDIT.

5. Hit 2 to see matrix B.

   This matrix contains the expected counts.

Casio fx-9750GII: Chi-square test of homogeneity and independence

1. Navigate to STAT (MENU button, then hit the 2 button or select STAT).

2. Choose the TEST option (F3 button).

3. Choose the CHI option (F3 button).

4. Choose the 2WAY option (F2 button).

5. Enter the data into a matrix:

   - Hit ▷MAT (F2 button).
   - Navigate to a matrix you would like to use (e.g. Mat C) and hit EXE.
   - Specify the matrix dimensions: m is for rows, n is for columns.
   - Enter the data.
   - Return to the test page by hitting EXIT twice.

6. Enter the Observed matrix that was used by hitting MAT (F1 button) and the matrix letter (e.g. C).

7. Enter the Expected matrix where the expected values will be stored (e.g. D).

8. Hit the EXE button, which returns
   x^2 chi-square test statistic
   p   p-value
   df   degrees of freedom

9. To see the expected values of the matrix, go to ▷MAT (F6 button) and select the corresponding matrix.

		Congress		Total
	Obama	Democrats	Republicans	
Approve	842	736	541	2119
Disapprove	616	646	842	2104
Total	1458	1382	1383	4223

Table 6.21: Pew Research poll results of a March 2012 poll.

⊙ **Guided Practice 6.56** Use Table 6.21 and a calculator to find the expected values and the X^2 statistic, *df*, and p-value for the corresponding chi-square test.[33]

---

[33]First create a 2 × 3 matrix ith the data. The final summaries should be $X^2 = 106.4$, p-value = $8.06 \times 10^{-24} \approx 0$, and *df* = 2. Below is the matrix of expected values:

	Obama	Congr. Dem.	Congr. Rep.
Approve	731.59	693.45	693.96
Disapprove	726.41	688.55	689.04

6.5   Exercises

6.5.1   Inference for a single proportion

6.1   Vegetarian college students. Suppose that 8% of college students are vegetarians. Determine if the following statements are true or false, and explain your reasoning.

(a) The distribution of the sample proportions of vegetarians in random samples of size 60 is approximately normal since $n \geq 30$.

(b) The distribution of the sample proportions of vegetarian college students in random samples of size 50 is right skewed.

(c) A random sample of 125 college students where 12% are vegetarians would be considered unusual.

(d) A random sample of 250 college students where 12% are vegetarians would be considered unusual.

(e) The standard error would be reduced by one-half if we increased the sample size from 125 to 250.

6.2   Young Americans, Part I. About 77% of young adults think they can achieve the American dream. Determine if the following statements are true or false, and explain your reasoning.[34]

(a) The distribution of sample proportions of young Americans who think they can achieve the American dream in samples of size 20 is left skewed.

(b) The distribution of sample proportions of young Americans who think they can achieve the American dream in random samples of size 40 is approximately normal since $n \geq 30$.

(c) A random sample of 60 young Americans where 85% think they can achieve the American dream would be considered unusual.

(d) A random sample of 120 young Americans where 85% think they can achieve the American dream would be considered unusual.

6.3   Orange tabbies. Suppose that 90% of orange tabby cats are male. Determine if the following statements are true or false, and explain your reasoning.

(a) The distribution of sample proportions of random samples of size 30 is left skewed.

(b) Using a sample size that is 4 times as large will reduce the standard error of the sample proportion by one-half.

(c) The distribution of sample proportions of random samples of size 140 is approximately normal.

(d) The distribution of sample proportions of random samples of size 280 is approximately normal.

6.4   Young Americans, Part II. About 25% of young Americans have delayed starting a family due to the continued economic slump. Determine if the following statements are true or false, and explain your reasoning.[35]

(a) The distribution of sample proportions of young Americans who have delayed starting a family due to the continued economic slump in random samples of size 12 is right skewed.

(b) In order for the distribution of sample proportions of young Americans who have delayed starting a family due to the continued economic slump to be approximately normal, we need random samples where the sample size is at least 40.

(c) A random sample of 50 young Americans where 20% have delayed starting a family due to the continued economic slump would be considered unusual.

(d) A random sample of 150 young Americans where 20% have delayed starting a family due to the continued economic slump would be considered unusual.

(e) Tripling the sample size will reduce the standard error of the sample proportion by one-third.

---

[34]A. Vaughn. "Poll finds young adults optimistic, but not about money". In: *Los Angeles Times* (2011).
[35]Demos.org. "The State of Young America: The Poll". In: (2011).

6.5 Prop 19 in California. In a 2010 Survey USA poll, 70% of the 119 respondents between the ages of 18 and 34 said they would vote in the 2010 general election for Prop 19, which would change California law to legalize marijuana and allow it to be regulated and taxed. At a 95% confidence level, this sample has an 8% margin of error. Based on this information, determine if the following statements are true or false, and explain your reasoning.[36]

(a) We are 95% confident that between 62% and 78% of the California voters in this sample support Prop 19.

(b) We are 95% confident that between 62% and 78% of all California voters between the ages of 18 and 34 support Prop 19.

(c) If we considered many random samples of 119 California voters between the ages of 18 and 34, and we calculated 95% confidence intervals for each, 95% of them will include the true population proportion of 18-34 year old Californians who support Prop 19.

(d) In order to decrease the margin of error to 4%, we would need to quadruple (multiply by 4) the sample size.

(e) Based on this confidence interval, there is sufficient evidence to conclude that a majority of California voters between the ages of 18 and 34 support Prop 19.

6.6 2010 Healthcare Law. On June 28, 2012 the U.S. Supreme Court upheld the much debated 2010 healthcare law, declaring it constitutional. A Gallup poll released the day after this decision indicates that 46% of 1,012 Americans agree with this decision. At a 95% confidence level, this sample has a 3% margin of error. Based on this information, determine if the following statements are true or false, and explain your reasoning.[37]

(a) We are 95% confident that between 43% and 49% of Americans in this sample support the decision of the U.S. Supreme Court on the 2010 healthcare law.

(b) We are 95% confident that between 43% and 49% of Americans support the decision of the U.S. Supreme Court on the 2010 healthcare law.

(c) If we considered many random samples of 1,012 Americans, and we calculated the sample proportions of those who support the decision of the U.S. Supreme Court, 95% of those sample proportions will be between 43% and 49%.

(d) The margin of error at a 90% confidence level would be higher than 3%.

6.7 Fireworks on July 4th. In late June 2012, Survey USA published results of a survey stating that 56% of the 600 randomly sampled Kansas residents planned to set off fireworks on July 4th. Determine the margin of error for the 56% point estimate using a 95% confidence level.[38]

6.8 Elderly drivers. In January 2011, The Marist Poll published a report stating that 66% of adults nationally think licensed drivers should be required to retake their road test once they reach 65 years of age. It was also reported that interviews were conducted on 1,018 American adults, and that the margin of error was 3% using a 95% confidence level.[39]

(a) Verify the margin of error reported by The Marist Poll.

(b) Based on a 95% confidence interval, does the poll provide convincing evidence that *more than* 70% of the population think that licensed drivers should be required to retake their road test once they turn 65?

---

[36]Survey USA, Election Poll #16804, data collected July 8-11, 2010.
[37]Gallup, Americans Issue Split Decision on Healthcare Ruling, data collected June 28, 2012.
[38]Survey USA, News Poll #19333, data collected on June 27, 2012.
[39]Marist Poll, Road Rules: Re-Testing Drivers at Age 65?, March 4, 2011.

6.9   Life after college. We are interested in estimating the proportion of graduates at a mid-sized university who found a job within one year of completing their undergraduate degree. Suppose we conduct a survey and find out that 348 of the 400 randomly sampled graduates found jobs. The graduating class under consideration included over 4500 students.

(a) Describe the population parameter of interest. What is the value of the point estimate of this parameter?

(b) Check if the conditions for constructing a confidence interval based on these data are met.

(c) Calculate a 95% confidence interval for the proportion of graduates who found a job within one year of completing their undergraduate degree at this university, and interpret it in the context of the data.

(d) What does "95% confidence" mean?

(e) Now calculate a 99% confidence interval for the same parameter and interpret it in the context of the data.

(f) Compare the widths of the 95% and 99% confidence intervals. Which one is wider? Explain.

6.10   Life rating in Greece. Greece has faced a severe economic crisis since the end of 2009. A Gallup poll surveyed 1,000 randomly sampled Greeks in 2011 and found that 25% of them said they would rate their lives poorly enough to be considered "suffering".[40]

(a) Describe the population parameter of interest. What is the value of the point estimate of this parameter?

(b) Check if the conditions required for constructing a confidence interval based on these data are met.

(c) Construct a 95% confidence interval for the proportion of Greeks who are "suffering".

(d) Without doing any calculations, describe what would happen to the confidence interval if we decided to use a higher confidence level.

(e) Without doing any calculations, describe what would happen to the confidence interval if we used a larger sample.

6.11   Study abroad. A survey on 1,509 high school seniors who took the SAT and who completed an optional web survey between April 25 and April 30, 2007 shows that 55% of high school seniors are fairly certain that they will participate in a study abroad program in college.[41]

(a) Is this sample a representative sample from the population of all high school seniors in the US? Explain your reasoning.

(b) Let's suppose the conditions for inference are met. Even if your answer to part (a) indicated that this approach would not be reliable, this analysis may still be interesting to carry out (though not report). Construct a 90% confidence interval for the proportion of high school seniors (of those who took the SAT) who are fairly certain they will participate in a study abroad program in college, and interpret this interval in context.

(c) What does "90% confidence" mean?

(d) Based on this interval, would it be appropriate to claim that the majority of high school seniors are fairly certain that they will participate in a study abroad program in college?

---

[40]Gallup World, More Than One in 10 "Suffering" Worldwide, data collected throughout 2011.

[41]studentPOLL, College-Bound Students' Interests in Study Abroad and Other International Learning Activities, January 2008.

6.12 Legalization of marijuana, Part I. The 2010 General Social Survey asked 1,259 US residents: "Do you think the use of marijuana should be made legal, or not?" 48% of the respondents said it should be made legal.[42]

(a) Is 48% a sample statistic or a population parameter? Explain.

(b) Construct a 95% confidence interval for the proportion of US residents who think marijuana should be made legal, and interpret it in the context of the data.

(c) A critic points out that this 95% confidence interval is only accurate if the statistic follows a normal distribution, or if the normal model is a good approximation. Is this true for these data? Explain.

(d) A news piece on this survey's findings states, "Majority of Americans think marijuana should be legalized." Based on your confidence interval, is this news piece's statement justified?

6.13 Public option, Part I. A *Washington Post* article from 2009 reported that "support for a government-run health-care plan to compete with private insurers has rebounded from its summertime lows and wins clear majority support from the public." More specifically, the article says "seven in 10 Democrats back the plan, while almost nine in 10 Republicans oppose it. Independents divide 52 percent against, 42 percent in favor of the legislation." (6% responded with "other".) There were 819 Democrats, 566 Republicans and 783 Independents surveyed.[43]

(a) A political pundit on TV claims that a majority of Independents oppose the health care public option plan. Do these data provide strong evidence to support this statement?

(b) Would you expect a confidence interval for the proportion of Independents who oppose the public option plan to include 0.5? Explain.

6.14 The Civil War. A national survey conducted in 2011 among a simple random sample of 1,507 adults shows that 56% of Americans think the Civil War is still relevant to American politics and political life.[44]

(a) Conduct a hypothesis test to determine if these data provide strong evidence that the majority of the Americans think the Civil War is still relevant.

(b) Interpret the p-value in this context.

(c) Calculate a 90% confidence interval for the proportion of Americans who think the Civil War is still relevant. Interpret the interval in this context, and comment on whether or not the confidence interval agrees with the conclusion of the hypothesis test.

6.15 Browsing on the mobile device. A 2012 survey of 2,254 American adults indicates that 17% of cell phone owners do their browsing on their phone rather than a computer or other device.[45]

(a) According to an online article, a report from a mobile research company indicates that 38 percent of Chinese mobile web users only access the internet through their cell phones.[46] Conduct a hypothesis test to determine if these data provide strong evidence that the proportion of Americans who only use their cell phones to access the internet is different than the Chinese proportion of 38%.

(b) Interpret the p-value in this context.

(c) Calculate a 95% confidence interval for the proportion of Americans who access the internet on their cell phones, and interpret the interval in this context.

[42]National Opinion Research Center, General Social Survey, 2010.

[43]D. Balz and J. Cohen. "Most support public option for health insurance, poll finds". In: *The Washington Post* (2009).

[44]Pew Research Center Publications, Civil War at 150: Still Relevant, Still Divisive, data collected between March 30 - April 3, 2011.

[45]Pew Internet, Cell Internet Use 2012, data collected between March 15 - April 13, 2012.

[46]S. Chang. "The Chinese Love to Use Feature Phone to Access the Internet". In: *M.I.C Gadget* (2012).

6.16 Is college worth it? Part I. Among a simple random sample of 331 American adults who do not have a four-year college degree and are not currently enrolled in school, 48% said they decided not to go to college because they could not afford school.[47]

(a) A newspaper article states that only a minority of the Americans who decide not to go to college do so because they cannot afford it and uses the point estimate from this survey as evidence. Conduct a hypothesis test to determine if these data provide strong evidence supporting this statement.

(b) Would you expect a confidence interval for the proportion of American adults who decide not to go to college because they cannot afford it to include 0.5? Explain.

6.17 Taste test. Some people claim that they can tell the difference between a diet soda and a regular soda in the first sip. A researcher wanting to test this claim randomly sampled 80 such people. He then filled 80 plain white cups with soda, half diet and half regular through random assignment, and asked each person to take one sip from their cup and identify the soda as diet or regular. 53 participants correctly identified the soda.

(a) Do these data provide strong evidence that these people are able to detect the difference between diet and regular soda, in other words, are the results significantly better than just random guessing?

(b) Interpret the p-value in this context.

6.18 Is college worth it? Part II. Exercise 6.16 presents the results of a poll where 48% of 331 Americans who decide to not go to college do so because they cannot afford it.

(a) Calculate a 90% confidence interval for the proportion of Americans who decide to not go to college because they cannot afford it, and interpret the interval in context.

(b) Suppose we wanted the margin of error for the 90% confidence level to be about 1.5%. How large of a survey would you recommend?

6.19 College smokers. We are interested in estimating the proportion of students at a university who smoke. Out of a random sample of 200 students from this university, 40 students smoke.

(a) Calculate a 95% confidence interval for the proportion of students at this university who smoke, and interpret this interval in context. (Reminder: check conditions)

(b) If we wanted the margin of error to be no larger than 2% at a 95% confidence level for the proportion of students who smoke, how big of a sample would we need?

6.20 Legalize Marijuana, Part II. As discussed in Exercise 6.12, the 2010 General Social Survey reported a sample where about 48% of US residents thought marijuana should be made legal. If we wanted to limit the margin of error of a 95% confidence interval to 2%, about how many Americans would we need to survey ?

6.21 Public option, Part II. Exercise 6.13 presents the results of a poll evaluating support for the health care public option in 2009, reporting that 52% of Independents in the sample opposed the public option. If we wanted to estimate this number to within 1% with 90% confidence, what would be an appropriate sample size?

---

[47]Pew Research Center Publications, Is College Worth It?, data collected between March 15-29, 2011.

6.22 Acetaminophen and liver damage. It is believed that large doses of acetaminophen (the active ingredient in over the counter pain relievers like Tylenol) may cause damage to the liver. A researcher wants to conduct a study to estimate the proportion of acetaminophen users who have liver damage. For participating in this study, he will pay each subject $20 and provide a free medical consultation if the patient has liver damage.

(a) If he wants to limit the margin of error of his 98% confidence interval to 2%, what is the minimum amount of money he needs to set aside to pay his subjects?

(b) The amount you calculated in part (a) is substantially over his budget so he decides to use fewer subjects. How will this affect the width of his confidence interval?

6.5.2 Difference of two proportions

6.23 Social experiment, Part I. A "social experiment" conducted by a TV program questioned what people do when they see a very obviously bruised woman getting picked on by her boyfriend. On two different occasions at the same restaurant, the same couple was depicted. In one scenario the woman was dressed "provocatively" and in the other scenario the woman was dressed "conservatively". The table below shows how many restaurant diners were present under each scenario, and whether or not they intervened.

		Scenario		
		Provocative	Conservative	Total
Intervene	Yes	5	15	20
	No	15	10	25
	Total	20	25	45

Explain why the sampling distribution of the difference between the proportions of interventions under provocative and conservative scenarios does not follow an approximately normal distribution.

6.24 Heart transplant success. The Stanford University Heart Transplant Study was conducted to determine whether an experimental heart transplant program increased lifespan. Each patient entering the program was officially designated a heart transplant candidate, meaning that he was gravely ill and might benefit from a new heart. Patients were randomly assigned into treatment and control groups. Patients in the treatment group received a transplant, and those in the control group did not. The table below displays how many patients survived and died in each group.[48]

	control	treatment
alive	4	24
dead	30	45

A hypothesis test would reject the conclusion that the survival rate is the same in each group, and so we might like to calculate a confidence interval. Explain why we cannot construct such an interval using the normal approximation. What might go wrong if we constructed the confidence interval despite this problem?

---

[48]B. Turnbull et al. "Survivorship of Heart Transplant Data". In: *Journal of the American Statistical Association* 69 (1974), pp. 74–80.

6.25   Gender and color preference. A 2001 study asked 1,924 male and 3,666 female under-graduate college students their favorite color. A 95% confidence interval for the difference between the proportions of males and females whose favorite color is black ($p_{male} - p_{female}$) was calculated to be (0.02, 0.06). Based on this information, determine if the following statements are true or false, and explain your reasoning for each statement you identify as false.[49]

(a) We are 95% confident that the true proportion of males whose favorite color is black is 2% lower to 6% higher than the true proportion of females whose favorite color is black.

(b) We are 95% confident that the true proportion of males whose favorite color is black is 2% to 6% higher than the true proportion of females whose favorite color is black.

(c) 95% of random samples will produce 95% confidence intervals that include the true difference between the population proportions of males and females whose favorite color is black.

(d) We can conclude that there is a significant difference between the proportions of males and females whose favorite color is black and that the difference between the two sample proportions is too large to plausibly be due to chance.

(e) The 95% confidence interval for ($p_{female} - p_{male}$) cannot be calculated with only the information given in this exercise.

6.26   The Daily Show. A 2010 Pew Research foundation poll indicates that among 1,099 college graduates, 33% watch The Daily Show. Meanwhile, 22% of the 1,110 people with a high school degree but no college degree in the poll watch The Daily Show. A 95% confidence interval for ($p_{college\ grad} - p_{HS\ or\ less}$), where p is the proportion of those who watch The Daily Show, is (0.07, 0.15). Based on this information, determine if the following statements are true or false, and explain your reasoning if you identify the statement as false.[50]

(a) At the 5% significance level, the data provide convincing evidence of a difference between the proportions of college graduates and those with a high school degree or less who watch The Daily Show.

(b) We are 95% confident that 7% less to 15% more college graduates watch The Daily Show than those with a high school degree or less.

(c) 95% of random samples of 1,099 college graduates and 1,110 people with a high school degree or less will yield differences in sample proportions between 7% and 15%.

(d) A 90% confidence interval for ($p_{college\ grad} - p_{HS\ or\ less}$) would be wider.

(e) A 95% confidence interval for ($p_{HS\ or\ less} - p_{college\ grad}$) is (-0.15,-0.07).

6.27   Public Option, Part III. Exercise 6.13 presents the results of a poll evaluating support for the health care public option plan in 2009. 70% of 819 Democrats and 42% of 783 Independents support the public option.

(a) Calculate a 95% confidence interval for the difference between ($p_D - p_I$) and interpret it in this context. We have already checked conditions for you.

(b) True or false: If we had picked a random Democrat and a random Independent at the time of this poll, it is more likely that the Democrat would support the public option than the Independent.

---

[49]L Ellis and C Ficek. "Color preferences according to gender and sexual orientation". In: *Personality and Individual Differences* 31.8 (2001), pp. 1375–1379.

[50]The Pew Research Center, Americans Spending More Time Following the News, data collected June 8-28, 2010.

6.28 Sleep deprivation, CA vs. OR, Part I. According to a report on sleep deprivation by the Centers for Disease Control and Prevention, the proportion of California residents who reported insufficient rest or sleep during each of the preceding 30 days is 8.0%, while this proportion is 8.8% for Oregon residents. These data are based on simple random samples of 11,545 California and 4,691 Oregon residents. Calculate a 95% confidence interval for the difference between the proportions of Californians and Oregonians who are sleep deprived and interpret it in context of the data.[51]

6.29 Offshore drilling, Part I. A 2010 survey asked 827 randomly sampled registered voters in California "Do you support? Or do you oppose? Drilling for oil and natural gas off the Coast of California? Or do you not know enough to say?" Below is the distribution of responses, separated based on whether or not the respondent graduated from college.[52]

(a) What percent of college graduates and what percent of the non-college graduates in this sample do not know enough to have an opinion on drilling for oil and natural gas off the Coast of California?

(b) Conduct a hypothesis test to determine if the data provide strong evidence that the proportion of college graduates who do not have an opinion on this issue is different than that of non-college graduates.

	College Grad	
	Yes	No
Support	154	132
Oppose	180	126
Do not know	104	131
Total	438	389

6.30 Sleep deprivation, CA vs. OR, Part II. Exercise 6.28 provides data on sleep deprivation rates of Californians and Oregonians. The proportion of California residents who reported insufficient rest or sleep during each of the preceding 30 days is 8.0%, while this proportion is 8.8% for Oregon residents. These data are based on simple random samples of 11,545 California and 4,691 Oregon residents.

(a) Conduct a hypothesis test to determine if these data provide strong evidence the rate of sleep deprivation is different for the two states. (Reminder: check conditions)

(b) It is possible the conclusion of the test in part (a) is incorrect. If this is the case, what type of error was made?

6.31 Offshore drilling, Part II. Results of a poll evaluating support for drilling for oil and natural gas off the coast of California were introduced in Exercise 6.29.

	College Grad	
	Yes	No
Support	154	132
Oppose	180	126
Do not know	104	131
Total	438	389

(a) What percent of college graduates and what percent of the non-college graduates in this sample support drilling for oil and natural gas off the Coast of California?

(b) Conduct a hypothesis test to determine if the data provide strong evidence that the proportion of college graduates who support off-shore drilling in California is different than that of non-college graduates.

---

[51]CDC, Perceived Insufficient Rest or Sleep Among Adults — United States, 2008.
[52]Survey USA, Election Poll #16804, data collected July 8-11, 2010.

6.32 Full body scan, Part I. A news article reports that "Americans have differing views on two potentially inconvenient and invasive practices that airports could implement to uncover potential terrorist attacks." This news piece was based on a survey conducted among a random sample of 1,137 adults nationwide, interviewed by telephone November 7-10, 2010, where one of the questions on the survey was "Some airports are now using 'full-body' digital x-ray machines to electronically screen passengers in airport security lines. Do you think these new x-ray machines should or should not be used at airports?" Below is a summary of responses based on party affiliation.[53]

| | | Party Affiliation | | |
		Republican	Democrat	Independent
Answer	Should	264	299	351
	Should not	38	55	77
	Don't know/No answer	16	15	22
	Total	318	369	450

(a) Conduct an appropriate hypothesis test evaluating whether there is a difference in the proportion of Republicans and Democrats who think the full-body scans should be applied in airports. Assume that all relevant conditions are met.

(b) The conclusion of the test in part (a) may be incorrect, meaning a testing error was made. If an error was made, was it a Type 1 or a Type 2 Error? Explain.

6.33 Sleep deprived transportation workers. The National Sleep Foundation conducted a survey on the sleep habits of randomly sampled transportation workers and a control sample of non-transportation workers. The results of the survey are shown below.[54]

| | | Transportation Professionals | | | |
	Control	Pilots	Truck Drivers	Train Operators	Bux/Taxi/Limo Drivers
Less than 6 hours of sleep	35	19	35	29	21
6 to 8 hours of sleep	193	132	117	119	131
More than 8 hours	64	51	51	32	58
Total	292	202	203	180	210

Conduct a hypothesis test to evaluate if these data provide evidence of a difference between the proportions of truck drivers and non-transportation workers (the control group) who get less than 6 hours of sleep per day, i.e. are considered sleep deprived.

[53]S. Condon. "Poll: 4 in 5 Support Full-Body Airport Scanners". In: *CBS News* (2010).
[54]National Sleep Foundation, 2012 Sleep in America Poll: Transportation Workers' Sleep, 2012.

6.34 Prenatal vitamins and Autism. Researchers studying the link between prenatal vitamin use and autism surveyed the mothers of a random sample of children aged 24 - 60 months with autism and conducted another separate random sample for children with typical development. The table below shows the number of mothers in each group who did and did not use prenatal vitamins during the three months before pregnancy (periconceptional period).[55]

		Autism		
		Autism	Typical development	Total
Periconceptional	No vitamin	111	70	181
prenatal vitamin	Vitamin	143	159	302
	Total	254	229	483

(a) State appropriate hypotheses to test for independence of use of prenatal vitamins during the three months before pregnancy and autism.

(b) Complete the hypothesis test and state an appropriate conclusion. (Reminder: verify any necessary conditions for the test.)

(c) A New York Times article reporting on this study was titled "Prenatal Vitamins May Ward Off Autism". Do you find the title of this article to be appropriate? Explain your answer. Additionally, propose an alternative title.[56]

6.35 HIV in sub-Saharan Africa. In July 2008 the US National Institutes of Health announced that it was stopping a clinical study early because of unexpected results. The study population consisted of HIV-infected women in sub-Saharan Africa who had been given single dose Nevaripine (a treatment for HIV) while giving birth, to prevent transmission of HIV to the infant. The study was a randomized comparison of continued treatment of a woman (after successful childbirth) with Nevaripine vs. Lopinavir, a second drug used to treat HIV. 240 women participated in the study; 120 were randomized to each of the two treatments. Twenty-four weeks after starting the study treatment, each woman was tested to determine if the HIV infection was becoming worse (an outcome called *virologic failure*). Twenty-six of the 120 women treated with Nevaripine experienced virologic failure, while 10 of the 120 women treated with the other drug experienced virologic failure.[57]

(a) Create a two-way table presenting the results of this study.

(b) State appropriate hypotheses to test for independence of treatment and virologic failure.

(c) Complete the hypothesis test and state an appropriate conclusion. (Reminder: verify any necessary conditions for the test.)

6.36 Diabetes and unemployment. A 2012 Gallup poll surveyed Americans about their employment status and whether or not they have diabetes. The survey results indicate that 1.5% of the 47,774 employed (full or part time) and 2.5% of the 5,855 unemployed 18-29 year olds have diabetes.[58]

(a) Create a two-way table presenting the results of this study.

(b) State appropriate hypotheses to test for independence of incidence of diabetes and employment status.

(c) The sample difference is about 1%. If we completed the hypothesis test, we would find that the p-value is very small (about 0), meaning the difference is statistically significant. Use this result to explain the difference between statistically significant and practically significant findings.

---

[55]R.J. Schmidt et al. "Prenatal vitamins, one-carbon metabolism gene variants, and risk for autism". In: *Epidemiology* 22.4 (2011), p. 476.

[56]R.C. Rabin. "Patterns: Prenatal Vitamins May Ward Off Autism". In: *New York Times* (2011).

[57]S. Lockman et al. "Response to antiretroviral therapy after a single, peripartum dose of nevirapine". In: *Obstetrical & gynecological survey* 62.6 (2007), p. 361.

[58]Gallup Wellbeing, Employed Americans in Better Health Than the Unemployed, data collected Jan. 2, 2011 - May 21, 2012.

6.37 Active learning. A teacher wanting to increase the active learning component of her course is concerned about student reactions to changes she is planning to make. She conducts a survey in her class, asking students whether they believe more active learning in the classroom (hands on exercises) instead of traditional lecture will helps improve their learning. She does this at the beginning and end of the semester and wants to evaluate whether students' opinions have changed over the semester. Can she used the methods we learned in this chapter for this analysis? Explain your reasoning.

6.38 An apple a day keeps the doctor away. A physical education teacher at a high school wanting to increase awareness on issues of nutrition and health asked her students at the beginning of the semester whether they believed the expression "an apple a day keeps the doctor away", and 40% of the students responded yes. Throughout the semester she started each class with a brief discussion of a study highlighting positive effects of eating more fruits and vegetables. She conducted the same apple-a-day survey at the end of the semester, and this time 60% of the students responded yes. Can she used the methods we learned in this chapter for this analysis? Explain your reasoning.

6.5.3 Testing for goodness of fit using chi-square

6.39 True or false, Part I. Determine if the statements below are true or false. For each false statement, suggest an alternative wording to make it a true statement.

(a) The chi-square distribution, just like the normal distribution, has two parameters, mean and standard deviation.

(b) The chi-square distribution is always right skewed, regardless of the value of the degrees of freedom parameter.

(c) The chi-square statistic is always positive.

(d) As the degrees of freedom increases, the shape of the chi-square distribution becomes more skewed.

6.40 True or false, Part II. Determine if the statements below are true or false. For each false statement, suggest an alternative wording to make it a true statement.

(a) As the degrees of freedom increases, the mean of the chi-square distribution increases.

(b) If you found $X^2 = 10$ with $df = 5$ you would fail to reject H_0 at the 5% significance level.

(c) When finding the p-value of a chi-square test, we always shade the tail areas in both tails.

(d) As the degrees of freedom increases, the variability of the chi-square distribution decreases.

6.41 Open source textbook. A professor using an open source introductory statistics book predicts that 60% of the students will purchase a hard copy of the book, 25% will print it out from the web, and 15% will read it online. At the end of the semester he asks his students to complete a survey where they indicate what format of the book they used. Of the 126 students, 71 said they bought a hard copy of the book, 30 said they printed it out from the web, and 25 said they read it online.

(a) State the hypotheses for testing if the professor's predictions were inaccurate.

(b) How many students did the professor expect to buy the book, print the book, and read the book exclusively online?

(c) This is an appropriate setting for a chi-square test. List the conditions required for a test and verify they are satisfied.

(d) Calculate the chi-squared statistic, the degrees of freedom associated with it, and the p-value.

(e) Based on the p-value calculated in part (d), what is the conclusion of the hypothesis test? Interpret your conclusion in this context.

6.42 Evolution vs. creationism. A Gallup Poll released in December 2010 asked 1019 adults living in the Continental U.S. about their belief in the origin of humans. These results, along with results from a more comprehensive poll from 2001 (that we will assume to be exactly accurate), are summarized in the table below:[59]

Response	*Year*	
	2010	2001
Humans evolved, with God guiding (1)	38%	37%
Humans evolved, but God had no part in process (2)	16%	12%
God created humans in present form (3)	40%	45%
Other / No opinion (4)	6%	6%

(a) Calculate the actual number of respondents in 2010 that fall in each response category.

(b) State hypotheses for the following research question: have beliefs on the origin of human life changed since 2001?

(c) Calculate the expected number of respondents in each category under the condition that the null hypothesis from part (b) is true.

(d) Conduct a chi-square test and state your conclusion. (Reminder: verify conditions.)

6.43 Rock-paper-scissors. Rock-paper-scissors is a hand game played by two or more people where players choose to sign either rock, paper, or scissors with their hands. For your AP Statistics class project, you want to evaluate whether players choose between these three options randomly, or if certain options are favored above others. You ask two friends to play rock-paper-scissors and count the times each option is played. The following table summarizes the data:

Rock	Paper	Scissors
43	21	35

Use these data to evaluate whether players choose between these three options randomly, or if certain options are favored above others. Make sure to clearly outline each step of your analysis, and interpret your results in context of the data and the research question.

6.44 Barking deer. Microhabitat factors associated with forage and bed sites of barking deer in Hainan Island, China were examined from 2001 to 2002. In this region woods make up 4.8% of the land, cultivated grass plot makes up 14.7%, and deciduous forests makes up 39.6%. Of the 426 sites where the deer forage, 4 were categorized as woods, 16 as cultivated grassplot, and 61 as deciduous forests. The table below summarizes these data.[60]

Woods	Cultivated grassplot	Deciduous forests	Other	Total
4	16	67	345	426

(a) Write the hypotheses for testing if barking deer prefer to forage in certain habitats over others.

(b) What type of test can we use to answer this research question?

(c) Check if the assumptions and conditions required for this test are satisfied.

(d) Do these data provide convincing evidence that barking deer prefer to forage in certain habitats over others? Conduct an appropriate hypothesis test to answer this research question.

Photo by Shrikant Rao
(http://flic.kr/p/4Xjdkk)
CC BY 2.0 license

[59]Four in 10 Americans Believe in Strict Creationism, December 17, 2010, www.gallup.com/poll/145286/Four-Americans-Believe-Strict-Creationism.aspx.

[60]Liwei Teng et al. "Forage and bed sites characteristics of Indian muntjac (Muntiacus muntjak) in Hainan Island, China". In: *Ecological Research* 19.6 (2004), pp. 675–681.

6.5.4   Homogeneity and independence in two-way tables

6.45   Quitters. Does being part of a support group affect the ability of people to quit smoking? A county health department enrolled 300 smokers in a randomized experiment. 150 participants were assigned to a group that used a nicotine patch and met weekly with a support group; the other 150 received the patch and did not meet with a support group. At the end of the study, 40 of the participants in the patch plus support group had quit smoking while only 30 smokers had quit in the other group.

(a) Create a two-way table presenting the results of this study.

(b) Answer each of the following questions under the null hypothesis that being part of a support group does not affect the ability of people to quit smoking, and indicate whether the expected values are higher or lower than the observed values.

    i. How many subjects in the "patch + support" group would you expect to quit?

    ii. How many subjects in the "patch only" group would you expect to not quit?

6.46   Full body scan, Part II. The table below summarizes a data set we first encountered in Exercise 6.32 regarding views on full-body scans and political affiliation. The differences in each political group may be due to chance. Complete the following computations under the null hypothesis of independence between an individual's party affiliation and his support of full-body scans. It may be useful to first add on an extra column for row totals before proceeding with the computations.

		Party Affiliation		
		Republican	Democrat	Independent
	Should	264	299	351
Answer	Should not	38	55	77
	Don't know/No answer	16	15	22
	Total	318	369	450

(a) How many Republicans would you expect to not support the use of full-body scans?

(b) How many Democrats would you expect to support the use of full-body scans?

(c) How many Independents would you expect to not know or not answer?

6.47   Offshore drilling, Part III. The table below summarizes a data set we first encountered in Exercise 6.29 that examines the responses of a random sample of college graduates and non-graduates on the topic of oil drilling. Complete a chi-square test for these data to check whether there is a statistically significant difference in responses from college graduates and non-graduates.

	College Grad	
	Yes	No
Support	154	132
Oppose	180	126
Do not know	104	131
Total	438	389

6.48 Coffee and Depression. Researchers conducted a study investigating the relationship between caffeinated coffee consumption and risk of depression in women. They collected data on 50,739 women free of depression symptoms at the start of the study in the year 1996, and these women were followed through 2006. The researchers used questionnaires to collect data on caffeinated coffee consumption, asked each individual about physician-diagnosed depression, and also asked about the use of antidepressants. The table below shows the distribution of incidences of depression by amount of caffeinated coffee consumption.[61]

		Caffeinated coffee consumption					
		≤ 1 cup/week	2-6 cups/week	1 cup/day	2-3 cups/day	≥ 4 cups/day	Total
Clinical	Yes	670	373	905	564	95	2,607
depression	No	11,545	6,244	16,329	11,726	2,288	48,132
	Total	12,215	6,617	17,234	12,290	2,383	50,739

(a) What type of test is appropriate for evaluating if there is an association between coffee intake and depression?

(b) Write the hypotheses for the test you identified in part (a).

(c) Calculate the overall proportion of women who do and do not suffer from depression.

(d) Identify the expected count for the highlighted cell, and calculate the contribution of this cell to the test statistic, i.e. $(Observed - Expected)^2 / Expected$.

(e) The test statistic is $X^2 = 20.93$. What is the p-value?

(f) What is the conclusion of the hypothesis test?

(g) One of the authors of this study was quoted on the NYTimes as saying it was "too early to recommend that women load up on extra coffee" based on just this study.[62] Do you agree with this statement? Explain your reasoning.

6.49 Shipping holiday gifts. A December 2010 survey asked 500 randomly sampled Los Angeles residents which shipping carrier they prefer to use for shipping holiday gifts. The table below shows the distribution of responses by age group as well as the expected counts for each cell (shown in parentheses).

		Age						Total
		18-34		35-54		55+		
	USPS	72	(81)	97	(102)	76	(62)	245
	UPS	52	(53)	76	(68)	34	(41)	162
Shipping Method	FedEx	31	(21)	24	(27)	9	(16)	64
	Something else	7	(5)	6	(7)	3	(4)	16
	Not sure	3	(5)	6	(5)	4	(3)	13
	Total	165		209		126		500

(a) State the null and alternative hypotheses for testing for independence of age and preferred shipping method for holiday gifts among Los Angeles residents.

(b) Are the conditions for inference using a chi-square test satisfied?

---

[61]M. Lucas et al. "Coffee, caffeine, and risk of depression among women". In: *Archives of internal medicine* 171.17 (2011), p. 1571.

[62]A. O'Connor. "Coffee Drinking Linked to Less Depression in Women". In: *New York Times* (2011).

6.50 How's it going? The American National Election Studies (ANES) collects data on voter attitudes and intentions as well as demographic information. In this question we will focus on two variables from the 2012 ANES dataset:[63]

- region (levels: Northeast, North Central, South, and West), and

- whether the respondent feels things in this country are generally going in the right direction or things have pretty seriously gotten off on the wrong track.

To keep calculations simple we will work with a random sample of 500 respondents from the ANES dataset. The distribution of responses are as follows:

	Right Direction	Wrong Track	Total
Northeast	29	54	83
North Central	44	77	121
South	62	131	193
West	36	67	103
Total	171	329	500

(a) Region: According to the 2010 Census, 18% of US residents live in the Northeast, 22% live in the North Central region, 37% live in the South, and 23% live in the West. Evaluate whether the ANES sample is representative of the population distribution of US residents. Make sure to clearly state the hypotheses, check conditions, calculate the appropriate test statistic and the p-value, and make your conclusion in context of the data. Also comment on what your conclusion says about whether or not this sample can be considered to be representative.

(b) Region and direction:

   (i) We would like to evaluate the relationship between region and feeling about the country's direction. What is the response variable and what is the explanatory variable?

   (ii) What are the hypotheses for evaluating this relationship?

   (iii) Complete the hypothesis test and interpret your results in context of the data and the research question.

---

[63]The American National Election Studies (ANES). The ANES 2012 Time Series Study [dataset]. Stanford University and the University of Michigan [producers].

Chapter 7

Inference for numerical data

Chapter 5 introduced a framework for statistical inference based on confidence intervals and hypotheses. Chapter 6 summarized inference procedures for categorical data (counts and proportions). In this chapter, we focus on inference procedures for numerical data and we encounter several new point estimates and scenarios. In each case, the inference ideas remain the same:

1. Determine which point estimate or test statistic is useful.

2. Identify an appropriate distribution for the point estimate or test statistic.

3. Apply the ideas from Chapter 5 using the distribution from step 2.

Each section in Chapter 7 explores a new situation: a single mean (7.1), the mean of differences (7.2), the difference between means (7.3); and the comparison of means across multiple groups (7.4).

7.1 Inference for a single mean with the t-distribution

When certain conditions are satisfied, the sampling distribution associated with a sample mean or difference of two sample means is nearly normal. However, this becomes more complex when the sample size is small, where *small* here typically means a sample size smaller than 30 observations. For this reason, we'll use a new distribution called the t-distribution that will often work for both small and large samples of numerical data.

7.1.1 Using the z-distribution for inference when μ is unknown and σ is known

We have seen in Section 4.2 that the distribution of a sample mean is normal if the population is normal or if the sample size is at least 30. In these problems, we used the population mean and population standard deviation to find a Z-score. However, in the case of inference, the parameters will be unknown. In rare circumstances we may know the standard deviation of a population, even though we do not know its mean. For example, in some industrial process, the mean may be known to shift over time, while the standard deviation of the process remains the same. In these cases, we can use the normal model as the basis for our inference procedures. We use $\bar{x}$ as our point estimate for μ and the SD formula calculated in Section 4.2: $SD = \frac{\sigma}{\sqrt{n}}$.

$$\text{CI: } \bar{x} \ \pm \ Z^* \frac{\sigma}{\sqrt{n}} \qquad\qquad Z = \frac{\bar{x} - \text{null value}}{\frac{\sigma}{\sqrt{n}}}$$

What happens if we do not know the population standard deviation σ, as is usually the case? The best we can do is use the sample standard deviation, denoted by s, to estimate the population standard deviation.

$$SE = \frac{s}{\sqrt{n}}$$

However, when we do this we run into a problem: when carrying out our inference procedures we will be trying to estimate *two* quantities: both the mean and the standard deviation. Looking at the SD and SE formulas, we can make some important observations that will give us a hint as to what will happen when we use s instead of σ.

- For a given population, σ is a fixed number and does not vary.
- s, the standard deviation of a sample, will vary from one sample to the next and will not be exactly equal to σ.
- The larger the sample size n, the better the estimate s will tend to be for σ.

For this reason, the normal model still works well when the sample size is larger than about 30. For smaller sample sizes, we run into a problem: our estimate of s, which is used to compute the standard error, isn't as reliable and tends to add more variability to our estimate of the mean. It is this extra variability that leads us to a new distribution: the t-distribution.

7.1.2   Introducing the t-distribution

When we use the sample standard deviation s in place of the population standard deviation σ to standardize the sample mean, we get an entirely new distribution - one that is similar to the normal distribution, but has greater spread. This distribution is known as the t-distribution. A t-distribution, shown as a solid line in Figure 7.1, has a bell shape. However, its tails are thicker than the normal model's. This means observations are more likely to fall beyond two standard deviations from the mean than under the normal distribution.[1] These extra thick tails are exactly the correction we need to resolve the problem of a poorly estimated standard deviation.

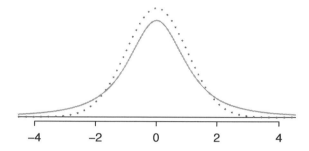

Figure 7.1: Comparison of a t-distribution (solid line) and a normal distribution (dotted line).

The t-distribution, always centered at zero, has a single parameter: degrees of freedom. The **degrees of freedom (df)** describe the precise form of the bell-shaped t-distribution. Several t-distributions are shown in Figure 7.2. When there are more degrees of freedom, the t-distribution looks very much like the standard normal distribution.

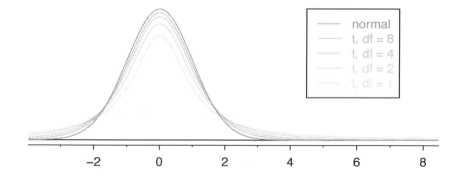

Figure 7.2: The larger the degrees of freedom, the more closely the t-distribution resembles the standard normal model.

Degrees of freedom (df)

The degrees of freedom describe the shape of the t-distribution. The larger the degrees of freedom, the more closely the distribution approximates the normal model.

---

[1] The standard deviation of the t-distribution is actually a little more than 1. However, it is useful to always think of the t-distribution as having a standard deviation of 1 in all of our applications.

When the degrees of freedom is about 30 or more, the t-distribution is nearly indistinguishable from the normal distribution. In Section 7.1.3, we relate degrees of freedom to sample size.

We will find it very useful to become familiar with the t-distribution, because it plays a very similar role to the normal distribution during inference for numerical data. We use a t-**table**, partially shown in Table 7.3, in place of the normal probability table for numerical data when the population standard deviation is unknown, especially when the sample size is small. A larger table is presented in Appendix B.3.

	one tail	0.100	0.050	0.025	0.010	0.005
df	1	3.078	6.314	12.71	31.82	63.66
	2	1.886	2.920	4.303	6.965	9.925
	3	1.638	2.353	3.182	4.541	5.841
	$\vdots$	$\vdots$	$\vdots$	$\vdots$	$\vdots$	$\vdots$
	17	1.333	1.740	2.110	2.567	2.898
	18	**1.330**	**1.734**	**2.101**	**2.552**	**2.878**
	19	1.328	1.729	2.093	2.539	2.861
	20	1.325	1.725	2.086	2.528	2.845
	$\vdots$	$\vdots$	$\vdots$	$\vdots$	$\vdots$	$\vdots$
	1000	1.282	1.646	1.962	2.330	2.581
	∞	1.282	1.645	1.960	2.326	2.576
Confidence level C		80%	90%	95%	98%	99%

Table 7.3: An abbreviated look at the t-table. Each row represents a different t-distribution. The columns describe the cutoffs for specific tail areas. The row with $df = 18$ has been **highlighted**.

Each row in the t-table represents a t-distribution with different degrees of freedom. The columns correspond to tail probabilities. For instance, if we know we are working with the t-distribution with $df = 18$, we can examine row 18, which is **highlighted** in Table 7.3. If we want the value in this row that identifies the cutoff for an upper tail of 10%, we can look in the column where *one tail* is 0.100. This cutoff is 1.33. If we had wanted the cutoff for the lower 10%, we would use -1.33. Just like the normal distribution, all t-distributions are symmetric.

● **Example 7.1** What proportion of the t-distribution with 18 degrees of freedom falls below -2.10?

Just like a normal probability problem, we first draw the picture in Figure 7.4 and shade the area below -2.10. To find this area, we identify the appropriate row: $df = 18$. Then we identify the column containing the absolute value of -2.10; it is the third column. Because we are looking for just one tail, we examine the top line of the table, which shows that a one tail area for a value in the third row corresponds to 0.025. About 2.5% of the distribution falls below -2.10. In the next example we encounter a case where the exact T value is not listed in the table.

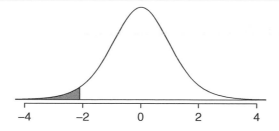

Figure 7.4: The *t*-distribution with 18 degrees of freedom. The area below -2.10 has been shaded.

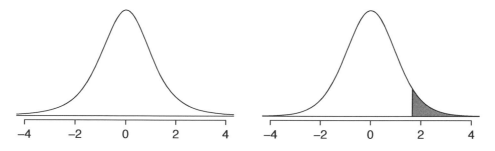

Figure 7.5: Left: The *t*-distribution with 20 degrees of freedom, with the area above 1.65 shaded. Right: The *t*-distribution with 2 degrees of freedom, with the area further than 3 units from 0 shaded.

● **Example 7.2** For the *t*-distribution with 18 degrees of freedom, what percent of the curve is contained between -1.330 and +1.330?

Using row $df = 18$, we find 1.330 in the table. The area in each tail is 0.100 for a total of 0.200, which leaves 0.800 in the middle between -1.33 and +1.33. This corresponds to a confidence level of 80%.

● **Example 7.3** For the *t*-distribution with 3 degrees of freedom, as shown in the left panel of Figure 7.5, what should the value of $t^\star$ be so that 95% of the area of the curve falls between $-t^\star$ and $+t^\star$?

We can look at the column in the *t*-table that says 95% at the bottom row for the confidence level and trace it up to row $df = 3$ to find that $t^\star = 3.182$.

● **Example 7.4** A *t*-distribution with 20 degrees of freedom is shown in the right panel of Figure 7.5. Estimate the proportion of the distribution falling above 1.65.

We identify the row in the *t*-table using the degrees of freedom: $df = 20$. Then we look for 1.65; it is not listed. It falls between the first and second columns. Since these values bound 1.65, their tail areas will bound the tail area corresponding to 1.65. We identify the one tail area of the first and second columns, 0.050 and 0.10, and we conclude that between 5% and 10% of the distribution is more than 1.65 standard deviations above the mean. If we like, we can identify the precise area using statistical software: 0.0573.

When the desired degrees of freedom is not listed on the table, choose a conservative value: round the degrees of freedom down, i.e. move *up* to the previous row listed. Another options is to use a calculator or statistical software to get a precise value.

7.1.3   The *t*-distribution and the standard error of a mean

When estimating the mean and standard deviation from a small sample, the *t*-distribution is a more accurate tool than the normal model. This is true for both small and large samples.

TIP: When to use the *t*-distribution

Use the *t*-distribution for inference of the sample mean when observations are independent and nearly normal. You may relax the nearly normal condition as the sample size increases. For example, the data distribution may be moderately skewed when the sample size is at least 30.

To proceed with the *t*-distribution for inference about a single mean, we must check two conditions.

Independence of observations. We verify this condition just as we did before. We collect a simple random sample from less than 10% of the population, or if it was an experiment or random process, we carefully check to the best of our abilities that the observations were independent.

n $\geq$ 30 or observations come from a nearly normal distribution. We can easily check if the sample size is at least 30. If it is not, then this second condition requires more care. We often (i) take a look at a graph of the data, such as a dot plot or box plot, for obvious departures from the normal model, and (ii) consider whether any previous experiences alert us that the data may not be nearly normal.

When examining a sample mean and estimated standard deviation from a sample of n independent and nearly normal observations, we use a *t*-distribution with $n - 1$ degrees of freedom (*df*). For example, if the sample size was 19, then we would use the *t*-distribution with $df = 19 - 1 = 18$ degrees of freedom and proceed exactly as we did in Chapter 5, except that *now we use the t-table*.

The *t*-distribution and the SE of a mean

In general, when the population mean is uknown, the population standard deviation will also be unknown. When this is the case, we estimate the population standard deviation with the sample standard deviation and we use SE instead of SD.

$$SE_{\bar{x}} = \frac{s}{\sqrt{n}}$$

When we use the sample standard deviation, we use the *t*-distribution with $df = n - 1$ degrees of freedom instead of the normal distribution.

7.1.4   The normality condition

When the sample size n is at least 30, the Central Limit Theorem tells us that we do not have to worry too much about skew in the data. When this is not true, we need verify that the observations come from a nearly normal distribution. In some cases, this may be known, such as if the population is the heights of adults.

What do we do, though, if the population is not known to be approximately normal AND the sample size is small? We must look at the distribution of the data and check for excessive skew.

> **Caution: Checking the normality condition**
> We should exercise caution when verifying the normality condition for small samples. It is important to not only examine the data but also think about where the data come from. For example, ask: would I expect this distribution to be symmetric, and am I confident that outliers are rare?

You may relax the normality condition as the sample size goes up. If the sample size is 10 or more, slight skew is not problematic. Once the sample size hits about 30, then moderate skew is reasonable. Data with strong skew or outliers require a more cautious analysis.

7.1.5   One sample t-intervals

Dolphins are at the top of the oceanic food chain, which causes dangerous substances such as mercury to concentrate in their organs and muscles. This is an important problem for both dolphins and other animals, like humans, who occasionally eat them.

Figure 7.6: A Risso's dolphin.

Photo by Mike Baird (www.bairdphotos.com). CC BY 2.0 license.

Here we identify a confidence interval for the average mercury content in dolphin muscle using a sample of 19 Risso's dolphins from the Taiji area in Japan.[2] The data are

---

[2]Taiji was featured in the movie *The Cove*, and it is a significant source of dolphin and whale meat in Japan. Thousands of dolphins pass through the Taiji area annually, and we will assume these 19 dolphins represent a simple random sample from those dolphins. Data reference: Endo T and Haraguchi K. 2009. High mercury levels in hair samples from residents of Taiji, a Japanese whaling town. Marine Pollution Bulletin 60(5):743-747.

summarized in Table 7.7. The minimum and maximum observed values can be used to evaluate whether or not there are obvious outliers or skew.

n	$\bar{x}$	s	minimum	maximum
19	4.4	2.3	1.7	9.2

Table 7.7: Summary of mercury content in the muscle of 19 Risso's dolphins from the Taiji area. Measurements are in μg/wet g (micrograms of mercury per wet gram of muscle).

● **Example 7.5** Are the independence and normality conditions satisfied for this data set?

The observations are a simple random sample and consist of less than 10% of the population, therefore independence is reasonable. To check the normality condition of the population, we would like to graph the data from the sample. However, we do not have all of the data. Instead, we will have to look at the summary statistics provided in Table 7.7. These summary statistics do not suggest any skew or outliers; all observations are within 2.5 standard deviations of the mean. Based on this evidence, the normality assumption seems reasonable.

In the normal model, we used $z^\star$ and the standard deviation to determine the width of a confidence interval. We revise the confidence interval formula slightly when using the t-distribution:

$$\bar{x} \pm t^\star_{df} SE$$

$t^\star_{df}$

Multiplication factor for t-interval

The sample mean is computed just as before: $\bar{x} = 4.4$. In place of the standard deviation of $\bar{x}$, we use the standard error of $\bar{x}$: $SE_{\bar{x}} = s/\sqrt{n} = 0.528$.

The value $t^\star_{df}$ is a cutoff we obtain based on the confidence level and the t-distribution with df degrees of freedom. Before determining this cutoff, we will first need the degrees of freedom.

Degrees of freedom for a single sample

If the sample has n observations and we are examining a single mean, then we use the t-distribution with $df = n - 1$ degrees of freedom.

In our current example, we should use the t-distribution with $df = 19 - 1 = 18$ degrees of freedom. Then identifying $t^\star_{18}$ is similar to how we found $z^\star$.

- For a 95% confidence interval, we want to find the cutoff $t^\star_{18}$ such that 95% of the t-distribution is between $-t^\star_{18}$ and $t^\star_{18}$.

- We look in the t-table on page 314, find the column with 95% along the bottom row and then the row with 18 degrees of freedom: $t^\star_{18} = 2.10$.

Generally the value of $t^\star_{df}$ is slightly larger than what we would get under the normal model with $z^\star$.

Finally, we can substitute all our values into the confidence interval equation to create the 95% confidence interval for the average mercury content in muscles from Risso's dolphins that pass through the Taiji area:

$$\bar{x} \ \pm \ t^{\star}_{18} SE$$
$$4.4 \ \pm \ 2.10 \times 0.528 \quad df = 18$$
$$(3.29 \ , \ 5.51)$$

We are 95% confident the true average mercury content of muscles in Risso's dolphins is between 3.29 and 5.51 μg/wet gram. This is above the Japanese regulation level of 0.4 μg/wet gram.

Finding a t-interval for the mean

Based on a sample of n independent and nearly normal observations, a confidence interval for the population mean is

$$\bar{x} \ \pm \ t^{\star}_{df} SE \qquad df = n - 1$$

where $\bar{x}$ is the sample mean, $t^{\star}_{df}$ corresponds to the confidence level and degrees of freedom, and SE is the standard error given by $\frac{s}{\sqrt{n}}$.

Constructing a confidence interval for a mean

1. State the name of the CI being used.

   - 1-sample t-interval.

2. Verify conditions.

   - A simple random sample.
   - Population is known to be normal OR $n \geq 30$ OR graph of sample is approximately symmetric with no outliers, making the assumption that the population is normal a reasonable one.

3. Plug in the numbers and write the interval in the form

   $$\text{point estimate} \ \pm \ t^{\star} \times \text{SE of estimate}$$

   - The point estimate is $\bar{x}$
   - $df = n - 1$
   - Plug in a critical value $t^{\star}$ using the t-table at row$= n - 1$
   - Use $SE = \frac{s}{\sqrt{n}}$
   - Evaluate the CI and write in the form (_ , _)

4. Interpret the interval: "We are [XX]% confident that the true average of [...] is between [...] and [...]."

5. State the conclusion to the original question.

⊙ **Guided Practice 7.6**   The FDA's webpage provides some data on mercury content of fish.[3] Based on a sample of 15 croaker white fish (Pacific), a sample mean and standard deviation were computed as 0.287 and 0.069 ppm (parts per million), respectively. The 15 observations ranged from 0.18 to 0.41 ppm. We will assume these observations are independent. Construct an appropriate 95% confidence interval for the true average mercury content of croaker white fish (Pacific). Is there evidence that the average mercury content is greater than 0.275 ppm?[4]

7.1.6   Choosing a sample size when estimating a mean

Many companies are concerned about rising healthcare costs. A company may estimate certain health characteristics of its employees, such as blood pressure, to project its future cost obligations. However, it might be too expensive to measure the blood pressure of every employee at a large company, and the company may choose to take a sample instead.

● **Example 7.7**   Blood pressure oscillates with the beating of the heart, and the systolic pressure is defined as the peak pressure when a person is at rest. The average systolic blood pressure for people in the U.S. is about 130 mmHg with a standard deviation of about 25 mmHg. How large of a sample is necessary to estimate the average systolic blood pressure with a margin of error of 4 mmHg using a 95% confidence level?

$$ME_{95\%} = 1.96 \times \frac{\sigma_{employee}}{\sqrt{n}}$$

The challenge in this case is to find the sample size n so that this margin of error is less than or equal to $m = 4$, which we write as an inequality:-1mm

$$1.96 \times \frac{\sigma_{employee}}{\sqrt{n}} \leq 4$$

To proceed and solve for n, we substitute the best estimate we have for $\sigma_{employee}$: 25.

$$1.96 \times \frac{25}{\sqrt{n}} \leq 4$$

$$1.96 \times \frac{25}{4} \leq \sqrt{n}$$

$$\left(1.96 \times \frac{25}{4}\right)^2 \leq n$$

$$150.06 \leq n$$

$$n = 151$$

The minimum sample size that meets the condition is 151. We round up because the sample size must be an integer and it must be *greater than or equal to* 150.06.

---

[3]www.fda.gov/food/foodborneillnesscontaminants/metals/ucm115644.htm

[4]The interval called for in this problem is a 1-sample *t*-interval. We will assume that the sample was random. n is small, but there are no obvious outliers; all observations are within 2 standard deviations of the mean. If there is skew, it is not evident. Therefore we do not have reason to believe the mercury content in the population is not nearly normal in this type of fish. We can now identify and calculate the necessary quantities. The point estimate is the sample average, which is 0.287. The standard error: $SE = \frac{0.069}{\sqrt{15}} = 0.0178$. Degrees of freedom: $df = n - 1 = 14$. Using the *t*-table, we identify $t_{14}^\star = 2.145$. The confidence interval is given by: $0.287 \pm 2.145 \times 0.0178 \rightarrow (0.249, 0.325)$. We are 95% confident that the true average mercury content of croaker white fish (Pacific) is between 0.249 and 0.325 ppm. Because the interval contains 0.275 as well as value less than 0.275, we do not have evidence that the true *average* mercury content is greater than 0.275, even though our sample average was 0.287.

A potentially controversial part of Example 7.7 is the use of the U.S. standard deviation for the employee standard deviation. Usually the standard deviation is not known. In such cases, it is reasonable to review scientific literature or market research to make an educated guess about the standard deviation.

Margin of error for a sample mean

The margin of error for a sample mean is similar to the formula we add and subtract in a confidence interval:

$$ME = z^{\star} \frac{\sigma}{\sqrt{n}}$$

The value $z^{\star}$ is chosen to correspond to the desired confidence level, and σ is the standard deviation associated with the population.

Identify a sample size for a particular margin of error

To estimate the necessary sample size to achieve a margin of error of m, we require the margin of error ME to be less than or equal to m:

$$z^{\star} \frac{\sigma}{\sqrt{n}} \leq m$$

We solve for the sample size, n.

Sample size computations are helpful in planning data collection, and they require careful forethought.

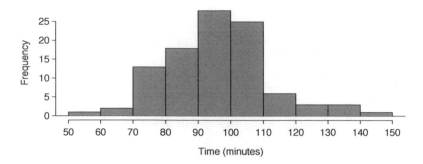

Figure 7.8: Histogram of `time` for a single sample of size 100.

7.1.7 Hypothesis testing for a mean

Is the typical US runner getting faster or slower over time? We consider this question in the context of the Cherry Blossom Run, comparing runners in 2006 and 2012. Technological advances in shoes, training, and diet might suggest runners would be faster in 2012. An opposing viewpoint might say that with the average body mass index on the rise, people tend to run slower. In fact, all of these components might be influencing run time.

The average time for all runners who finished the Cherry Blossom Run in 2006 was 93.29 minutes (93 minutes and about 17 seconds). We want to determine using data from 100 participants in the 2012 Cherry Blossom Run whether runners in this race are getting faster or slower, versus the other possibility that there has been no change.

⊙ **Guided Practice 7.8**   What are appropriate hypotheses for this context?[5]

⊙ **Guided Practice 7.9**   The data come from a simple random sample from less than 10% of all participants, so the observations are independent. However, should we be worried about skew in the data? A histogram of the times is shown in Figure 7.8. [6]

With independence satisfied and skew not a concern, we can proceed with performing a hypothesis test using the t-distribution.

⊙ **Guided Practice 7.10**   The sample mean and sample standard deviation are 95.61 and 15.78 minutes, respectively. Recall that the sample size is 100. What is the p-value for the test, and what is your conclusion?[7]

---

[5]H_0: The average 10 mile run time in 2012 was the same as in 2006 (93.29 minutes). $\mu = 93.29$. H_A: The average 10 mile run time for 2012 was *different* than 93.29 minutes. $\mu \neq 93.29$.

[6]Since the sample size 100 is greater than 30, we do not need to worry about slight skew in the data.

[7]With the conditions satisfied for the t-distribution, we can compute the standard error ($SE = 15.78/\sqrt{100} = 1.58$ and the T score: $T = \frac{95.61 - 93.29}{1.58} = 1.47$. For $df = 100 - 1 = 99$, we would find a p-value between 0.10 and 0.20 (two-sided!). Because the p-value is greater than 0.05, we do not reject the null hypothesis. That is, the data do not provide strong evidence that the average run time for the Cherry Blossom Run in 2012 is any different than the 2006 average.

Hypothesis test for a mean

1. State the name of the test being used.

   - 1-sample t-test.

2. Verify conditions.

   - Data come from a simple random sample.
   - Population is known to be normal OR $n \geq 30$ OR graph of data is approximately symmetric with no outliers, making the assumption that population is normal a reasonable one.

3. Write the hypotheses in plain language, then set them up in mathematical notation.

   - $H_0 : \mu = \mu_0$
   - $H_0 : \mu \neq$ or $<$ or $> \mu_0$

4. Identify the significance level α.

5. Calculate the test statistic and df: $T = \frac{\text{point estimate} - \text{null value}}{\text{SE of estimate}}$

   - The point estimate is $\bar{x}$
   - $SE = \frac{s}{\sqrt{n}}$
   - $df = n - 1$

6. Find the p-value, compare it to α, and state whether to reject or not reject the null hypothesis.

7. Write the conclusion in the context of the question.

⊙ **Guided Practice 7.11** Recall the example about the mercury content in croaker white fish (Pacific). Based on a sample of 15, a sample mean and standard deviation were computed as 0.287 and 0.069 ppm (parts per million), respectively. Carry out an appropriate test to determine if 0.25 is a reasonable value for the average mercury content.[8]

---

[8]We should carry out a 1-sample t-test. The conditions have already been checked. $H_0 : \mu = 0.25$; The true average mercury content is 0.25 ppm. $H_A : \mu \neq 0.25$; The true average mercury content is not equal to 0.25 ppm. Let $\alpha = 0.05$. $SE = \frac{0.069}{\sqrt{15}} = 0.0178$. $T = \frac{0.287 - 0.25}{0.0178} = 2.07$ $df = 15 - 1 = 14$. p-value $= 0.057 > 0.05$, so we do not reject the null hypothesis. We do not have sufficient evidence that the average mercury content in croaker white fish is not 0.25.

● **Example 7.12** Recall that the 95% confidence interval for the average mercuy content in croaker white fish was (0.249, 0.325). Discuss whether the conclusion of the test of hypothesis is consistent or inconsistent with the conclusion of the hypothesis test.

It is consistent because 0.25 is located (just barely) inside the confidence interval, so it is a reasonable value. Our hypothesis test did not reject the hypothesis that $\mu = 0.25$, implying that it is a plausible value. Note, though, that the hypothesis test did not *prove* that $\mu = .25$. A hypothesis cannot prove that the mean is a specific value. It can only find evidence that it is not a specific value. Note also that the p-value was close to the cutoff of 0.05. This is because the value 0.25 was close to edge of the confidence interval.

7.1.8 Calculator: the 1-sample t-test and t-interval

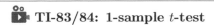 **TI-83/84: 1-sample t-test**

Use STAT, TESTS, T-Test.

1. Choose STAT.

2. Right arrow to TESTS.

3. Down arrow and choose 2:T-Test.

4. Choose Data if you have all the data or Stats if you have the mean and standard deviation.

5. Let μ_0 be the null or hypothesized value of μ.

   - If you choose Data, let List be L1 or the list in which you entered your data (don't forget to enter the data!) and let Freq be 1.

   - If you choose Stats, enter the mean, SD, and sample size.

6. Choose $\neq$, $<$, or $>$ to correspond to H$_A$.

7. Choose Calculate and hit ENTER, which returns:

t	t statistic	Sx	the sample standard deviation
p	p-value	n	the sample size
x̄	the sample mean		

Casio fx-9750GII: 1-sample *t*-test

1. Navigate to STAT (MENU button, then hit the 2 button or select STAT).

2. If necessary, enter the data into a list.

3. Choose the TEST option (F3 button).

4. Choose the t option (F2 button).

5. Choose the 1-S option (F1 button).

6. Choose either the Var option (F2) or enter the data in using the List option.

7. Specify the test details:

   - Specify the sidedness of the test using the F1, F2, and F3 keys.
   - Enter the null value, $\mu 0$.
   - If using the Var option, enter the summary statistics. If using List, specify the list and leave Freq values at 1.

8. Hit the EXE button, which returns

	alternative hypothesis	$\bar{x}$	sample mean
t	T statistic	sx	sample standard deviation
p	p-value	n	sample size

TI-83/84: 1-sample *t*-interval

Use STAT, TESTS, TInterval.

1. Choose STAT.

2. Right arrow to TESTS.

3. Down arrow and choose 8:TInterval.

4. Choose Data if you have all the data or Stats if you have the mean and standard deviation.

   - If you choose Data, let List be L1 or the list in which you entered your data (don't forget to enter the data!) and let Freq be 1.
   - If you choose Stats, enter the mean, SD, and sample size.

5. Let C-Level be the desired confidence level.

6. Choose Calculate and hit ENTER, which returns:

(__, __)	the confidence interval
$\bar{x}$	the sample mean
Sx	the sample SD
n	the sample size

▶ Casio fx-9750GII: 1-sample t-interval

1. Navigate to STAT (MENU button, then hit the 2 button or select STAT).

2. If necessary, enter the data into a list.

3. Choose the INTR option (F3 button), t (F2 button), and 1-S (F1 button).

4. Choose either the Var option (F2) or enter the data in using the List option.

5. Specify the interval details:

   - Confidence level of interest for C-Level.
   - If using the Var option, enter the summary statistics. If using List, specify the list and leave Freq value at 1.

6. Hit the EXE button, which returns

Left, Right	ends of the confidence interval
x̄	sample mean
sx	sample standard deviation
n	sample size

⊙ **Guided Practice 7.13**   The average time for all runners who finished the Cherry Blossom Run in 2006 was 93.29 minutes. In 2012, the average time for 100 randomly selected participants was 95.61, with a standard deviation of 15.78 minutes. Use a calculator to find the T statistic and p-value for the appropriate test to see if the average time for the participants in 2012 is different than it was in 2006.[9]

⊙ **Guided Practice 7.14**   Use a calculator to find a 95% confidence interval for the true improvement of students that use this SAT prep company.[10]

---

[9]Let μ_0 be 93.29. Choose $\neq$ to correspond to H_A. $T = 1.47$, $df = 99$, and p-value= 0.14.

[10]The interval is $(105.21, 166.59)$.

7.2 Inference for paired data

Are textbooks actually cheaper online? Here we compare the price of textbooks at UCLA's bookstore and prices at Amazon.com. Seventy-three UCLA courses were randomly sampled in Spring 2010, representing less than 10% of all UCLA courses.[11] A portion of this data set is shown in Table 7.9.

	dept	course	ucla	amazon	diff
1	Am Ind	C170	27.67	27.95	-0.28
2	Anthro	9	40.59	31.14	9.45
3	Anthro	135T	31.68	32.00	-0.32
4	Anthro	191HB	16.00	11.52	4.48
⋮	⋮	⋮	⋮	⋮	⋮
72	Wom Std	M144	23.76	18.72	5.04
73	Wom Std	285	27.70	18.22	9.48

Table 7.9: Six cases of the `textbooks` data set.

7.2.1 Paired observations and samples

Each textbook has two corresponding prices in the data set: one for the UCLA bookstore and one for Amazon. Therefore, each textbook price from the UCLA bookstore has a natural correspondence with a textbook price from Amazon. When two sets of observations have this special correspondence, they are said to be **paired**.

Paired data

Two sets of observations are *paired* if each observation in one set has a special correspondence or connection with exactly one observation in the other data set.

To analyze paired data, it is often useful to look at the difference in outcomes of each pair of observations. In the `textbook` data set, we look at the difference in prices, which is represented as the `diff` variable in the `textbooks` data. Here the differences are taken as

UCLA price − Amazon price

for each book. It is important that we always subtract using a consistent order; here Amazon prices are always subtracted from UCLA prices. If this difference is positive, the UCLA price is higher. If ths difference is negative, the Amazon price is higher. If this difference is zero, the two prices are equal. A histogram of these differences is shown in Figure 7.10. Using differences between paired observations is a common and useful way to analyze paired data.

⊙ **Guided Practice 7.15** The first difference shown in Table 7.9 is computed as $27.67 - 27.95 = -0.28$. Based on the table and on the histogram of differences in Figure 7.10, which store tends to have the higher prices in the sample?[12]

---

[11]When a class had multiple books, only the most expensive text was considered.

[12]Because the most of the differences are positive, UCLA prices tend to be higher than Amazon. Note that it is important to identify the order in which the differences are taken.

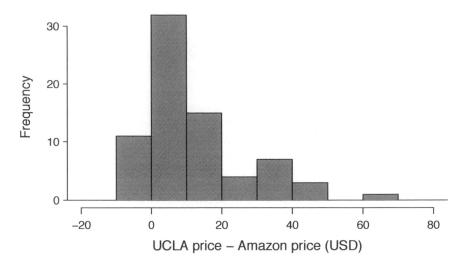

Figure 7.10: Histogram of the difference in price for each of the 73 books sampled. These data are strongly skewed.

7.2.2   Hypothesis testing for paired data

To analyze a paired data set, we use the exact same tools that we developed in the previous section. Now we apply them to the differences in the paired observations.

n_{diff}	$\bar{x}_{diff}$	s_{diff}
73	12.76	14.26

Table 7.11: Summary statistics for the price differences. There were 73 books, so there are 73 differences.

● **Example 7.16** Set up and implement a hypothesis test to determine whether, on average, there is a difference between Amazon's price for a book and the UCLA bookstore's price.

There are two scenarios: there is no difference or there is some difference in average prices. The *no difference* scenario is always the null hypothesis:

H_0: $\mu_{diff} = 0$. There is no difference in the average textbook price.

H_A: $\mu_{diff} \neq 0$. There is a difference in average prices.

The standard deviation of all of the differences in unknown, so we will use the standard deviation of the sample differences. The observations are based on a simple random sample from less than 10% of all books sold at the bookstore, so independence is reasonable; the distribution of differences, shown in Figure 7.10, is strongly skewed, but the sample size $n = 73$ is well over 30. Because all three conditions are reasonably satisfied, we can conclude the t-test is reasonable.

We compute the standard error associated with $\bar{x}_{diff}$ using the standard deviation

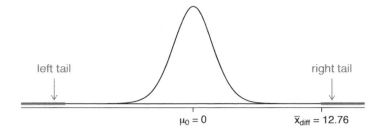

Figure 7.12: Sampling distribution for the mean difference in book prices, if the true average difference is zero.

of the differences ($s_{diff} = 14.26$) and the number of differences ($n_{diff} = 73$):

$$SE_{\bar{x}_{diff}} = \frac{s_{diff}}{\sqrt{n_{diff}}} = \frac{14.26}{\sqrt{73}} = 1.67$$

To visualize the p-value, the sampling distribution of $\bar{x}_{diff}$ is drawn as though H_0 is true, which is shown in Figure 7.12. The p-value is represented by the two (very) small tails.

To find the tail areas, we compute the test statistic, which is the T score of $\bar{x}_{diff}$ under the null condition that the actual mean difference is 0:

$$t = \frac{\bar{x}_{diff} - 0}{SE_{x_{diff}}} = \frac{12.76 - 0}{1.67} = 7.59 \qquad df = 72$$

This T score is so large it isn't even in the table, which ensures the single tail area will be 0.0002 or smaller. A calculator gives a tail area as 4.5×10^{-11}. Since the p-value corresponds to both tails in this case and the t-distribution is symmetric, the p-value can be estimated as twice the one-tail area:

$$\text{p-value} = 2 \times (\text{one tail area}) \approx 2 \times 4.5 \times 10^{-11} = 9 \times 10^{-11} \approx 0$$

Because the p-value is less than 0.05, we reject the null hypothesis. We have found convincing evidence that Amazon is, on average, cheaper than the UCLA bookstore for UCLA course textbooks.

Hypothesis test for paired data

1. State the name of the test being used.

   - Matched pairs t-test.

2. Verify conditions.

   - Paired data from a random sample or experiment.
   - Population of differences is known to be normal OR $n_{diff} \geq 30$ OR graph of sample differences is approximately symmetric with no outliers, making the assumption that population of differences is normal a reasonable one.

3. Write the hypotheses in plain language, then set them up in mathematical notation.

   - $H_0 : \mu_{diff} = 0$
   - $H_0 : \mu_{diff} \neq$ or $<$ or > 0

4. Identify the significance level α.

5. Calculate the test statistic and df: $t = \dfrac{\text{point estimate} - \text{null value}}{\text{SE of estimate}}$

   - The point estimate is $\bar{x}_{diff}$
   - Use $SE = \dfrac{s_{diff}}{\sqrt{n_{diff}}}$
   - $df = n_{diff} - 1$

6. Find the p-value and compare it to α to determine whether to reject or not reject H_0.

7. Write the conclusion in the context of the question.

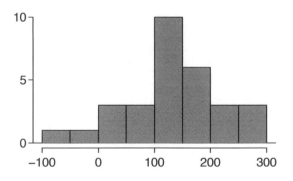

Figure 7.13: Sample distribution of: SAT score after course - SAT score before course. The distribution is approximately symmetric.

⊙ **Guided Practice 7.17** An SAT preparation company claims that its students' scores improve by over 100 points on average after their course. A consumer group would like to evaluate this claim, and they collect data on a random sample of 30 students who took the class. Each of these students took the SAT before and after taking the company's course, and so we have a difference in scores for each student. We will examine these differences $x_1 = 57$, $x_2 = 133$, ..., $x_{30} = 140$ as a sample to evaluate the company's claim. The distribution of the differences, shown in Figure 7.13, has mean 135.9 and standard deviation 82.2. Do these data provide convincing evidence to back up the company's claim? [13]

⊙ **Guided Practice 7.18** Because we rejected the null hypothesis, does this mean that taking the company's class improves student scores by more than 100 points on average?[14]

7.2.3 Confidence intervals for the mean of a difference μ_{diff}

In the previous examples, we carried out a matched pairs t-test, where the null hypothesis was that the true average of the paired differences is zero. Sometimes we want to estimate the true average of paired differences with a confidence interval, and we use a **matched pairs t-interval**. Consider again the table of data on the difference in price between UCLA and Amazon for each of the 73 books sampled.

n_{diff}	$\bar{x}_{diff}$	s_{diff}
73	12.76	14.26

Table 7.14: Summary statistics for the price differences. There were 73 books, so there are 73 differences.

We we construct a 95% confidence interval for the average price difference between books at the UCLA bookstore and books on Amazon. Conditions have already verified and the standard error computed in Example 7.16. To find the interval, identify $t^\star$. Since $df = 72$ is not on the t-table, round the df down to 60 to get a $t^\star$ of 2.00 for 95% confidence. Plugging in the $t^\star$, the point estimate, and the standard error into the confidence interval formula we get:

$$\text{point estimate} \pm t^\star SE \qquad\qquad df = n - 1$$
$$12.76 \pm 2.00 \times 1.67 \qquad\qquad df = 72$$
$$(9.42, 16.10)$$

---

[13]These are *paired data*, so we analyze the score differences with a matched pairs t-test. Conditions: This is a random sample from less than 10% of the company's students (assuming they have more than 300 former students), so the independence condition is reasonable. $n = 30 \geq 30$. This is a one-sided test. H_0: student scores do not improve by more than 100 after taking the company's course. $\mu_{diff} = 100$ H_A: students scores improve by more than 100 points on average after taking the company's course. $\mu_{diff} > 100$. Let

$$\alpha = 0.05 \qquad SE_{diff} = \frac{82.2}{\sqrt{30}} = 15.0 \qquad T = \frac{135.9 - 100}{15.0} = 2.4 \qquad \text{with } df = 29$$

p-value= $0.012 < \alpha$ so we reject the null hypothesis. The data provide convincing evidence to support the company's claim that student scores improve by more than 100 points following the class.

[14]This is an observational study, so we cannot make this causal conclusion. For instance, maybe SAT test takers tend to improve their score over time even if they don't take a special SAT class, or perhaps only the most motivated students take such SAT courses.

We are 95% confident that the UCLA bookstore is, on average, between $9.42 and $16.10 more expensive than Amazon for UCLA course books. This interval does not contain zero, so it is consistent with the fact that our earlier hypothesis test *rejected* the null hypothesis that the average difference was 0. Because our interval is entirely above 0, we have evidence that the true average difference is not 0. Unlike the test of hypothesis, though, the confidence interval tells us about how much more expensive the UCLA bookstore is.

Constructing a confidence interval for paired data

1. State the name of the CI being used.

   - Matched pairs t-interval

2. Verify conditions.

   - Paired data from a random sample or experiment.
   - Population of diferrences is known to be normal OR $n_{diff} \geq 30$ OR graph of sample differences is approximately symmetric with no outliers, making the assumption that the population of differences is normal a reasonable one.

3. Plug in the numbers and write the interval in the form

$$\text{point estimate} \pm t^\star \times \text{SE of estimate}$$

   - The point estimate of $\bar{x}_{diff}$
   - $df = n_{diff} - 1$
   - Plug in the critical value $t^\star$ using the t-table at row $n_{diff} - 1$
   - Use $SE = \frac{s_{diff}}{\sqrt{n_{diff}}}$

4. Evaluate the CI and write in the form (_ , _).

5. Interpret the interval: "We are [XX]% confident that the true mean of the differences in [...] is between [...] and [...]."

6. State the conclusion to the original question.

⊙ **Guided Practice 7.19** In the SAT preparation company example, we saw that $\bar{x}_{diff}$ was 135.9 and s_{diff} was 82.2. That is, the average change in students' scores after the class was a 135.9 point increase and the SD of the change or difference in their scores was 82.2 points. Construct a 95% confidence interval to estimate the true average change in score after taking the class. Is there evidence for the company's claim that students score an average of 100 points higher after the class?[15]

---

[15]Because this is a before and after scenario, we use a matched pairs t-interval. The conditions were verified in the previous section. The confidence interval is : $135.9 \pm 2.045(15.0) \rightarrow (105.2, 166.6)$. We can be 95% confident that the true *average* increase in scores after the prep class is between 105.2 and 166.6. Because the entire interval is above 100, there is evidence that on average students score more than 100 points higher after the course. Recall that this does not prove that the increase is *due to* the course.

⊙ **Guided Practice 7.20** The 95% confidence interval in the previous exercise was calculated as (105.2, 166.6). True or false: about 95% of the students that take the class saw an increase of at least 105.2 points.[16]

7.2.4 Calculator: the matched pairs t-test and t-interval

The matched pairs t-test and CI proceed the same way as the 1-sample t-test and confidence interval. Instead of using the data or the summary statistics from the single sample, make sure to use the data of *differences* or the summary statistics for the *differences*.

TI-83/84: matched pairs t-test

Use STAT, TESTS, T-Test.

1. Choose STAT.

2. Right arrow to TESTS.

3. Down arrow and choose 2:T-Test.

4. Choose Data if you have all the data or Stats if you have the mean and standard deviation.

5. Let μ_0 be the null or hypothesized value of μ_{diff}.

   - If you choose Data, let List be L3 or the list in which you entered the differences (don't forget to enter the differences!) and let Freq be 1.

   - If you choose Stats, enter the mean, SD, and sample size of the differences.

6. Choose $\neq$, <, or > to correspond to H_A.

7. Choose   Calculate   and   hit   ENTER,   which   returns:
t	t statistic
p	p-value
$\bar{x}$	the sample mean of the differences
Sx	the sample SD of the differences
n	the sample size of the differences

---

[16]False. This confidence interval estimates the *average* increase - not the increase of individuals. As can be seen in Figure 7.13, much greater than 5% saw an increase of less than 105.2 points. Some individuals even saw a *decrease* in their score as indicated by the negative differences.

TI-83/84: matched pairs t-interval

Use STAT, TESTS, TInterval.

1. Choose STAT.

2. Right arrow to TESTS.

3. Down arrow and choose 8:TInterval.

4. Choose Data if you have all the data or Stats if you have the mean and standard deviation.

   - If you choose Data, let List be L3 or the list in which you entered the differences (don't forget to enter the differences!) and let Freq be 1.

   - If you choose Stats, enter the mean, SD, and sample size of the differences.

5. Let C-Level be the desired confidence level.

6. Choose Calculate and hit ENTER, which returns:

(__,__)	the confidence interval for the mean of the differences
$\bar{x}$	the sample mean of the differences
Sx	the sample SD of the differences
n	the number of differences in the sample

Casio fx-9750GII: matched pairs t-test or confidence interval

1. Compute the paired differences of the observations.

2. Using the computed differences, follow the instructions for a 1-sample t-test or confidence interval.

---

	dept	ucla	amazon
1	Am Ind	27.67	27.95
2	Anthro	40.59	31.14
3	Anthro	31.68	32.00
4	Anthro	16.00	11.52
5	Art His	18.95	14.21
6	Art His	14.95	10.17
7	Asia Am	24.7	20.06

Table 7.15: A partial table of the textbooks data.

⊙ **Guided Practice 7.21**   Use the first 7 values of the 7.9 data set produced above and calculate the T score and p-value to test whether, on average, Amazon's textbook price is cheaper that UCLA's price.[17]

⊙ **Guided Practice 7.22**   Use the data from Table 7.15 to calculate a 95% confidence interval for the average difference in textbook price between Amazon and UCLA.[18]

7.3   Difference of two means using the *t*-distribution

It is also useful to be able to compare two means for small samples. For instance, a teacher might like to test the notion that two versions of an exam were equally difficult. She could do so by randomly assigning each version to students. If she found that the average scores on the exams were so different that we cannot write it off as chance, then she may want to award extra points to students who took the more difficult exam.

In a medical context, we might investigate whether embryonic stem cells can improve heart pumping capacity in individuals who have suffered a heart attack. We could look for evidence of greater heart health in the stem cell group against a control group.

In this section we use the *t*-distribution for the difference in sample means. We will again drop the minimum sample size condition and instead impose a strong condition on the distribution of the data.

7.3.1   Sampling distribution for the difference of two means

In this section we consider a difference in two population means, $\mu_1 - \mu_2$, under the condition that the data are not paired. The methods are similar in theory but different in the details. Just as with a single sample, we identify conditions to ensure a point estimate of the difference $\bar{x}_1 - \bar{x}_2$ is nearly normal. Next we introduce a formula for the standard deviation of $\bar{x}_1 - \bar{x}_2$, which allows us to apply our general tools from Section 5.

We apply these methods to two examples: participants in the 2012 Cherry Blossom Run and newborn infants. This section is motivated by questions like "Is there convincing evidence that newborns from mothers who smoke have a different average birth weight than newborns from mothers who don't smoke?"

We start by looking at the population mean and standard deviation for the run times of men and women participants in the 2009 Cherry Blossom Run. Table 7.16 summarizes these values.

	men	women
μ	87.65	102.13
σ	12.5	15.2

Table 7.16: Summary of the run time of participants in the 2009 Cherry Blossom Run.

---

[17]Create a list of the differences, and use the data or list option to perform the test. Let μ_0 be 0, and select the appropriate list. Freq should be 1, and the test sidedness should be $>$. $T = 3.076$ and p-value= 0.0109.

[18]Choose a C-Level of 0.95, and the final result should be (0.80354, 7.0507).

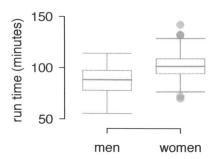

Figure 7.17: Side-by-side box plots for the sample of 2009 Cherry Blossom Run participants.

The two populations (men and women) are independent of one-another, so the data are not paired.[19] If we take two separate random samples of men and women from this race, what is the expected value for the difference in their average times? Not surprisingly, the expected value of $\bar{x}_w - \bar{x}_m$ is $\mu_1 - \mu_2$. We can quantify the variability in the point estimate, using the following formula for its standard deviation:

$$SD_{\bar{x}_w - \bar{x}_m} = \sqrt{(SD_{\bar{x}_w})^2 + (SD_{\bar{x}_m})^2}$$
$$= \sqrt{\left(\frac{\sigma_{\bar{x}_w}}{\sqrt{n_w}}\right)^2 + \left(\frac{\sigma_{\bar{x}_m}}{\sqrt{n_m}}\right)^2}$$
$$= \sqrt{\frac{\sigma_w^2}{n_w} + \frac{\sigma_m^2}{n_m}}$$

⊙ **Guided Practice 7.23**   Let's say we take a random sample of 55 women and a random sample of 45 men. Use the SD formula for the difference of two means to compute the SD for the difference in the average run time for males and females.[20]

Distribution of a difference of sample means

The sample difference of two means, $\bar{x}_1 - \bar{x}_2$, is nearly normal with mean $\mu_1 - \mu_2$ and standard deviation

$$SD_{\bar{x}_1 - \bar{x}_2} = \sqrt{\frac{\sigma_1^2}{n_1} + \frac{\sigma_2^2}{n_2}} \tag{7.24}$$

when each sample mean is nearly normal and all observations are independent. Recall that each sample mean will be nearly normal if the population is normal or if the sample size is at least 30.

---

[19]Probability theory guarantees that the difference of two independent normal random variables is also normal. Because each sample mean is nearly normal and observations in the samples are independent, we are assured the difference is also nearly normal.

[20]$\sqrt{\frac{15.2^2}{55} + \frac{12.5^2}{45}} = 2.77$

7.3.2 Point estimates and standard errors for differences of means

In the example of two exam versions, the teacher would like to evaluate whether there is convincing evidence that the difference in average scores between the two exams is not due to chance.

It will be useful to extend the t-distribution method from Section 7.1 to apply to a difference of means:

$$\bar{x}_1 - \bar{x}_2 \quad \text{as a point estimate for} \quad \mu_1 - \mu_2$$

First, we verify the small sample conditions (independence and nearly normal data) for each sample separately, then we verify that the samples are also independent. For instance, if the teacher believes students in her class are independent, the exam scores are nearly normal, and the students taking each version of the exam were independent, then we can use the t-distribution for inference on the point estimate $\bar{x}_1 - \bar{x}_2$.

The formula for the standard error of $\bar{x}_1 - \bar{x}_2$, introduced in Section 7.3.1, also applies to small samples:

$$SE_{\bar{x}_1 - \bar{x}_2} = \sqrt{SE_{\bar{x}_1}^2 + SE_{\bar{x}_2}^2} = \sqrt{\frac{s_1^2}{n_1} + \frac{s_2^2}{n_2}} \tag{7.25}$$

Because we will use the t-distribution, we will need to identify the appropriate degrees of freedom. This can be done using a calculator or computer software. An alternative technique is to use the smaller of $n_1 - 1$ and $n_2 - 1$. [21]

Using the t-distribution for a difference in means

The t-distribution can be used for inference when working with the standardized difference of two means if (1) each sample meets the conditions for using the t-distribution and (2) the samples are independent. We estimate the standard error of the difference of two means using Equation (7.25).

7.3.3 Hypothesis testing for the difference of two means

Summary statistics for each exam version are shown in Table 7.18. The teacher would like to evaluate whether this difference is so large that it provides convincing evidence that Version B was more difficult (on average) than Version A.

Version	n	$\bar{x}$	s	min	max
A	30	79.4	14	45	100
B	27	74.1	20	32	100

Table 7.18: Summary statistics of scores for each exam version.

---

[21] This technique for degrees of freedom is conservative with respect to a Type 1 Error; it is more difficult to reject the null hypothesis using this *df* method.

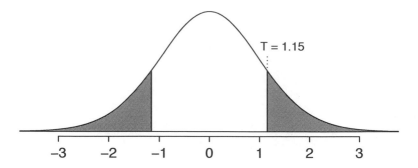

Figure 7.19: The t-distribution with 26 degrees of freedom. The shaded right tail represents values with $T \geq 1.15$. Because it is a two-sided test, we also shade the corresponding lower tail.

⊙ **Guided Practice 7.26**   Construct a two-sided hypothesis test to evaluate whether the observed difference in sample means, $\bar{x}_A - \bar{x}_B = 5.3$, might be due to chance.[22]

⊙ **Guided Practice 7.27**   To evaluate the hypotheses in Guided Practice 7.26 using the t-distribution, we must first verify assumptions. (a) Does it seem reasonable that the scores are independent within each group? (b) What about the normality condition for each group? (c) Do you think scores from the two groups would be independent of each other (i.e. the two samples are independent)?[23]

After verifying the conditions for each sample and confirming the samples are independent of each other, we are ready to conduct the test using the t-distribution. In this case, we are estimating the true difference in average test scores using the sample data, so the point estimate is $\bar{x}_A - \bar{x}_B = 5.3$. The standard error of the estimate can be calculated using Equation (7.25):

$$SE = \sqrt{\frac{s_A^2}{n_A} + \frac{s_B^2}{n_B}} = \sqrt{\frac{14^2}{30} + \frac{20^2}{27}} = 4.62$$

Finally, we construct the test statistic:

$$T = \frac{\text{point estimate} - \text{null value}}{SE} = \frac{(79.4 - 74.1) - 0}{4.62} = 1.15$$

If we have a calculator or computer handy, we can identify the degrees of freedom as 45.97. Otherwise we use the smaller of $n_1 - 1$ and $n_2 - 1$: $df = 26$.

---

[22]Because the teacher did not expect one exam to be more difficult prior to examining the test results, she should use a two-sided hypothesis test. H_0: the exams are equally difficult, on average. $\mu_A - \mu_B = 0$. H_A: one exam was more difficult than the other, on average. $\mu_A - \mu_B \neq 0$.

[23](a) It is probably reasonable to conclude the scores are independent. (b) The summary statistics suggest the data are roughly symmetric about the mean, and it doesn't seem unreasonable to suggest the data might be normal. Note that since these samples are each nearing 30, moderate skew in the data would be acceptable. (c) It seems reasonable to suppose that the samples are independent since the exams were handed out randomly.

⊙ **Guided Practice 7.28** Identify the p-value, shown in Figure 7.19. Use $df = 26$.[24]

In Guided Practice 7.28, we could have used $df = 45.97$. However, this value is not listed in the table. In such cases, we use the next lower degrees of freedom (unless the computer also provides the p-value). For example, we could have used $df = 45$ but not $df = 46$. As before, we provide a summary of the steps to perform when carrying out such a test.

Hypothesis test for the difference of two means

1. State the name of the test being used.

   - 2-sample t-test.

2. Verify conditions.

   - 2 independent random samples OR 2 randomly allocated treatments.
   - Both populations known to be normal OR $n_1 \geq 30$ and $n_2 \geq 30$ OR graphs of both samples are approximately symmetric with no outliers, making the assumption that the populations are normal a reasonable one.

3. Write the hypotheses in plain language, then set them up in mathematical notation.

   - $H_0 : \mu_1 = \mu_2$ or $\mu_1 - \mu_2 = 0$
   - $H_0 : \mu_1 \neq$ or $<$ or $> \mu_2$

4. Identify the significance level α.

5. Calculate the test statistic and df: $T = \frac{\text{point estimate} - \text{null value}}{\text{SE of estimate}}$

   - The point estimate is $\bar{x}_1 - \bar{x}_2$
   - $SE = \sqrt{\frac{s_1^2}{n_1} + \frac{s_2^2}{n_2}}$
   - Find and record the df from a calculator.

6. Find the p-value and compare it to α to determine whether to reject or not reject H_0.

7. Write the conclusion in the context of the question.

---

[24]We examine row $df = 26$ in the t-table. Because this value is smaller than the value in the left column, the p-value is larger than 0.200 (two tails!). Because the p-value is so large, we do not reject the null hypothesis. That is, the data do not convincingly show that one exam version is more difficult than the other, and the teacher should not be convinced that she should add points to the Version B exam scores.

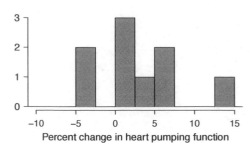 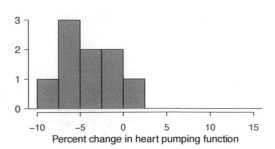

Figure 7.21: Histograms for both the embryonic stem cell group and the control group. Higher values are associated with greater improvement. We don't see any evidence of skew in these data; however, it is worth noting that skew would be difficult to detect with such a small sample.

	n	$\bar{x}$	s
ESCs	9	3.50	5.17
control	9	-4.33	2.76

Table 7.20: Summary statistics for the embryonic stem cell data set.

● **Example 7.29**   Do embryonic stem cells (ESCs) help improve heart function following a heart attack? Table 7.20 contains summary statistics for an experiment to test ESCs in sheep that had a heart attack. Each of these sheep was randomly assigned to the ESC or control group, and the change in their hearts' pumping capacity was measured. A positive value generally corresponds to increased pumping capacity, which suggests a stronger recovery. The sample data is graphed in Figure 7.21. Use the given information and an appropriate an appopriate statistical test to answer the research question.

We will carry out a 2-sample t-test. The first condition is met because it is stated that there were two randomly allocated treatments. For the second condition, we must look at a graphs of the data. The data are very limited, so we can only check for obvious outliers in the raw data in Figure 7.21. Since the distributions are (very) roughly symmetric, we will assume the populations are approximately normal.

H_0: $\mu_{esc} - \mu_{control} = 0$. The stem cells do not improve heart pumping function.
H_A: $\mu_{esc} - \mu_{control} > 0$. The stem cells do improve heart pumping function.

Let $\alpha = 0.05$. Now we compute the sample difference, the standard error for that point estimate, and the test statistic:

$$\bar{x}_{esc} - \bar{x}_{control} = 7.83 \quad SE = \sqrt{\frac{5.17^2}{9} + \frac{2.76^2}{9}} = 1.95 \quad T = \frac{7.83 - 0}{1.95} = 4.01$$

Using a calculator, $df = 12.2$ and p-value $= 8.4 \times 10^{-4}$. The p-value is much less than 0.05, so we reject the null hypothesis. The data provide convincing evidence that embryonic stem cells improve the heart's pumping function in sheep that have suffered a heart attack.

7.3.4 Confidence intervals for $\mu_1 - \mu_2$

The results from the previous section provided evidence that ESCs actually help improve the pumping function of the heart. But how large is this improvement? To answer this question, we can use a confidence interval.

Confidence intervals take the form

$$\text{point estimate} \pm \text{critical value} \times SE$$

Using the point estimate and the SE calculated in the previous section, we get the general form of a confidence interval for a difference in means, $\mu_1 - \mu_2$.

$$(\bar{x}_1 - \bar{x}_2) \pm t^\star \sqrt{\frac{s_1^2}{n_1} + \frac{s_2^2}{n_2}}$$

⊙ **Guided Practice 7.30** In Example 7.29, you found that the point estimate, $\bar{x}_{esc} - \bar{x}_{control} = 7.83$, has a standard error of 1.95. Using $df = 8$, create a 99% confidence interval for the improvement due to ESCs.[25]

---

[25]We know the point estimate, 7.83, and the standard error, 1.95. We also verified the conditions for using the t-distribution in Example 7.29. Thus, we only need identify $t_8^\star$ to create a 99% confidence interval: $t_8^\star = 3.36$. The 99% confidence interval for the improvement from ESCs is given by

$$\text{point estimate} \pm t^\star SE$$
$$7.83 \pm 3.36 \times 1.95 \quad df = 8$$
$$(1.33 , 14.43)$$

That is, we are 99% confident that the true improvement in heart pumping function is somewhere between 1.33% and 14.43%.

Constructing a confidence interval for the difference of two means

1. State the name of the CI being used.

    - 2-sample t-interval.

2. Verify conditions.

    - 2 independent random samples OR 2 randomly allocated treatments.
    - Both populations are known to be normal OR n_1 and $n_2 \geq 30$ OR graphs of both samples are approximately symmetric with no outliers, i.e. the assumption that the populations are normal is reasonable.

3. Plug in the numbers and write the interval in the form

$$\text{point estimate } \pm t^\star \times \text{SE of estimate}$$

    - The point estimate is $\bar{x}_1 - \bar{x}_2$
    - Find and record the df using a calculator
    - Find the critical value $t^\star$ using the t-table at row $= df$ (round down to nearest integer)
    - Use $SE = \sqrt{\frac{s_1^2}{n_1} + \frac{s_2^2}{n_2}}$

4. Evaluate the CI and write in the form (_ , _).

5. Interpret the interval: "We are [XX]% confident that the true difference in the mean of [...] is between [...] and [...]."

6. State the conclusion to the original question.

An instructor decided to run two slight variations of the same exam. Prior to passing out the exams, she shuffled the exams together to ensure each student received a random version. Summary statistics for how students performed on these two exams are shown in Table 7.22. Anticipating complaints from students who took Version B, she would like to evaluate whether the difference observed in the groups is so large that it provides convincing evidence that Version B was more difficult (on average) than Version A.

Version	n	$\bar{x}$	s	min	max
A	30	79.4	14	45	100
B	27	74.1	20	32	100

Table 7.22: Summary statistics of scores for each exam version.

● **Example 7.31** Construct a 90% confidence interval for the difference in average scores. At this confidence level, is there evidence that one test was more difficult than the other?

We have two randomly allocated treatments (tests) and the scores for both groups do not show excessive skew, so we can assume that the population distributions are approximately normal. The point estimate is $\bar{x}_A - \bar{x}_B = 5.3$. The standard error of the estimate can be calculated as

$$SE = \sqrt{\frac{14^2}{30} + \frac{20^2}{27}} = 4.62$$

A calculator gives the degrees of freedom as 45.97. The confidence interval is given by $5.3 \pm 1.684(4.62) \rightarrow (-2.5, 13.1)$. Because the interval contains both positive and negative values the data do not convincingly show that one exam version is more difficult than the other, and the teacher should not be convinced that she should add points to the Version B exam scores.

7.3.5 Calculator: the 2-sample t-test and t-interval

▶ TI-83/84: 2-sample t-test

Use STAT, TESTS, 2-SampTTest.

1. Choose STAT.

2. Right arrow to TESTS.

3. Choose 4:2-SampTTest.

4. Choose Data if you have all the data or Stats if you have the means and standard deviations.

   - If you choose Data, let List1 be L1 or the list that contains sample 1 and let List2 be L2 or the list that contains sample 2 (don't forget to enter the data!). Let Freq1 and Freq2 be 1.
   - If you choose Stats, enter the mean, SD, and sample size for sample 1 and for sample 2

5. Choose $\neq$, $<$, or $>$ to correspond to H$_A$.

6. Let Pooled be NO.

7. Choose Calculate and hit ENTER, which returns:

t	t statistic	Sx1	SD of sample 1
p	p-value	Sx2	SD of sample 2
df	degrees of freedom	n1	size of sample 1
$\bar{x}_1$	mean of sample 1	n2	size of sample 2
$\bar{x}_2$	mean of sample 2		

📽 Casio fx-9750GII: 2-sample t-test

1. Navigate to STAT (MENU button, then hit the 2 button or select STAT).

2. If necessary, enter the data into a list.

3. Choose the TEST option (F3 button).

4. Choose the t option (F2 button).

5. Choose the 2-S option (F2 button).

6. Choose either the Var option (F2) or enter the data in using the List option.

7. Specify the test details:

   - Specify the sidedness of the test using the F1, F2, and F3 keys.
   - If using the Var option, enter the summary statistics for each group. If using List, specify the lists and leave Freq values at 1.
   - Choose whether to pool the data or not.

8. Hit the EXE button, which returns

$\mu1 - \mu2$	alt. hypothesis	$\bar{x}1, \bar{x}2$	sample means
t	t statistic	sx1, sx2	sample standard deviations
p	p-value	n1, n2	sample sizes
df	degrees of freedom		

📽 TI-83/84: 2-sample t-interval

Use STAT, TESTS, 2-SampTInt.

1. Choose STAT.

2. Right arrow to TESTS.

3. Down arrow and choose 0:2-SampTTInt.

4. Choose Data if you have all the data or Stats if you have the means and standard deviations.

   - If you choose Data, let List1 be L1 or the list that contains sample 1 and let List2 be L2 or the list that contains sample 2 (don't forget to enter the data!). Let Freq1 and Freq2 be 1.
   - If you choose Stats, enter the mean, SD, and sample size for sample 1 and for sample 2.

5. Let C-Level be the desired confidence level and let Pooled be No.

6. Choose Calculate and hit ENTER, which returns:

(__,__)	the confidence interval	Sx1	SD of sample 1
df	degrees of freedom	Sx2	SD of sample 2
$\bar{x}_1$	mean of sample 1	n1	size of sample 1
$\bar{x}_2$	mean of sample 2	n2	size of sample 2

	n	$\bar{x}$	s
ESCs	9	3.50	5.17
control	9	-4.33	2.76

Table 7.23: Summary statistics for the embryonic stem cell data set.

Casio fx-9750GII: 2-sample t-interval

1. Navigate to STAT (MENU button, then hit the 2 button or select STAT).

2. If necessary, enter the data into a list.

3. Choose the INTR option (F4 button).

4. Choose the t option (F2 button).

5. Choose the 2-S option (F2 button).

6. Choose either the Var option (F2) or enter the data in using the List option.

7. Specify the test details:

   - Confidence level of interest for C-Level.
   - If using the Var option, enter the summary statistics for each group. If using List, specify the lists and leave Freq values at 1.
   - Choose whether to pool the data or not.

8. Hit the EXE button, which returns

Left, Right	ends of the confidence interval
df	degrees of freedom
$\bar{x}1$, $\bar{x}2$	sample means
sx1, sx2	sample standard deviations
n1, n2	sample sizes

⊙ **Guided Practice 7.32** Use the data from the ESC experiment shown in Table 7.23 and a calculator to find the appropriate degrees of freedom and construct a 90% confidence interval.[26]

⊙ **Guided Practice 7.33** Use the data from the ESC example and a calculator to find an appropriate statistic, degrees of freedom, and p-value for a two-sided hypothesis test.[27]

---

[26]The interval is $(4.3543, 11.307)$ with $df = 12.2$.

[27]$T = 4.008$, $df = 12.2$, and p-value$= 0.00168$.

7.4 Comparing many means with ANOVA (special topic)

Sometimes we want to compare means across many groups. We might initially think to do pairwise comparisons; for example, if there were three groups, we might be tempted to compare the first mean with the second, then with the third, and then finally compare the second and third means for a total of three comparisons. However, this strategy can be treacherous. If we have many groups and do many comparisons, it is likely that we will eventually find a difference just by chance, even if there is no difference in the populations.

In this section, we will learn a new method called **analysis of variance (ANOVA)** and a new test statistic called F. ANOVA uses a single hypothesis test to check whether the means across many groups are equal:

H_0: The mean outcome is the same across all groups. In statistical notation, $\mu_1 = \mu_2 = \cdots = \mu_k$ where μ_i represents the mean of the outcome for observations in category i.

H_A: At least one mean is different.

Generally we must check three conditions on the data before performing ANOVA:

- the observations are independent within and across groups,

- the data within each group are nearly normal, and

- the variability across the groups is about equal.

When these three conditions are met, we may perform an ANOVA to determine whether the data provide strong evidence against the null hypothesis that all the μ_i are equal.

● **Example 7.34** College departments commonly run multiple lectures of the same introductory course each semester because of high demand. Consider a statistics department that runs three lectures of an introductory statistics course. We might like to determine whether there are statistically significant differences in first exam scores in these three classes (A, B, and C). Describe appropriate hypotheses to determine whether there are any differences between the three classes.

The hypotheses may be written in the following form:

H_0: The average score is identical in all lectures. Any observed difference is due to chance. Notationally, we write $\mu_A = \mu_B = \mu_C$.

H_A: The average score varies by class. We would reject the null hypothesis in favor of the alternative hypothesis if there were larger differences among the class averages than what we might expect from chance alone.

Strong evidence favoring the alternative hypothesis in ANOVA is described by unusually large differences among the group means. We will soon learn that assessing the variability of the group means relative to the variability among individual observations within each group is key to ANOVA's success.

● **Example 7.35** Examine Figure 7.24. Compare groups I, II, and III. Can you visually determine if the differences in the group centers is due to chance or not? Now compare groups IV, V, and VI. Do these differences appear to be due to chance?

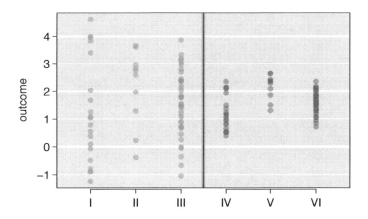

Figure 7.24: Side-by-side dot plot for the outcomes for six groups.

Any real difference in the means of groups I, II, and III is difficult to discern, because the data within each group are very volatile relative to any differences in the average outcome. On the other hand, it appears there are differences in the centers of groups IV, V, and VI. For instance, group V appears to have a higher mean than that of the other two groups. Investigating groups IV, V, and VI, we see the differences in the groups' centers are noticeable because those differences are large *relative to the variability in the individual observations within each group.*

7.4.1 Is batting performance related to player position in MLB?

We would like to discern whether there are real differences between the batting performance of baseball players according to their position: outfielder (OF), infielder (IF), designated hitter (DH), and catcher (C). We will use a data set called `bat10`, which includes batting records of 327 Major League Baseball (MLB) players from the 2010 season. Six of the 327 cases represented in `bat10` are shown in Table 7.25, and descriptions for each variable are provided in Table 7.26. The measure we will use for the player batting performance (the outcome variable) is on-base percentage (OBP). The on-base percentage roughly represents the fraction of the time a player successfully gets on base or hits a home run.

	name	team	position	AB	H	HR	RBI	AVG	OBP
1	I Suzuki	SEA	OF	680	214	6	43	0.315	0.359
2	D Jeter	NYY	IF	663	179	10	67	0.270	0.340
3	M Young	TEX	IF	656	186	21	91	0.284	0.330
⋮	⋮	⋮	⋮	⋮	⋮	⋮	⋮		
325	B Molina	SF	C	202	52	3	17	0.257	0.312
326	J Thole	NYM	C	202	56	3	17	0.277	0.357
327	C Heisey	CIN	OF	201	51	8	21	0.254	0.324

Table 7.25: Six cases from the `bat10` data matrix.

variable	description
name	Player name
team	The abbreviated name of the player's team
position	The player's primary field position (OF, IF, DH, C)
AB	Number of opportunities at bat
H	Number of hits
HR	Number of home runs
RBI	Number of runs batted in
AVG	Batting average, which is equal to H/AB
OBP	On-base percentage, which is roughly equal to the fraction of times a player gets on base or hits a home run

Table 7.26: Variables and their descriptions for the bat10 data set.

⊙ **Guided Practice 7.36** The null hypothesis under consideration is the following: $\mu_{OF} = \mu_{IF} = \mu_{DH} = \mu_C$. Write the null and corresponding alternative hypotheses in plain language.[28]

● **Example 7.37** The player positions have been divided into four groups: outfield (OF), infield (IF), designated hitter (DH), and catcher (C). What would be an appropriate point estimate of the on-base percentage by outfielders, μ_{OF}?

A good estimate of the on-base percentage by outfielders would be the sample average of OBP for just those players whose position is outfield: $\bar{x}_{OF} = 0.334$.

Table 7.27 provides summary statistics for each group. A side-by-side box plot for the on-base percentage is shown in Figure 7.28. Notice that the variability appears to be approximately constant across groups; nearly constant variance across groups is an important assumption that must be satisfied before we consider the ANOVA approach.

	OF	IF	DH	C
Sample size (n_i)	120	154	14	39
Sample mean ($\bar{x}_i$)	0.334	0.332	0.348	0.323
Sample SD (s_i)	0.029	0.037	0.036	0.045

Table 7.27: Summary statistics of on-base percentage, split by player position.

● **Example 7.38** The largest difference between the sample means is between the designated hitter and the catcher positions. Consider again the original hypotheses:

H_0: $\mu_{OF} = \mu_{IF} = \mu_{DH} = \mu_C$

H_A: The average on-base percentage (μ_i) varies across some (or all) groups.

Why might it be inappropriate to run the test by simply estimating whether the difference of μ_{DH} and μ_C is statistically significant at a 0.05 significance level?

The primary issue here is that we are inspecting the data before picking the groups that will be compared. It is inappropriate to examine all data by eye (informal

---

[28]H_0: The average on-base percentage is equal across the four positions.
H_A: The average on-base percentage varies across some (or all) groups.

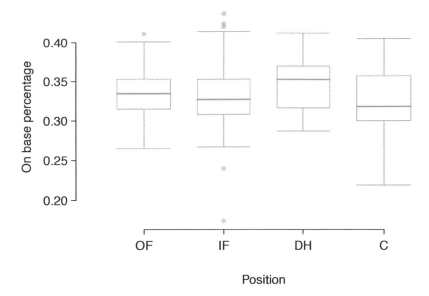

Figure 7.28: Side-by-side box plot of the on-base percentage for 327 players across four groups. There is one prominent outlier visible in the infield group, but with 154 observations in the infield group, this outlier is not a concern.

testing) and only afterwards decide which parts to formally test. This is called **data snooping** or **data fishing**. Naturally we would pick the groups with the large differences for the formal test, leading to an inflation in the Type 1 Error rate. To understand this better, let's consider a slightly different problem.

Suppose we are to measure the aptitude for students in 20 classes in a large elementary school at the beginning of the year. In this school, all students are randomly assigned to classrooms, so any differences we observe between the classes at the start of the year are completely due to chance. However, with so many groups, we will probably observe a few groups that look rather different from each other. If we select only these classes that look so different, we will probably make the wrong conclusion that the assignment wasn't random. While we might only formally test differences for a few pairs of classes, we informally evaluated the other classes by eye before choosing the most extreme cases for a comparison.

For additional information on the ideas expressed in Example 7.38, we recommend reading about the **prosecutor's fallacy**.[29]

In the next section we will learn how to use the F statistic and ANOVA to test whether observed differences in sample means could have happened just by chance even if there was no difference in the respective population means.

---

[29]See, for example, andrewgelman.com/2007/05/18/the_prosecutors.

7.4.2 Analysis of variance (ANOVA) and the F test

The method of analysis of variance in this context focuses on answering one question: is the variability in the sample means so large that it seems unlikely to be from chance alone? This question is different from earlier testing procedures since we will *simultaneously* consider many groups, and evaluate whether their sample means differ more than we would expect from natural variation. We call this variability the **mean square between groups** (MSG), and it has an associated degrees of freedom, $df_G = k - 1$ when there are k groups. The MSG can be thought of as a scaled variance formula for means. If the null hypothesis is true, any variation in the sample means is due to chance and shouldn't be too large. Details of MSG calculations are provided in the footnote,[30] however, we typically use software for these computations.

The mean square between the groups is, on its own, quite useless in a hypothesis test. We need a benchmark value for how much variability should be expected among the sample means if the null hypothesis is true. To this end, we compute a pooled variance estimate, often abbreviated as the **mean square error** (MSE), which has an associated degrees of freedom value $df_E = n - k$. It is helpful to think of MSE as a measure of the variability within the groups. Details of the computations of the MSE are provided in the footnote[31] for interested readers.

When the null hypothesis is true, any differences among the sample means are only due to chance, and the MSG and MSE should be about equal. As a test statistic for ANOVA, we examine the fraction of MSG and MSE:

$$F = \frac{MSG}{MSE} \tag{7.39}$$

The MSG represents a measure of the between-group variability, and MSE measures the variability within each of the groups.

⊙ **Guided Practice 7.40** For the baseball data, $MSG = 0.00252$ and $MSE = 0.00127$. Identify the degrees of freedom associated with MSG and MSE and verify the F statistic is approximately 1.994.[32]

---

[30]Let $\bar{x}$ represent the mean of outcomes across all groups. Then the mean square between groups is computed as

$$MSG = \frac{1}{df_G}SSG = \frac{1}{k-1}\sum_{i=1}^{k} n_i\left(\bar{x}_i - \bar{x}\right)^2$$

where SSG is called the **sum of squares between groups** and n_i is the sample size of group i.

[31]Let $\bar{x}$ represent the mean of outcomes across all groups. Then the **sum of squares total** (SST) is computed as

$$SST = \sum_{i=1}^{n}\left(x_i - \bar{x}\right)^2$$

where the sum is over all observations in the data set. Then we compute the **sum of squared errors** (SSE) in one of two equivalent ways:

$$SSE = SST - SSG$$
$$= (n_1 - 1)s_1^2 + (n_2 - 1)s_2^2 + \cdots + (n_k - 1)s_k^2$$

where s_i^2 is the sample variance (square of the standard deviation) of the residuals in group i. Then the MSE is the standardized form of SSE: $MSE = \frac{1}{df_E}SSE$.

[32]There are $k = 4$ groups, so $df_G = k - 1 = 3$. There are $n = n_1 + n_2 + n_3 + n_4 = 327$ total observations, so $df_E = n - k = 323$. Then the F statistic is computed as the ratio of MSG and MSE: $F = \frac{MSG}{MSE} = \frac{0.00252}{0.00127} = 1.984 \approx 1.994$. ($F = 1.994$ was computed by using values for MSG and MSE that were not rounded.)

We can use the F statistic to evaluate the hypotheses in what is called an **F test**. A p-value can be computed from the F statistic using an F distribution, which has two associated parameters: df_1 and df_2. For the F statistic in ANOVA, $df_1 = df_G$ and $df_2 = df_E$. An F distribution with 3 and 323 degrees of freedom, corresponding to the F statistic for the baseball hypothesis test, is shown in Figure 7.29.

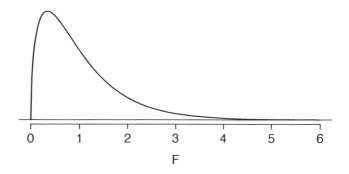

Figure 7.29: An F distribution with $df_1 = 3$ and $df_2 = 323$.

The larger the observed variability in the sample means (MSG) relative to the within-group observations (MSE), the larger F will be and the stronger the evidence against the null hypothesis. Because larger values of F represent stronger evidence against the null hypothesis, we use the upper tail of the distribution to compute a p-value.

The F statistic and the F test

Analysis of variance (ANOVA) is used to test whether the mean outcome differs across 2 or more groups. ANOVA uses a test statistic F, which represents a standardized ratio of variability in the sample means relative to the variability within the groups. If H_0 is true and the model assumptions are satisfied, the statistic F follows an F distribution with parameters $df_1 = k - 1$ and $df_2 = n - k$. The upper tail of the F distribution is used to represent the p-value.

⊙ **Guided Practice 7.41** The test statistic for the baseball example is $F = 1.994$. Shade the area corresponding to the p-value in Figure 7.29. [33]

● **Example 7.42** The p-value corresponding to the shaded area in the solution of Guided Practice 7.41 is equal to about 0.115. Does this provide strong evidence against the null hypothesis?

The p-value is larger than 0.05, indicating the evidence is not strong enough to reject the null hypothesis at a significance level of 0.05. That is, the data do not provide strong evidence that the average on-base percentage varies by player's primary field position.

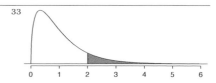

7.4.3   Reading an ANOVA table from software

The calculations required to perform an ANOVA by hand are tedious and prone to human error. For these reasons, it is common to use statistical software to calculate the F statistic and p-value.

An ANOVA can be summarized in a table very similar to that of a regression summary, which we will see in Chapter 8. Table 7.30 shows an ANOVA summary to test whether the mean of on-base percentage varies by player positions in the MLB. Many of these values should look familiar; in particular, the F test statistic and p-value can be retrieved from the last columns.

	Df	Sum Sq	Mean Sq	F value	Pr(>F)
position	3	0.0076	0.0025	1.9943	0.1147
Residuals	323	0.4080	0.0013		

$$s_{pooled} = 0.036 \text{ on } df = 323$$

Table 7.30: ANOVA summary for testing whether the average on-base percentage differs across player positions.

7.4.4   Graphical diagnostics for an ANOVA analysis

There are three conditions we must check for an ANOVA analysis: all observations must be independent, the data in each group must be nearly normal, and the variance within each group must be approximately equal.

Independence. If the data are a simple random sample from less than 10% of the population, this condition is satisfied. For processes and experiments, carefully consider whether the data may be independent (e.g. no pairing). For example, in the MLB data, the data were not sampled. However, there are not obvious reasons why independence would not hold for most or all observations.

Approximately normal. As with one- and two-sample testing for means, the normality assumption is especially important when the sample size is quite small. The normal probability plots for each group of the MLB data are shown in Figure 7.31; there is some deviation from normality for infielders, but this isn't a substantial concern since there are about 150 observations in that group and the outliers are not extreme. Sometimes in ANOVA there are so many groups or so few observations per group that checking normality for each group isn't reasonable. See the footnote[34] for guidance on how to handle such instances.

Constant variance. The last assumption is that the variance in the groups is about equal from one group to the next. This assumption can be checked by examining a side-by-side box plot of the outcomes across the groups, as in Figure 7.28 on page 349. In this case, the variability is similar in the four groups but not identical. We see in Table 7.27 on page 348 that the standard deviation varies a bit from one group to the next. Whether these differences are from natural variation is unclear, so we should report this uncertainty with the final results.

---

[34]First calculate the **residuals** of the baseball data, which are calculated by taking the observed values and subtracting the corresponding group means. For example, an outfielder with OBP of 0.405 would have a residual of $0.405 - \bar{x}_{OF} = 0.071$. Then to check the normality condition, create a normal probability plot using all the residuals simultaneously.

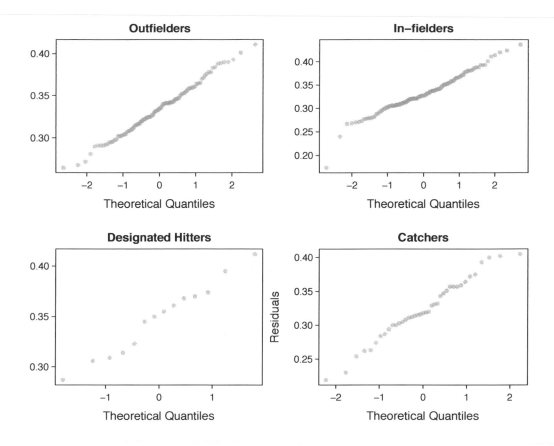

Figure 7.31: Normal probability plot of OBP for each field position.

Caution: Diagnostics for an ANOVA analysis
Independence is always important to an ANOVA analysis. The normality condition is very important when the sample sizes for each group are relatively small. The constant variance condition is especially important when the sample sizes differ between groups.

7.4.5 Multiple comparisons and controlling Type 1 Error rate

When we reject the null hypothesis in an ANOVA analysis, we might wonder, which of these groups have different means? To answer this question, we compare the means of each possible pair of groups. For instance, if there are three groups and there is strong evidence that there are some differences in the group means, there are three comparisons to make: group 1 to group 2, group 1 to group 3, and group 2 to group 3. These comparisons can be accomplished using a two-sample t-test, but we use a modified significance level and a pooled estimate of the standard deviation across groups. Usually this pooled standard deviation can be found in the ANOVA table, e.g. along the bottom of Table 7.30.

Class i	A	B	C
n_i	58	55	51
$\bar{x}_i$	75.1	72.0	78.9
s_i	13.9	13.8	13.1

Table 7.32: Summary statistics for the first midterm scores in three different lectures of the same course.

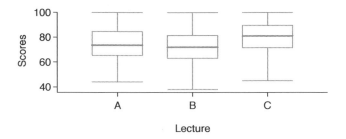

Figure 7.33: Side-by-side box plot for the first midterm scores in three different lectures of the same course.

● **Example 7.43** Example 7.34 on page 346 discussed three statistics lectures, all taught during the same semester. Table 7.32 shows summary statistics for these three courses, and a side-by-side box plot of the data is shown in Figure 7.33. We would like to conduct an ANOVA for these data. Do you see any deviations from the three conditions for ANOVA?

In this case (like many others) it is difficult to check independence in a rigorous way. Instead, the best we can do is use common sense to consider reasons the assumption of independence may not hold. For instance, the independence assumption may not be reasonable if there is a star teaching assistant that only half of the students may access; such a scenario would divide a class into two subgroups. No such situations were evident for these particular data, and we believe that independence is acceptable.

The distributions in the side-by-side box plot appear to be roughly symmetric and show no noticeable outliers.

The box plots show approximately equal variability, which can be verified in Table 7.32, supporting the constant variance assumption.

⊙ **Guided Practice 7.44** An ANOVA was conducted for the midterm data, and summary results are shown in Table 7.34. What should we conclude?[35]

	Df	Sum Sq	Mean Sq	F value	Pr(>F)
lecture	2	1290.11	645.06	3.48	0.0330
Residuals	161	29810.13	185.16		

$s_{pooled} = 13.61$ on $df = 161$

Table 7.34: ANOVA summary table for the midterm data.

---

[35]The p-value of the test is 0.0330, less than the default significance level of 0.05. Therefore, we reject the null hypothesis and conclude that the difference in the average midterm scores are not due to chance.

There is strong evidence that the different means in each of the three classes is not simply due to chance. We might wonder, which of the classes are actually different? As discussed in earlier chapters, a two-sample t-test could be used to test for differences in each possible pair of groups. However, one pitfall was discussed in Example 7.38 on page 348: when we run so many tests, the Type 1 Error rate increases. This issue is resolved by using a modified significance level.

Multiple comparisons and the Bonferroni correction for α

The scenario of testing many pairs of groups is called **multiple comparisons**. The **Bonferroni correction** suggests that a more stringent significance level is more appropriate for these tests:

$$\alpha^* = \alpha/K$$

where K is the number of comparisons being considered (formally or informally). If there are k groups, then usually all possible pairs are compared and $K = \frac{k(k-1)}{2}$.

● **Example 7.45** In Guided Practice 7.44, you found strong evidence of differences in the average midterm grades between the three lectures. Complete the three possible pairwise comparisons using the Bonferroni correction and report any differences.

We use a modified significance level of $\alpha^* = 0.05/3 = 0.0167$. Additionally, we use the pooled estimate of the standard deviation: $s_{pooled} = 13.61$ on $df = 161$, which is provided in the ANOVA summary table.

Lecture A versus Lecture B: The estimated difference and standard error are, respectively,

$$\bar{x}_A - \bar{x}_B = 75.1 - 72 = 3.1 \qquad SE = \sqrt{\frac{13.61^2}{58} + \frac{13.61^2}{55}} = 2.56$$

This results in a T score of 1.21 on $df = 161$ (we use the df associated with s_{pooled}). Statistical software was used to precisely identify the two-tailed p-value since the modified significance of 0.0167 is not found in the t-table. The p-value (0.228) is larger than $\alpha^* = 0.0167$, so there is not strong evidence of a difference in the means of lectures A and B.

Lecture A versus Lecture C: The estimated difference and standard error are 3.8 and 2.61, respectively. This results in a T score of 1.46 on $df = 161$ and a two-tailed p-value of 0.1462. This p-value is larger than α^*, so there is not strong evidence of a difference in the means of lectures A and C.

Lecture B versus Lecture C: The estimated difference and standard error are 6.9 and 2.65, respectively. This results in a T score of 2.60 on $df = 161$ and a two-tailed p-value of 0.0102. This p-value is smaller than α^*. Here we find strong evidence of a difference in the means of lectures B and C.

We might summarize the findings of the analysis from Example 7.45 using the following notation:

$$\mu_A \overset{?}{=} \mu_B \qquad\qquad \mu_A \overset{?}{=} \mu_C \qquad\qquad \mu_B \neq \mu_C$$

The midterm mean in lecture A is not statistically distinguishable from those of lectures B or C. However, there is strong evidence that lectures B and C are different. In the first two pairwise comparisons, we did not have sufficient evidence to reject the null hypothesis. Recall that failing to reject H_0 does not imply H_0 is true.

Caution: Sometimes an ANOVA will reject the null but no groups will have statistically significant differences
It is possible to reject the null hypothesis using ANOVA and then to not subsequently identify differences in the pairwise comparisons. However, *this does not invalidate the ANOVA conclusion*. It only means we have not been able to successfully identify which groups differ in their means.

The ANOVA procedure examines the big picture: it considers all groups simultaneously to decipher whether there is evidence that some difference exists. Even if the test indicates that there is strong evidence of differences in group means, identifying with high confidence a specific difference as statistically significant is more difficult.

Consider the following analogy: we observe a Wall Street firm that makes large quantities of money based on predicting mergers. Mergers are generally difficult to predict, and if the prediction success rate is extremely high, that may be considered sufficiently strong evidence to warrant investigation by the Securities and Exchange Commission (SEC). While the SEC may be quite certain that there is insider trading taking place at the firm, the evidence against any single trader may not be very strong. It is only when the SEC considers all the data that they identify the pattern. This is effectively the strategy of ANOVA: stand back and consider all the groups simultaneously.

7.5 Exercises

7.5.1 Inference for a single mean with the t-distribution

7.1 Identify the critical t. An independent random sample is selected from an approximately normal population with unknown standard deviation. Find the degrees of freedom and the critical t-value ($t^\star$) for the given sample size and confidence level.

(a) $n = 6$, CL = 90%

(b) $n = 21$, CL = 98%

(c) $n = 29$, CL = 95%

(d) $n = 12$, CL = 99%

7.2 t-distribution. The figure on the right shows three unimodal and symmetric curves: the standard normal (z) distribution, the t-distribution with 5 degrees of freedom, and the t-distribution with 1 degree of freedom. Determine which is which, and explain your reasoning.

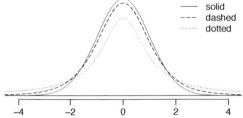

7.3 Find the p-value, Part I. An independent random sample is selected from an approximately normal population with an unknown standard deviation. Find the p-value for the given set of hypotheses and T test statistic. Also determine if the null hypothesis would be rejected at $\alpha = 0.05$.

(a) $H_A : \mu > \mu_0$, $n = 11$, $T = 1.91$

(b) $H_A : \mu < \mu_0$, $n = 17$, $T = -3.45$

(c) $H_A : \mu \neq \mu_0$, $n = 7$, $T = 0.83$

(d) $H_A : \mu > \mu_0$, $n = 28$, $T = 2.13$

7.4 Find the p-value, Part II. An independent random sample is selected from an approximately normal population with an unknown standard deviation. Find the p-value for the given set of hypotheses and T test statistic. Also determine if the null hypothesis would be rejected at $\alpha = 0.01$.

(a) $H_A : \mu > 0.5$, $n = 26$, $T = 2.485$

(b) $H_A : \mu < 3$, $n = 18$, $T = 0.5$

7.5 Working backwards, Part I. A 95% confidence interval for a population mean, μ, is given as (18.985, 21.015). This confidence interval is based on a simple random sample of 36 observations. Calculate the sample mean and standard deviation. Assume that all conditions necessary for inference are satisfied. Use the t-distribution in any calculations.

7.6 Working backwards, Part II. A 90% confidence interval for a population mean is (65, 77). The population distribution is approximately normal and the population standard deviation is unknown. This confidence interval is based on a simple random sample of 25 observations. Calculate the sample mean, the margin of error, and the sample standard deviation.

7.7 Sleep habits of New Yorkers. New York is known as "the city that never sleeps".
A random sample of 25 New Yorkers were asked how much sleep they get per night. Statistical
summaries of these data are shown below. Do these data provide strong evidence that New Yorkers
sleep less than 8 hours a night on average?

n	$\bar{x}$	s	min	max
25	7.73	0.77	6.17	9.78

(a) Write the hypotheses in symbols and in words.

(b) Check conditions, then calculate the test statistic, T, and the associated degrees of freedom.

(c) Find and interpret the p-value in this context. Drawing a picture may be helpful.

(d) What is the conclusion of the hypothesis test?

(e) If you were to construct a 90% confidence interval that corresponded to this hypothesis test,
would you expect 8 hours to be in the interval?

7.8 Fuel efficiency of Prius. Fueleconomy.gov, the official US government source for fuel economy information, allows users to share gas mileage information on their vehicles. The histogram
below shows the distribution of gas mileage in miles per gallon (MPG) from 14 users who drive a
2012 Toyota Prius. The sample mean is 53.3 MPG and the standard deviation is 5.2 MPG. Note
that these data are user estimates and since the source data cannot be verified, the accuracy of
these estimates are not guaranteed.[36]

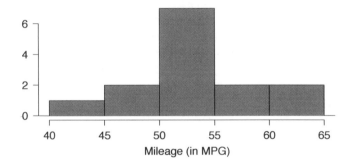

(a) We would like to use these data to evaluate the average gas mileage of all 2012 Prius drivers.
Do you think this is reasonable? Why or why not?

(b) The EPA claims that a 2012 Prius gets 50 MPG (city and highway mileage combined). Do
these data provide strong evidence against this estimate for drivers who participate on fuele-
conomy.gov? Note any assumptions you must make as you proceed with the test.

(c) Calculate a 95% confidence interval for the average gas mileage of a 2012 Prius by drivers who
participate on fueleconomy.gov.

7.9 Find the mean. You are given the following hypotheses:

$$H_0 : \mu = 60$$
$$H_A : \mu < 60$$

We know that the sample standard deviation is 8 and the sample size is 20. For what sample
mean would the p-value be equal to 0.05? Assume that all conditions necessary for inference are
satisfied.

7.10 $t^\star$ vs. $z^\star$. For a given confidence level, $t^\star_{df}$ is larger than $z^\star$. Explain how $t^\star_{df}$ being slightly
larger than $z^\star$ affects the width of the confidence interval.

---

[36]Fuelecomy.gov, Shared MPG Estimates: Toyota Prius 2012.

7.11 Play the piano. Georgianna claims that in a small city renowned for its music school, the average child takes at least 5 years of piano lessons. We have a random sample of 20 children from the city, with a mean of 4.6 years of piano lessons and a standard deviation of 2.2 years.

(a) Evaluate Georgianna's claim using a hypothesis test.

(b) Construct a 95% confidence interval for the number of years students in this city take piano lessons, and interpret it in context of the data.

(c) Do your results from the hypothesis test and the confidence interval agree? Explain your reasoning.

7.12 Auto exhaust and lead exposure. Researchers interested in lead exposure due to car exhaust sampled the blood of 52 police officers subjected to constant inhalation of automobile exhaust fumes while working traffic enforcement in a primarily urban environment. The blood samples of these officers had an average lead concentration of 124.32 μg/l and a SD of 37.74 μg/l; a previous study of individuals from a nearby suburb, with no history of exposure, found an average blood level concentration of 35 μg/l.[37]

(a) Write down the hypotheses that would be appropriate for testing if the police officers appear to have been exposed to a higher concentration of lead.

(b) Explicitly state and check all conditions necessary for inference on these data.

(c) Test the hypothesis that the downtown police officers have a higher lead exposure than the group in the previous study. Interpret your results in context.

(d) Based on your preceding result, without performing a calculation, would a 99% confidence interval for the average blood concentration level of police officers contain 35 μg/l?

7.13 Car insurance savings. A market researcher wants to evaluate car insurance savings at a competing company. Based on past studies he is assuming that the standard deviation of savings is $100. He wants to collect data such that he can get a margin of error of no more than $10 at a 95% confidence level. How large of a sample should he collect?

7.14 SAT scores. SAT scores of students at an Ivy League college are distributed with a standard deviation of 250 points. Two statistics students, Raina and Luke, want to estimate the average SAT score of students at this college as part of a class project. They want their margin of error to be no more than 25 points.

(a) Raina wants to use a 90% confidence interval. How large a sample should she collect?

(b) Luke wants to use a 99% confidence interval. Without calculating the actual sample size, determine whether his sample should be larger or smaller than Raina's, and explain your reasoning.

(c) Calculate the minimum required sample size for Luke.

7.5.2 Inference for paired data

7.15 Air quality. Air quality measurements were collected in a random sample of 25 country capitals in 2013, and then again in the same cities in 2014. We would like to use these data to compare average air quality between the two years.

(a) Should we use a one-sided or a two-sided test? Explain your reasoning.

(b) Should we use a paired or non-paired test? Explain your reasoning.

(c) Should we use a t-test or a z-test? Explain your reasoning.

---

[37] WI Mortada et al. "Study of lead exposure from automobile exhaust as a risk for nephrotoxicity among traffic policemen." In: *American journal of nephrology* 21.4 (2000), pp. 274–279.

7.16   True / False: paired. Determine if the following statements are true or false. If false, explain.

(a) In a paired analysis we first take the difference of each pair of observation, and then we do inference on these differences.

(b) Two data sets of different sizes cannot be analyzed as paired data.

(c) Each observation in one data set has a natural correspondence with exactly one observation from the other data set.

(d) Each observation in one data set is subtracted from the average of the other data set's observations.

7.17   Paired or not, Part I? In each of the following scenarios, determine if the data are paired.

(a) Compare pre- (beginning of semester) and post-test (end of semester) scores of students.

(b) Assess gender-related salary gap by comparing salaries of randomly sampled men and women.

(c) Compare artery thicknesses at the beginning of a study and after 2 years of taking Vitamin E for the same group of patients.

(d) Assess effectiveness of a diet regimen by comparing the before and after weights of subjects.

7.18   Paired or not, Part II? In each of the following scenarios, determine if the data are paired.

(a) We would like to know if Intel's stock and Southwest Airlines' stock have similar rates of return. To find out, we take a random sample of 50 days, and record Intel's and Southwest's stock on those same days.

(b) We randomly sample 50 items from Target stores and note the price for each. Then we visit Walmart and collect the price for each of those same 50 items.

(c) A school board would like to determine whether there is a difference in average SAT scores for students at one high school versus another high school in the district. To check, they take a simple random sample of 100 students from each high school.

7.19   Global warming, Part I. Is there strong evidence of global warming? Let's consider a small scale example, comparing how temperatures have changed in the US from 1968 to 2008. The daily high temperature reading on January 1 was collected in 1968 and 2008 for 51 randomly selected locations in the continental US. Then the difference between the two readings (temperature in 2008 - temperature in 1968) was calculated for each of the 51 different locations. The average of these 51 values was 1.1 degrees with a standard deviation of 4.9 degrees. We are interested in determining whether these data provide strong evidence of temperature warming in the continental US.

(a) Is there a relationship between the observations collected in 1968 and 2008? Or are the observations in the two groups independent? Explain.

(b) Write hypotheses for this research in symbols and in words.

(c) Check the conditions required to complete this test.

(d) Calculate the test statistic and find the p-value.

(e) What do you conclude? Interpret your conclusion in context.

(f) What type of error might we have made? Explain in context what the error means.

(g) Based on the results of this hypothesis test, would you expect a confidence interval for the average difference between the temperature measurements from 1968 and 2008 to include 0? Explain your reasoning.

7.20 High School and Beyond, Part I. The National Center of Education Statistics conducted a survey of high school seniors, collecting test data on reading, writing, and several other subjects. Here we examine a simple random sample of 200 students from this survey. Side-by-side box plots of reading and writing scores as well as a histogram of the differences in scores are shown below.

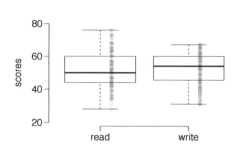
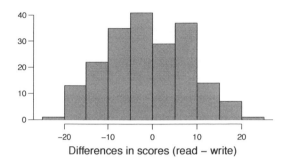

(a) Is there a clear difference in the average reading and writing scores?

(b) Are the reading and writing scores of each student independent of each other?

(c) Create hypotheses appropriate for the following research question: is there an evident difference in the average scores of students in the reading and writing exam?

(d) Check the conditions required to complete this test.

(e) The average observed difference in scores is $\bar{x}_{read-write} = -0.545$, and the standard deviation of the differences is 8.887 points. Do these data provide convincing evidence of a difference between the average scores on the two exams?

(f) What type of error might we have made? Explain what the error means in the context of the application.

(g) Based on the results of this hypothesis test, would you expect a confidence interval for the average difference between the reading and writing scores to include 0? Explain your reasoning.

7.21 Global warming, Part II. We considered the differences between the temperature readings in January 1 of 1968 and 2008 at 51 locations in the continental US in Exercise 7.19. The mean and standard deviation of the reported differences are 1.1 degrees and 4.9 degrees.

(a) Calculate a 90% confidence interval for the average difference between the temperature measurements between 1968 and 2008.

(b) Interpret this interval in context.

(c) Does the confidence interval provide convincing evidence that the temperature was higher in 2008 than in 1968 in the continental US? Explain.

7.22 High school and beyond, Part II. We considered the differences between the reading and writing scores of a random sample of 200 students who took the High School and Beyond Survey in Exercise 7.21. The mean and standard deviation of the differences are $\bar{x}_{read-write} = -0.545$ and 8.887 points.

(a) Calculate a 95% confidence interval for the average difference between the reading and writing scores of all students.

(b) Interpret this interval in context.

(c) Does the confidence interval provide convincing evidence that there is a real difference in the average scores? Explain.

**7.23	Gifted children.** Researchers collected a simple random sample of 36 children who had been identified as gifted in a large city. The following histograms show the distributions of the IQ scores of mothers and fathers of these children. Also provided are some sample statistics.[38]

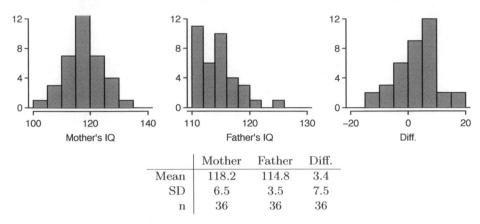

	Mother	Father	Diff.
Mean	118.2	114.8	3.4
SD	6.5	3.5	7.5
n	36	36	36

(a) Are the IQs of mothers and the IQs of fathers in this data set related? Explain.

(b) Conduct a hypothesis test to evaluate if the scores are equal on average. Make sure to clearly state your hypotheses, check the relevant conditions, and state your conclusion in the context of the data.

**7.24	Sample size and pairing.** Determine if the following statement is true or false, and if false, explain your reasoning: If comparing means of two groups with equal sample sizes, always use a paired test.

---

[38]F.A. Graybill and H.K. Iyer. *Regression Analysis: Concepts and Applications.* Duxbury Press, 1994, pp. 511–516.

7.5.3   Difference of two means using the *t*-distribution

7.25   Cleveland vs. Sacramento. Average income varies from one region of the country to another, and it often reflects both lifestyles and regional living expenses. Suppose a new graduate is considering a job in two locations, Cleveland, OH and Sacramento, CA, and he wants to see whether the average income in one of these cities is higher than the other. He would like to conduct a hypothesis test based on two small samples from the 2000 Census, but he first must consider whether the conditions are met to implement the test. Below are histograms for each city. Should he move forward with the hypothesis test? Explain your reasoning.

	Cleveland, OH
Mean	$ 35,749
SD	$ 39,421
n	21

	Sacramento, CA
Mean	$ 35,500
SD	$ 41,512
n	17

7.26   Oscar winners. The first Oscar awards for best actor and best actress were given out in 1929. The histograms below show the age distribution for all of the best actor and best actress winners from 1929 to 2012. Summary statistics for these distributions are also provided. Is a hypothesis test appropriate for evaluating whether the difference in the average ages of best actors and actresses might be due to chance? Explain your reasoning.[39]

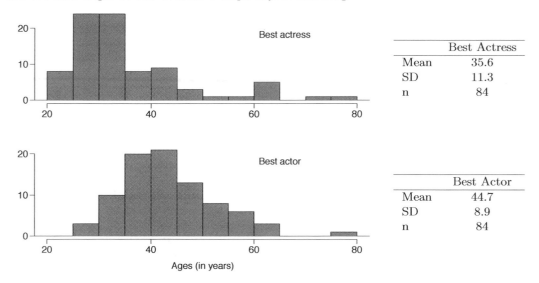

	Best Actress
Mean	35.6
SD	11.3
n	84

	Best Actor
Mean	44.7
SD	8.9
n	84

---

[39]Oscar winners from 1929 – 2012, data up to 2009 from the Journal of Statistics Education data archive and more current data from wikipedia.org.

7.27 Friday the 13th, Part I. In the early 1990's, researchers in the UK collected data on traffic flow, number of shoppers, and traffic accident related emergency room admissions on Friday the 13th and the previous Friday, Friday the 6th. The histograms below show the distribution of number of cars passing by a specific intersection on Friday the 6th and Friday the 13th for many such date pairs. Also given are some sample statistics, where the difference is the number of cars on the 6th minus the number of cars on the 13th.[40]

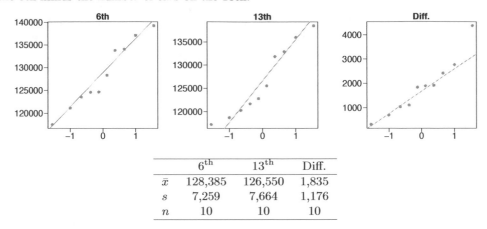

	6th	13th	Diff.
$\bar{x}$	128,385	126,550	1,835
s	7,259	7,664	1,176
n	10	10	10

(a) Are there any underlying structures in these data that should be considered in an analysis? Explain.

(b) What are the hypotheses for evaluating whether the number of people out on Friday the 6th is different than the number out on Friday the 13th?

(c) Check conditions to carry out the hypothesis test from part (b).

(d) Calculate the test statistic and the p-value.

(e) What is the conclusion of the hypothesis test?

(f) Interpret the p-value in this context.

(g) What type of error might have been made in the conclusion of your test? Explain.

7.28 Diamonds, Part I. Prices of diamonds are determined by what is known as the 4 Cs: cut, clarity, color, and carat weight. The prices of diamonds go up as the carat weight increases, but the increase is not smooth. For example, the difference between the size of a 0.99 carat diamond and a 1 carat diamond is undetectable to the naked human eye, but the price of a 1 carat diamond tends to be much higher than the price of a 0.99 diamond. In this question we use two random samples of diamonds, 0.99 carats and 1 carat, each sample of size 23, and compare the average prices of the diamonds. In order to be able to compare equivalent units, we first divide the price for each diamond by 100 times its weight in carats. That is, for a 0.99 carat diamond, we divide the price by 99. For a 1 carat diamond, we divide the price by 100. The distributions and some sample statistics are shown below.[41]

Conduct a hypothesis test to evaluate if there is a difference between the average standardized prices of 0.99 and 1 carat diamonds. Make sure to state your hypotheses clearly, check relevant conditions, and interpret your results in context of the data.

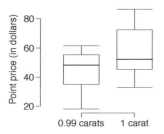

	0.99 carats	1 carat
Mean	$ 44.51	$ 56.81
SD	$ 13.32	$ 16.13
n	23	23

[40]T.J. Scanlon et al. "Is Friday the 13th Bad For Your Health?" In: *BMJ* 307 (1993), pp. 1584–1586.
[41]H. Wickham. *ggplot2: elegant graphics for data analysis*. Springer New York, 2009.

7.29   Friday the 13th, Part II. The Friday the 13th study reported in Exercise 7.27 also provides data on traffic accident related emergency room admissions. The distributions of these counts from Friday the 6th and Friday the 13th are shown below for six such paired dates along with summary statistics. You may assume that conditions for inference are met.

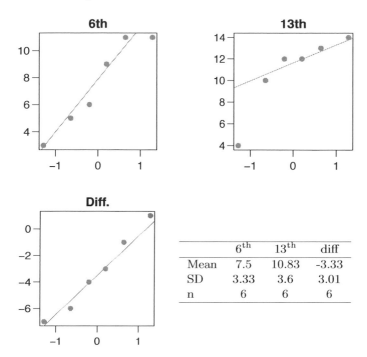

	6th	13th	diff
Mean	7.5	10.83	-3.33
SD	3.33	3.6	3.01
n	6	6	6

(a) Conduct a hypothesis test to evaluate if there is a difference between the average numbers of traffic accident related emergency room admissions between Friday the 6th and Friday the 13th.

(b) Calculate a 95% confidence interval for the difference between the average numbers of traffic accident related emergency room admissions between Friday the 6th and Friday the 13th.

(c) The conclusion of the original study states, "Friday 13th is unlucky for some. The risk of hospital admission as a result of a transport accident may be increased by as much as 52%. Staying at home is recommended." Do you agree with this statement? Explain your reasoning.

7.30   Diamonds, Part II. In Exercise 7.28, we discussed diamond prices (standardized by weight) for diamonds with weights 0.99 carats and 1 carat. See the table for summary statistics, and then construct a 95% confidence interval for the average difference between the standardized prices of 0.99 and 1 carat diamonds. You may assume the conditions for inference are met.

	0.99 carats	1 carat
Mean	$ 44.51	$ 56.81
SD	$ 13.32	$ 16.13
n	23	23

7.31 Chicken diet and weight, Part I. Chicken farming is a multi-billion dollar industry, and any methods that increase the growth rate of young chicks can reduce consumer costs while increasing company profits, possibly by millions of dollars. An experiment was conducted to measure and compare the effectiveness of various feed supplements on the growth rate of chickens. Newly hatched chicks were randomly allocated into six groups, and each group was given a different feed supplement. Below are some summary statistics from this data set along with box plots showing the distribution of weights by feed type.[42]

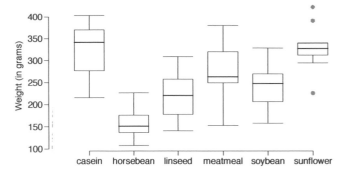

	Mean	SD	n
casein	323.58	64.43	12
horsebean	160.20	38.63	10
linseed	218.75	52.24	12
meatmeal	276.91	64.90	11
soybean	246.43	54.13	14
sunflower	328.92	48.84	12

(a) Describe the distributions of weights of chickens that were fed linseed and horsebean.

(b) Do these data provide strong evidence that the average weights of chickens that were fed linseed and horsebean are different? Use a 5% significance level.

(c) What type of error might we have committed? Explain.

(d) Would your conclusion change if we used $\alpha = 0.01$?

7.32 Fuel efficiency of manual and automatic cars, Part I. Each year the US Environmental Protection Agency (EPA) releases fuel economy data on cars manufactured in that year. Below are summary statistics on fuel efficiency (in miles/gallon) from random samples of cars with manual and automatic transmissions manufactured in 2012. Do these data provide strong evidence of a difference between the average fuel efficiency of cars with manual and automatic transmissions in terms of their average city mileage? Assume that conditions for inference are satisfied.[43]

	City MPG	
	Automatic	Manual
Mean	16.12	19.85
SD	3.58	4.51
n	26	26

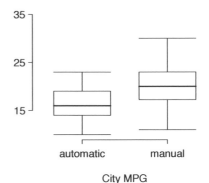

City MPG

7.33 Chicken diet and weight, Part II. Casein is a common weight gain supplement for humans. Does it have an effect on chickens? Using data provided in Exercise 7.31, test the hypothesis that the average weight of chickens that were fed casein is different than the average weight of chickens that were fed soybean. If your hypothesis test yields a statistically significant result, discuss whether or not the higher average weight of chickens can be attributed to the casein diet. Assume that conditions for inference are satisfied.

[42]Chicken Weights by Feed Type, from the `datasets` package in R..
[43]U.S. Department of Energy, Fuel Economy Data, 2012 Datafile.

7.34 Fuel efficiency of manual and automatic cars, Part II. The table provides summary statistics on highway fuel economy of cars manufactured in 2012 (from Exercise 7.32). Use these statistics to calculate a 98% confidence interval for the difference between average highway mileage of manual and automatic cars, and interpret this interval in the context of the data.[44]

	Hwy MPG	
	Automatic	Manual
Mean	22.92	27.88
SD	5.29	5.01
n	26	26

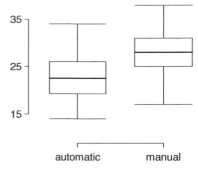

Hwy MPG

7.35 Gaming and distracted eating, Part I. A group of researchers are interested in the possible effects of distracting stimuli during eating, such as an increase or decrease in the amount of food consumption. To test this hypothesis, they monitored food intake for a group of 44 patients who were randomized into two equal groups. The treatment group ate lunch while playing solitaire, and the control group ate lunch without any added distractions. Patients in the treatment group ate 52.1 grams of biscuits, with a standard deviation of 45.1 grams, and patients in the control group ate 27.1 grams of biscuits, with a standard deviation of 26.4 grams. Do these data provide convincing evidence that the average food intake (measured in amount of biscuits consumed) is different for the patients in the treatment group? Assume that conditions for inference are satisfied.[45]

7.36 Gaming and distracted eating, Part II. The researchers from Exercise 7.35 also investigated the effects of being distracted by a game on how much people eat. The 22 patients in the treatment group who ate their lunch while playing solitaire were asked to do a serial-order recall of the food lunch items they ate. The average number of items recalled by the patients in this group was 4.9, with a standard deviation of 1.8. The average number of items recalled by the patients in the control group (no distraction) was 6.1, with a standard deviation of 1.8. Do these data provide strong evidence that the average number of food items recalled by the patients in the treatment and control groups are different?

[44]U.S. Department of Energy, Fuel Economy Data, 2012 Datafile.

[45]R.E. Oldham-Cooper et al. "Playing a computer game during lunch affects fullness, memory for lunch, and later snack intake". In: *The American Journal of Clinical Nutrition* 93.2 (2011), p. 308.

7.37 Prison isolation experiment, Part I. Subjects from Central Prison in Raleigh, NC, volunteered for an experiment involving an "isolation" experience. The goal of the experiment was to find a treatment that reduces subjects' psychopathic deviant T-scores. This score measures a person's need for control or their rebellion against control, and it is part of a commonly used mental health test called the Minnesota Multiphasic Personality Inventory (MMPI) test. The experiment had three treatment groups:

(1) Four hours of sensory restriction plus a 15 minute "therapeutic" tape advising that professional help is available.

(2) Four hours of sensory restriction plus a 15 minute "emotionally neutral" tape on training hunting dogs.

(3) Four hours of sensory restriction but no taped message.

Forty-two subjects were randomly assigned to these treatment groups, and an MMPI test was administered before and after the treatment. Distributions of the differences between pre and post treatment scores (pre - post) are shown below, along with some sample statistics. Use this information to independently test the effectiveness of each treatment. Make sure to clearly state your hypotheses, check conditions, and interpret results in the context of the data.[46]

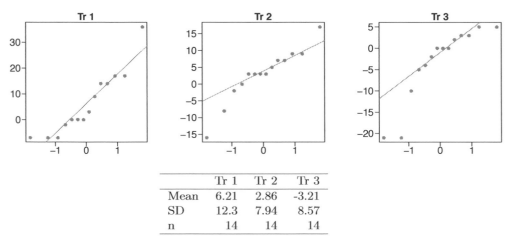

	Tr 1	Tr 2	Tr 3
Mean	6.21	2.86	-3.21
SD	12.3	7.94	8.57
n	14	14	14

7.38 True / False: comparing means. Determine if the following statements are true or false, and explain your reasoning for statements you identify as false.

(a) When comparing means of two samples where $n_1 = 20$ and $n_2 = 40$, we can use the normal model for the difference in means since $n_2 \geq 30$.

(b) As the degrees of freedom increases, the t-distribution approaches normality.

(c) We use a pooled standard error for calculating the standard error of the difference between means when sample sizes of groups are equal to each other.

---

[46]Prison isolation experiment.

7.5.4 Comparing many means with ANOVA (special topic)

7.39 Fill in the blank. When doing an ANOVA, you observe large differences in means between groups. Within the ANOVA framework, this would most likely be interpreted as evidence strongly favoring the _____ hypothesis.

7.40 Which test? We would like to test if students who are in the social sciences, natural sciences, arts and humanities, and other fields spend the same amount of time studying for this course. What type of test should we use? Explain your reasoning.

7.41 Chicken diet and weight, Part III. In Exercises 7.31 and 7.33 we compared the effects of two types of feed at a time. A better analysis would first consider all feed types at once: casein, horsebean, linseed, meat meal, soybean, and sunflower. The ANOVA output below can be used to test for differences between the average weights of chicks on different diets.

	Df	Sum Sq	Mean Sq	F value	Pr(>F)
feed	5	231,129.16	46,225.83	15.36	0.0000
Residuals	65	195,556.02	3,008.55		

Conduct a hypothesis test to determine if these data provide convincing evidence that the average weight of chicks varies across some (or all) groups. Make sure to check relevant conditions. Figures and summary statistics are shown below.

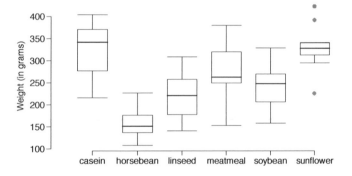

	Mean	SD	n
casein	323.58	64.43	12
horsebean	160.20	38.63	10
linseed	218.75	52.24	12
meatmeal	276.91	64.90	11
soybean	246.43	54.13	14
sunflower	328.92	48.84	12

7.42 Teaching descriptive statistics. A study compared five different methods for teaching descriptive statistics. The five methods were traditional lecture and discussion, programmed textbook instruction, programmed text with lectures, computer instruction, and computer instruction with lectures. 45 students were randomly assigned, 9 to each method. After completing the course, students took a 1-hour exam.

(a) What are the hypotheses for evaluating if the average test scores are different for the different teaching methods?

(b) What are the degrees of freedom associated with the F-test for evaluating these hypotheses?

(c) Suppose the p-value for this test is 0.0168. What is the conclusion?

7.43 Coffee, depression, and physical activity. Caffeine is the world's most widely used stimulant, with approximately 80% consumed in the form of coffee. Participants in a study investigating the relationship between coffee consumption and exercise were asked to report the number of hours they spent per week on moderate (e.g., brisk walking) and vigorous (e.g., strenuous sports and jogging) exercise. Based on these data the researchers estimated the total hours of metabolic equivalent tasks (MET) per week, a value always greater than 0. The table below gives summary statistics of MET for women in this study based on the amount of coffee consumed.[47]

| | Caffeinated coffee consumption | | | | | |
	$\leq$ 1 cup/week	2-6 cups/week	1 cup/day	2-3 cups/day	$\geq$ 4 cups/day	Total
Mean	18.7	19.6	19.3	18.9	17.5	
SD	21.1	25.5	22.5	22.0	22.0	
n	12,215	6,617	17,234	12,290	2,383	50,739

(a) Write the hypotheses for evaluating if the average physical activity level varies among the different levels of coffee consumption.

(b) Check conditions and describe any assumptions you must make to proceed with the test.

(c) Below is part of the output associated with this test. Fill in the empty cells.

	Df	Sum Sq	Mean Sq	F value	Pr(>F)
coffee					0.0003
Residuals		25,564,819			
Total		25,575,327			

(d) What is the conclusion of the test?

7.44 Student performance across discussion sections. A professor who teaches a large introductory statistics class (197 students) with eight discussion sections would like to test if student performance differs by discussion section, where each discussion section has a different teaching assistant. The summary table below shows the average final exam score for each discussion section as well as the standard deviation of scores and the number of students in each section.

	Sec 1	Sec 2	Sec 3	Sec 4	Sec 5	Sec 6	Sec 7	Sec 8
n_i	33	19	10	29	33	10	32	31
$\bar{x}_i$	92.94	91.11	91.80	92.45	89.30	88.30	90.12	93.35
s_i	4.21	5.58	3.43	5.92	9.32	7.27	6.93	4.57

The ANOVA output below can be used to test for differences between the average scores from the different discussion sections.

	Df	Sum Sq	Mean Sq	F value	Pr(>F)
section	7	525.01	75.00	1.87	0.0767
Residuals	189	7584.11	40.13		

Conduct a hypothesis test to determine if these data provide convincing evidence that the average score varies across some (or all) groups. Check conditions and describe any assumptions you must make to proceed with the test.

---

[47]M. Lucas et al. "Coffee, caffeine, and risk of depression among women". In: *Archives of internal medicine* 171.17 (2011), p. 1571.

7.45 GPA and major. Undergraduate students taking an introductory statistics course at Duke University conducted a survey about GPA and major. The side-by-side box plots show the distribution of GPA among three groups of majors. Also provided is the ANOVA output.

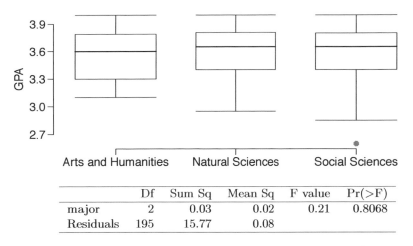

	Df	Sum Sq	Mean Sq	F value	Pr(>F)
major	2	0.03	0.02	0.21	0.8068
Residuals	195	15.77	0.08		

(a) Write the hypotheses for testing for a difference between average GPA across majors.

(b) What is the conclusion of the hypothesis test?

(c) How many students answered these questions on the survey, i.e. what is the sample size?

7.46 Work hours and education. The General Social Survey collects data on demographics, education, and work, among many other characteristics of US residents.[48] Using ANOVA, we can consider educational attainment levels for all 1,172 respondents at once. Below are the distributions of hours worked by educational attainment and relevant summary statistics that will be helpful in carrying out this analysis.

	Educational attainment					
	Less than HS	HS	Jr Coll	Bachelor's	Graduate	Total
Mean	38.67	39.6	41.39	42.55	40.85	40.45
SD	15.81	14.97	18.1	13.62	15.51	15.17
n	121	546	97	253	155	1,172

(a) Write hypotheses for evaluating whether the average number of hours worked varies across the five groups.

(b) Check conditions and describe any assumptions you must make to proceed with the test.

(c) Below is part of the output associated with this test. Fill in the empty cells.

	Df	Sum Sq	Mean Sq	F value	Pr(>F)
degree			501.54		0.0682
Residuals		267,382			
Total					

(d) What is the conclusion of the test?

7.47 True / False: ANOVA, Part I. Determine if the following statements are true or false in ANOVA, and explain your reasoning for statements you identify as false.

(a) As the number of groups increases, the modified significance level for pairwise tests increases as well.

(b) As the total sample size increases, the degrees of freedom for the residuals increases as well.

(c) The constant variance condition can be somewhat relaxed when the sample sizes are relatively consistent across groups.

(d) The independence assumption can be relaxed when the total sample size is large.

---

[48]National Opinion Research Center, General Social Survey, 2010.

7.48   Child care hours. The China Health and Nutrition Survey aims to examine the effects of the health, nutrition, and family planning policies and programs implemented by national and local governments.[49] It, for example, collects information on number of hours Chinese parents spend taking care of their children under age 6. The side-by-side box plots below show the distribution of this variable by educational attainment of the parent. Also provided below is the ANOVA output for comparing average hours across educational attainment categories.

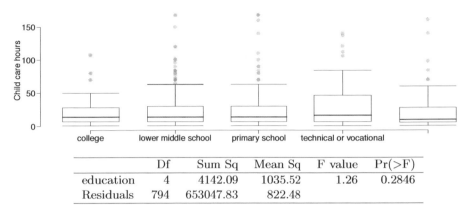

	Df	Sum Sq	Mean Sq	F value	Pr(>F)
education	4	4142.09	1035.52	1.26	0.2846
Residuals	794	653047.83	822.48		

(a) Write the hypotheses for testing for a difference between the average number of hours spent on child care across educational attainment levels.

(b) What is the conclusion of the hypothesis test?

7.49   Prison isolation experiment, Part II. Exercise 7.37 introduced an experiment that was conducted with the goal of identifying a treatment that reduces subjects' psychopathic deviant T-scores, where this score measures a person's need for control or his rebellion against control. In Exercise 7.37 you evaluated the success of each treatment individually. An alternative analysis involves comparing the success of treatments. The relevant ANOVA output is given below.

	Df	Sum Sq	Mean Sq	F value	Pr(>F)
treatment	2	639.48	319.74	3.33	0.0461
Residuals	39	3740.43	95.91		

$$s_{pooled} = 9.793 \text{ on } df = 39$$

(a) What are the hypotheses?

(b) What is the conclusion of the test? Use a 5% significance level.

(c) If in part (b) you determined that the test is significant, conduct pairwise tests to determine which groups are different from each other. If you did not reject the null hypothesis in part (b), recheck your solution.

7.50   True / False: ANOVA, Part II. Determine if the following statements are true or false, and explain your reasoning for statements you identify as false.

   If the null hypothesis that the means of four groups are all the same is rejected using ANOVA at a 5% significance level, then ...

(a) we can then conclude that all the means are different from one another.

(b) the standardized variability between groups is higher than the standardized variability within groups.

(c) the pairwise analysis will identify at least one pair of means that are significantly different.

(d) the appropriate α to be used in pairwise comparisons is 0.05 / 4 = 0.0125 since there are four groups.

---

[49]UNC Carolina Population Center, China Health and Nutrition Survey, 2006.

Chapter 8

Introduction to linear regression

Linear regression is a very powerful statistical technique. Many people have some familiarity with regression just from reading the news, where graphs with straight lines are overlaid on scatterplots. Linear models can be used for prediction or to evaluate whether there is a linear relationship between two numerical variables.

Figure 8.1 shows two variables whose relationship can be modeled perfectly with a straight line. The equation for the line is

$$y = 5 + 57.49x$$

Imagine what a perfect linear relationship would mean: you would know the exact value of y just by knowing the value of x. This is unrealistic in almost any natural process. For example, if we took family income x, this value would provide some useful information about how much financial support y a college may offer a prospective student. However, there would still be variability in financial support, even when comparing students whose families have similar financial backgrounds.

Linear regression assumes that the relationship between two variables, x and y, can be modeled by a straight line:

$$y = \beta_0 + \beta_1 x \tag{8.1}$$

β_0, β_1
Linear model
parameters

where β_0 and β_1 represent two model parameters (β is the Greek letter *beta*). (This use of β has nothing to do with the β we used to describe the probability of a Type 2 Error.) These parameters are estimated using data, and we write their point estimates as b_0 and b_1. When we use x to predict y, we usually call x the explanatory or **predictor** variable, and we call y the response.

It is rare for all of the data to fall on a straight line, as seen in the three scatterplots in Figure 8.2. In each case, the data fall around a straight line, even if none of the observations fall exactly on the line. The first plot shows a relatively strong downward linear trend, where the remaining variability in the data around the line is minor relative to the strength of the relationship between x and y. The second plot shows an upward trend that, while evident, is not as strong as the first. The last plot shows a very weak downward trend in the data, so slight we can hardly notice it. In each of these examples, we will have some uncertainty regarding our estimates of the model parameters, β_0 and β_1. For instance, we

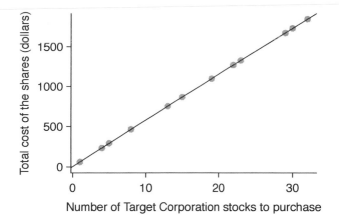

Figure 8.1: Requests from twelve separate buyers were simultaneously placed with a trading company to purchase Target Corporation stock (ticker **TGT**, April 26th, 2012), and the total cost of the shares were reported. Because the cost is computed using a linear formula, the linear fit is perfect.

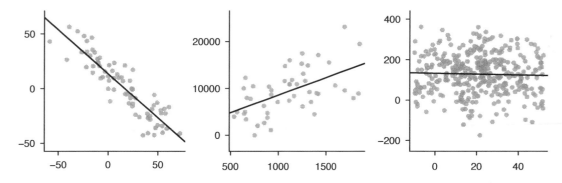

Figure 8.2: Three data sets where a linear model may be useful even though the data do not all fall exactly on the line.

might wonder, should we move the line up or down a little, or should we tilt it more or less? As we move forward in this chapter, we will learn different criteria for line-fitting, and we will also learn about the uncertainty associated with estimates of model parameters.

We will also see examples in this chapter where fitting a straight line to the data, even if there is a clear relationship between the variables, is not helpful. One such case is shown in Figure 8.3 where there is a very strong relationship between the variables even though the trend is not linear. We will discuss nonlinear trends in this chapter and the next, but the details of fitting nonlinear models are saved for a later course.

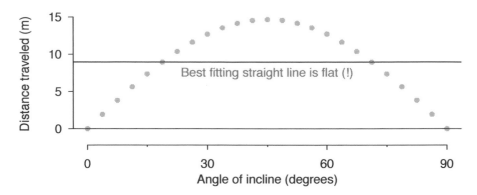

Figure 8.3: A linear model is not useful in this nonlinear case. These data are from an introductory physics experiment.

8.1   Line fitting, residuals, and correlation

It is helpful to think deeply about the line fitting process. In this section, we examine criteria for identifying a linear model and introduce a new statistic, *correlation*.

8.1.1   Beginning with straight lines

Scatterplots were introduced in Chapter 2 as a graphical technique to present two numerical variables simultaneously. Such plots permit the relationship between the variables to be examined with ease. Figure 8.4 shows a scatterplot for the head length and total length of 104 brushtail possums from Australia. Each point represents a single possum from the data.

The head and total length variables are associated. Possums with an above average total length also tend to have above average head lengths. While the relationship is not perfectly linear, it could be helpful to partially explain the connection between these variables with a straight line.

Straight lines should only be used when the data appear to have a linear relationship, such as the case shown in the left panel of Figure 8.6. The right panel of Figure 8.6 shows a case where a curved line would be more useful in understanding the relationship between the two variables.

Caution: Watch out for curved trends

We only consider models based on straight lines in this chapter. If data show a nonlinear trend, like that in the right panel of Figure 8.6, more advanced techniques should be used.

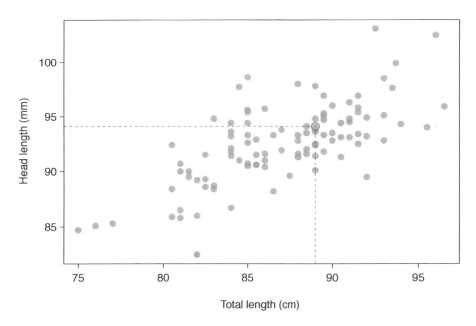

Figure 8.4: A scatterplot showing head length against total length for 104 brushtail possums. A point representing a possum with head length 94.1mm and total length 89cm is highlighted.

Figure 8.5: The common brushtail possum of Australia.

Photo by Peter Firminger on Flickr: http://flic.kr/p/6aPTn CC BY 2.0 license.

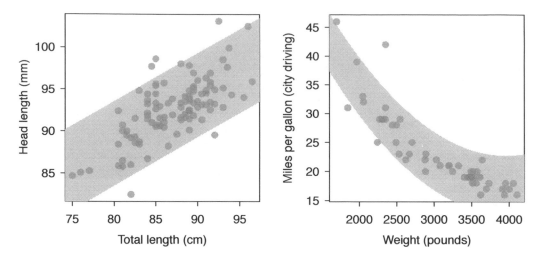

Figure 8.6: The figure on the left shows head length versus total length, and reveals that many of the points could be captured by a straight band. On the right, we see that a curved band is more appropriate in the scatterplot for `weight` and `mpgCity` from the `cars` data set.

8.1.2 Fitting a line by eye

We want to describe the relationship between the head length and total length variables in the possum data set using a line. In this example, we will use the total length as the predictor variable, x, to predict a possum's head length, y. We could fit the linear relationship by eye, as in Figure 8.7. The equation for this line is

$$\hat{y} = 41 + 0.59x \tag{8.2}$$

We can use this line to discuss properties of possums. For instance, the equation predicts a possum with a total length of 80 cm will have a head length of

$$\hat{y} = 41 + 0.59 \times 80$$
$$= 88.2$$

A "hat" on y is used to signify that this is an estimate. This estimate may be viewed as an average: the equation predicts that possums with a total length of 80 cm will have an average head length of 88.2 mm. Absent further information about an 80 cm possum, the prediction for head length that uses the average is a reasonable estimate.

8.1.3 Residuals

Residuals are the leftover variation in the data after accounting for the model fit:

$$\text{Data} = \text{Fit} + \text{Residual}$$

Each observation will have a residual. If an observation is above the regression line, then its residual, the vertical distance from the observation to the line, is positive. Observations below the line have negative residuals. One goal in picking the right linear model is for these residuals to be as small as possible.

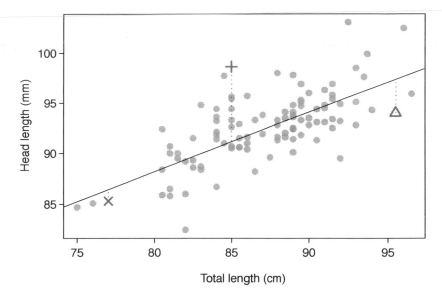

Figure 8.7: A reasonable linear model was fit to represent the relationship between head length and total length.

Three observations are noted specially in Figure 8.7. The observation marked by an "×" has a small, negative residual of about -1; the observation marked by "+" has a large residual of about +7; and the observation marked by "△" has a moderate residual of about -4. The size of a residual is usually discussed in terms of its absolute value. For example, the residual for "△" is larger than that of "×" because $|-4|$ is larger than $|-1|$.

Residual: difference between observed and expected

The residual of the ith observation (x_i, y_i) is the difference of the observed response (y_i) and the response we would predict based on the model fit $(\hat{y}_i)$:

$$\text{residual}_i = y_i - \hat{y}_i$$

We typically identify $\hat{y}_i$ by plugging x_i into the model.

● **Example 8.3** The linear fit shown in Figure 8.7 is given as $\hat{y} = 41 + 0.59x$. Based on this line, formally compute the residual of the observation $(77.0, 85.3)$. This observation is denoted by "×" on the plot. Check it against the earlier visual estimate, -1.

We first compute the predicted value of point "×" based on the model:

$$\hat{y}_\times = 41 + 0.59x_\times = 41 + 0.59 \times 77.0 = 86.4$$

Next we compute the difference of the actual head length and the predicted head length:

$$residual_\times = y_\times - \hat{y}_\times = 85.3 - 86.4 = -1.1$$

This is very close to the visual estimate of -1.

⊙ **Guided Practice 8.4**   If a model underestimates an observation, will the residual be positive or negative? What about if it overestimates the observation?[1]

⊙ **Guided Practice 8.5**   Compute the residuals for the observations $(85.0, 98.6)$ ("+" in the figure) and $(95.5, 94.0)$ ("△") using the linear relationship $\hat{y} = 41 + 0.59x$. [2]

Residuals are helpful in evaluating how well a linear model fits a data set. We often display them in a **residual plot** such as the one shown in Figure 8.8 for the regression line in Figure 8.7. The residuals are plotted at their original horizontal locations but with the vertical coordinate as the residual. For instance, the point $(85.0, 98.6)_+$ had a residual of 7.45, so in the residual plot it is placed at $(85.0, 7.45)$. Creating a residual plot is sort of like tipping the scatterplot over so the regression line is horizontal.

From the residual plot, we can better estimate the **standard deviation of the residuals**, often denoted by the letter s. The standard deviation of the residuals tells us the average size of the residuals. As such, it is a measure of the average deviation between the y values and the regression line. In other words, it tells us the average prediction error using the linear model.

● **Example 8.6**   Estimate the standard deviation of the residuals for predicting head length from total length using the regression line. Also, interpret the quantity in context.

To estimate this graphically, we use the residual plot. The approximate 68, 95 rule for standard deviations applies. Approximately 2/3 of the points are within $\pm$ 2.5 and approximately 95% of the points are within $\pm$ 5, so 2.5 is a good estimate for the standard deviation of the residuals. On average, the prediction of head length is off by about 2.5 cm.

Standard deviation of the residuals

The standard deviation of the residuals, often denoted by the letter s, tells us the average error in the predictions using the regression model. It can be estimated from a residual plot.

---

[1]If a model underestimates an observation, then the model estimate is below the actual. The residual, which is the actual observation value minus the model estimate, must then be positive. The opposite is true when the model overestimates the observation: the residual is negative.

[2](+) First compute the predicted value based on the model:

$$\hat{y}_+ = 41 + 0.59x_+ = 41 + 0.59 \times 85.0 = 91.15$$

Then the residual is given by

$$residual_+ = y_+ - \hat{y}_+ = 98.6 - 91.15 = 7.45$$

This was close to the earlier estimate of 7.

(△) $\hat{y}_△ = 41 + 0.59x_△ = 97.3$. $residual_△ = y_△ - \hat{y}_△ = -3.3$, close to the estimate of -4.

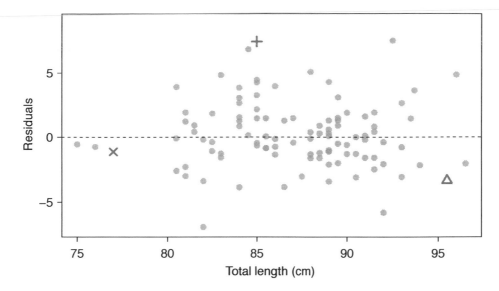

Figure 8.8: Residual plot for the model in Figure 8.7.

● **Example 8.7** One purpose of residual plots is to identify characteristics or patterns still apparent in data after fitting a model. Figure 8.9 shows three scatterplots with linear models in the first row and residual plots in the second row. Can you identify any patterns remaining in the residuals?

In the first data set (first column), the residuals show no obvious patterns. The residuals appear to be scattered randomly around the dashed line that represents 0.

The second data set shows a pattern in the residuals. There is some curvature in the scatterplot, which is more obvious in the residual plot. We should not use a straight line to model these data. Instead, a more advanced technique should be used.

The last plot shows very little upwards trend, and the residuals also show no obvious patterns. It is reasonable to try to fit a linear model to the data. However, it is unclear whether there is statistically significant evidence that the slope parameter is different from zero. The point estimate of the slope parameter, labeled b_1, is not zero, but we might wonder if this could just be due to chance. We will address this sort of scenario in Section 8.4.

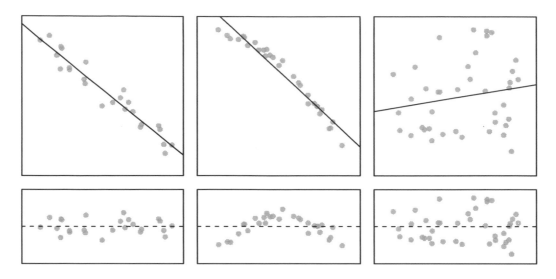

Figure 8.9: Sample data with their best fitting lines (top row) and their corresponding residual plots (bottom row).

8.1.4   Describing linear relationships with correlation

r
correlation

> **Correlation coefficient, r, measures the strength of a linear relationship**
>
> **Correlation**, which always takes values between -1 and 1, describes the strength of the linear relationship between two variables. It can be strong, moderate, or weak.

We can compute the correlation coefficient (or just correlation for short) using a formula, just as we did with the sample mean and standard deviation. However, this formula is rather complex,[3] so we generally perform the calculations on a computer or calculator. Figure 8.10 shows eight plots and their corresponding correlations. Only when the relationship is perfectly linear is the correlation either -1 or 1. If the relationship is strong and positive, the correlation will be near +1. If it is strong and negative, it will be near -1. If there is no apparent linear relationship between the variables, then the correlation will be near zero.

The correlation is intended to quantify the strength of a linear trend. Nonlinear trends, even when strong, sometimes produce correlations that do not reflect the strength of the relationship; see three such examples in Figure 8.11.

⊙ **Guided Practice 8.8**   It appears no straight line would fit any of the datasets represented in Figure 8.11. Try drawing nonlinear curves on each plot. Once you create a curve for each, describe what is important in your fit.[4]

---

[3]Formally, we can compute the correlation for observations (x_1, y_1), (x_2, y_2), ..., (x_n, y_n) using the formula $r = \frac{1}{n-1} \sum_{i=1}^{n} \frac{x_i - \bar{x}}{s_x} \frac{y_i - \bar{y}}{s_y}$, where $\bar{x}$, $\bar{y}$, s_x, and s_y are the sample means and standard deviations for each variable.

[4]We'll leave it to you to draw the lines. In general, the lines you draw should be close to most points and reflect overall trends in the data.

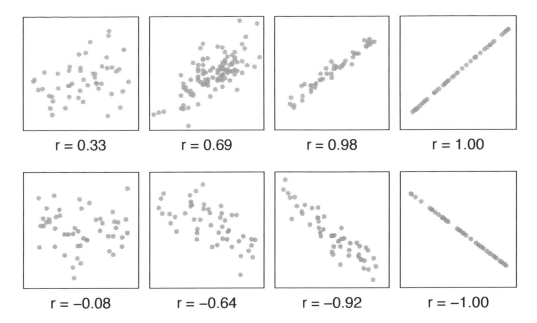

Figure 8.10: Sample scatterplots and their correlations. The first row shows variables with a positive relationship, represented by the trend up and to the right. The second row shows variables with a negative trend, where a large value in one variable is associated with a low value in the other.

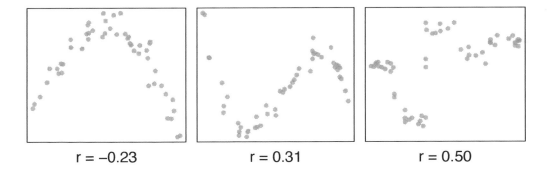

Figure 8.11: Sample scatterplots and their correlations. In each case, there is a strong relationship between the variables. However, the correlation is not very strong, and the relationship is not linear.

● **Example 8.9** Take a look at Figure 8.7 on page 379. How would this correlation change if head length were measured in cm rather than mm? What if head length were measure in inches rather than mm?

Here, changing the units of y corresponds to multiplying all the y values by a certain number. This would change the mean and the standard deviation of y, but it would not change the correlation. To see this, imagine dividing every number on the vertical axes by 10. The units of y are now cm rather than mm, but the graph has remain exactly the same

---

Changing units of x and y does not affect r.

The correlation between two variables should not be dependent upon the units in which the variables are recorded. Adding a constant, subtracting a constant, or multiplying a *positive* constant to all values of x or y does not affect the correlation.

8.2   Fitting a line by least squares regression

Fitting linear models by eye is open to criticism since it is based on an individual preference. In this section, we use *least squares regression* as a more rigorous approach.

This section considers family income and gift aid data from a random sample of fifty students in the 2011 freshman class of Elmhurst College in Illinois.[5] Gift aid is financial aid that does not need to be paid back, as opposed to a loan. A scatterplot of the data is shown in Figure 8.12 along with two linear fits. The lines follow a negative trend in the data; students who have higher family incomes tended to have lower gift aid from the university.

⊙ **Guided Practice 8.10**   Is the correlation positive or negative in Figure 8.12?[6]

8.2.1   An objective measure for finding the best line

We begin by thinking about what we mean by "best". Mathematically, we want a line that has small residuals. Perhaps our criterion could minimize the sum of the residual magnitudes:

$$|y_1 - \hat{y}_1| + |y_2 - \hat{y}_2| + \cdots + |y_n - \hat{y}_n| \tag{8.11}$$

which we could accomplish with a computer program. The resulting dashed line shown in Figure 8.12 demonstrates this fit can be quite reasonable. However, a more common practice is to choose the line that minimizes the sum of the squared residuals:

$$(y_1 - \hat{y}_1)^2 + (y_2 - \hat{y}_2)^2 + \cdots + (y_n - \hat{y}_n)^2 \tag{8.12}$$

The line that minimizes this **least squares criterion** is represented as the solid line in Figure 8.12. This is commonly called the **least squares line**. The following are three possible reasons to choose Criterion (8.12) over Criterion (8.11):

---

[5]These data were sampled from a table of data for all freshman from the 2011 class at Elmhurst College that accompanied an article titled *What Students Really Pay to Go to College* published online by *The Chronicle of Higher Education*: chronicle.com/article/What-Students-Really-Pay-to-Go/131435

[6]Larger family incomes are associated with lower amounts of aid, so the correlation will be negative. Using a computer, the correlation can be computed: -0.499.

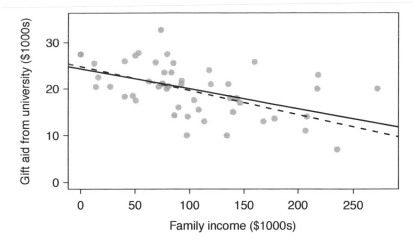

Figure 8.12: Gift aid and family income for a random sample of 50 freshman students from Elmhurst College. Two lines are fit to the data, the solid line being the *least squares line*.

1. It is the most commonly used method.

2. Computing the line based on Criterion (8.12) is much easier by hand and in most statistical software.

3. In many applications, a residual twice as large as another residual is more than twice as bad. For example, being off by 4 is usually more than twice as bad as being off by 2. Squaring the residuals accounts for this discrepancy.

The first two reasons are largely for tradition and convenience; the last reason explains why Criterion (8.12) is typically most helpful.[7]

8.2.2 Conditions for the least squares line

When fitting a least squares line, we generally require

Linearity. The data should show a linear trend. If there is a nonlinear trend (e.g. left panel of Figure 8.13), an advanced regression method from another book or later course should be applied.

Nearly normal residuals. Generally the residuals must be nearly normal. When this condition is found to be unreasonable, it is usually because of outliers or concerns about influential points, which we will discuss in greater depth in Section 8.3. An example of non-normal residuals is shown in the second panel of Figure 8.13.

Constant variability. The variability of points around the least squares line remains roughly constant. An example of non-constant variability is shown in the third panel of Figure 8.13.

---

[7]There are applications where Criterion (8.11) may be more useful, and there are plenty of other criteria we might consider. However, this book only applies the least squares criterion.

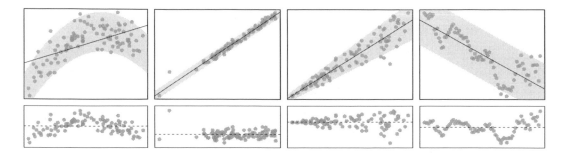

Figure 8.13: Four examples showing when the methods in this chapter are insufficient to apply to the data. In the left panel, a straight line does not fit the data. In the second panel, there are outliers; two points on the left are relatively distant from the rest of the data, and one of these points is very far away from the line. In the third panel, the variability of the data around the line increases with larger values of x. In the last panel, a time series data set is shown, where successive observations are highly correlated.

These conditions are best checked using a residual plot. If a residual plot has no pattern, such as a U-shape or the presence of outliers or non-constant variability in the residuals, then the conditions above may be considered to be satisfied.

TIP: Use a residual plot to determine if a linear model is appropriate
When a residual plot appears as a random cloud of points, a linear model is generally appropriate. If a residual plot has any type of pattern, a linear model is not appropriate.

Be cautious about applying regression to data collected sequentially in what is called a **time series**. Such data may have an underlying structure that should be considered in a model and analysis.

⊙ **Guided Practice 8.13**   Should we have concerns about applying least squares regression to the Elmhurst data in Figure 8.12?[8]

8.2.3   Finding the least squares line

For the Elmhurst data, we could write the equation of the least squares regression line as

$$\widehat{aid} = \beta_0 + \beta_1 \times family\_income$$

Here the equation is set up to predict gift aid based on a student's family income, which would be useful to students considering Elmhurst. These two values, β_0 and β_1, are the *parameters* of the regression line.

As in Chapters 4-6, the parameters are estimated using observed data. In practice, this estimation is done using a computer in the same way that other estimates, like a

---

[8]The trend appears to be linear, the data fall around the line with no obvious outliers, the variance is roughly constant. These are also not time series observations. Least squares regression can be applied to these data.

sample mean, can be estimated using a computer or calculator. However, we can also find the parameter estimates by applying two properties of the least squares line:

- The slope of the least squares line can be estimated by

$$b_1 = r\frac{s_y}{s_x} \tag{8.14}$$

  where r is the correlation between the two variables, and s_x and s_y are the sample standard deviations of the explanatory variable and response, respectively.

- If $\bar{x}$ is the mean of the horizontal variable (from the data) and $\bar{y}$ is the mean of the vertical variable, then the point $(\bar{x}, \bar{y})$ is on the least squares line. Plugging this point in for x and y in the least squares equation and solving for b_0 gives

$$\bar{y} = b_0 + b_1\bar{x} \qquad\qquad b_0 = \bar{y} - b_1\bar{x} \tag{8.15}$$

  When solving for the y-intercept, first find the slope, b_1, and plug the slope and the point $(\bar{x}, \bar{y})$ into the least squares equation.

We use b_0 and b_1 to represent the point estimates of the parameters β_0 and β_1.

b_0, b_1
Sample estimates of β_0, β_1

⊙ **Guided Practice 8.16** Table 8.14 shows the sample means for the family income and gift aid as \$101,800 and \$19,940, respectively. Plot the point $(101.8, 19.94)$ on Figure 8.12 on page 385 to verify it falls on the least squares line (the solid line).[9]

	family income, in \$1000s ("x")	gift aid, in \$1000s ("y")
mean	$\bar{x} = 101.8$	$\bar{y} = 19.94$
sd	$s_x = 63.2$	$s_y = 5.46$
		$r = -0.499$

Table 8.14: Summary statistics for family income and gift aid.

⊙ **Guided Practice 8.17** Using the summary statistics in Table 8.14, compute the slope and y-intercept for the regression line of gift aid against family income. Write the equation of the regression line.[10]

We mentioned earlier that a computer is usually used to compute the least squares line. A summary table based on computer output is shown in Table 8.15 for the Elmhurst data. The first column of numbers provides estimates for b_0 and b_1, respectively. Compare these to the result from Guided Practice 8.17.

---

[9]If you need help finding this location, draw a straight line up from the x-value of 100 (or thereabout). Then draw a horizontal line at 20 (or thereabout). These lines should intersect on the least squares line.

[10]Apply Equations (8.14) and (8.15) with the summary statistics from Table 8.14 to compute the slope and y-intercept:

$$b_1 = r\frac{s_y}{s_x} = (-0.499)\frac{5.46}{63.2} = -0.0431$$
$$b_0 = \bar{y} - b_1\bar{x} = 19.94 - (-0.0431)(101.8) = 24.3$$
$$\hat{y} = 24.3 - 0.0431x \qquad \text{or} \qquad \widehat{aid} = 24.3 - 0.0431\,\text{family\_income}$$

	Estimate	Std. Error	t value	Pr(>|t|)
(Intercept)	24.3193	1.2915	18.83	0.0000
family_income	-0.0431	0.0108	-3.98	0.0002

Table 8.15: Summary of least squares fit for the Elmhurst data. Compare the parameter estimates in the first column to the results of Guided Practice 8.17.

● **Example 8.18** Examine the second, third, and fourth columns in Table 8.15. Can you guess what they represent?

We'll describe the meaning of the columns using the second row, which corresponds to β_1. The first column provides the point estimate for β_1, as we calculated in an earlier example: -0.0431. The second column is a standard error for this point estimate: 0.0108. The third column is a T test statistic for the null hypothesis that $\beta_1 = 0$: $T = -3.98$. The last column is the p-value for the T test statistic for the null hypothesis $\beta_1 = 0$ and a two-sided alternative hypothesis: 0.0002. We will get into more of these details in Section 8.4.

● **Example 8.19** Suppose a high school senior is considering Elmhurst College. Can she simply use the linear equation that we have estimated to calculate her financial aid from the university?

She may use it as an estimate, though some qualifiers on this approach are important. First, the data all come from one freshman class, and the way aid is determined by the university may change from year to year. Second, the equation will provide an imperfect estimate. While the linear equation is good at capturing the trend in the data, no individual student's aid will be perfectly predicted.

8.2.4   Interpreting regression line parameter estimates

Interpreting parameters in a regression model is often one of the most important steps in the analysis.

● **Example 8.20** The slope and intercept estimates for the Elmhurst data are -0.0431 and 24.3. What do these numbers really mean?

Interpreting the slope parameter is helpful in almost any application. For each additional $1,000 of family income, we would expect a student to receive a net difference of $1,000 × (−0.0431) = −$43.10 in aid on average, i.e. $43.10 *less*. Note that a higher family income corresponds to less aid because the coefficient of family income is negative in the model. We must be cautious in this interpretation: while there is a real association, we cannot interpret a causal connection between the variables because these data are observational. That is, increasing a student's family income may not cause the student's aid to drop. (It would be reasonable to contact the college and ask if the relationship is causal, i.e. if Elmhurst College's aid decisions are partially based on students' family income.)

The estimated intercept $b_0 = 24.3$ (in $1000s) describes the average aid if a student's family had no income. The meaning of the intercept is relevant to this application since the family income for some students at Elmhurst is $0. In other applications, the intercept may have little or no practical value if there are no observations where x is near zero.

Interpreting parameters in a linear model

- The slope, b_1, describes the average increase or decrease in the y variable if the explanatory variable x is one unit larger.

- The y-intercept, b_0, describes the average or predicted outcome of y if $x = 0$. The linear model must be valid all the way to $x = 0$ for this to make sense, which in many applications is not the case.

8.2.5   Extrapolation is treacherous

When those blizzards hit the East Coast this winter, it proved to my satisfaction that global warming was a fraud. That snow was freezing cold. But in an alarming trend, temperatures this spring have risen. Consider this: On February 6th it was 10 degrees. Today it hit almost 80. At this rate, by August it will be 220 degrees. So clearly folks the climate debate rages on.

Stephen Colbert
April 6th, 2010 [11]

Linear models can be used to approximate the relationship between two variables. However, these models have real limitations. Linear regression is simply a modeling framework. The truth is almost always much more complex than our simple line. For example, we do not know how the data outside of our limited window will behave.

● **Example 8.21**   Use the model $\widehat{aid} = 24.3 - 0.0431 \times family\_income$ to estimate the aid of another freshman student whose family had income of \$1 million.

Recall that the units of family income are in \$1000s, so we want to calculate the aid for $family\_income = 1000$:

$$\widehat{aid} = 24.3 - 0.0431 \times family\_income$$
$$\widehat{aid} = 24.3 - 0.431(1000) = -18.8$$

The model predicts this student will have -\$18,800 in aid (!). Elmhurst College cannot (or at least does not) require any students to pay extra on top of tuition to attend.

Applying a model estimate to values outside of the realm of the original data is called **extrapolation**. Generally, a linear model is only an approximation of the real relationship between two variables. If we extrapolate, we are making an unreliable bet that the approximate linear relationship will be valid in places where it has not been analyzed.

---

[11]www.colbertnation.com/the-colbert-report-videos/269929

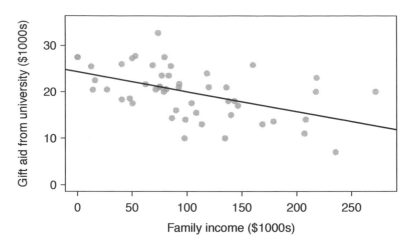

Figure 8.16: Gift aid and family income for a random sample of 50 freshman students from Elmhurst College, shown with the least squares regression line.

8.2.6   Using R^2 to describe the strength of a fit

We evaluated the strength of the linear relationship between two variables earlier using the correlation coefficient, r. However, it is more common to explain the strength of a linear fit using R^2, called **R-squared** or the **explained variance**. If provided with a linear model, we might like to describe how closely the data cluster around the linear fit.

The R^2 of a linear model describes the amount of variation in the response that is explained by the least squares line. For example, consider the Elmhurst data, shown in Figure 8.16. The variance of the response variable, aid received, is $s^2_{aid} = 29.8$. However, if we apply our least squares line, then this model reduces our uncertainty in predicting aid using a student's family income. The variability in the residuals describes how much variation remains after using the model: $s^2_{RES} = 22.4$. In short, there was a reduction of

$$\frac{s^2_{aid} - s^2_{RES}}{s^2_{aid}} = \frac{29.8 - 22.4}{29.8} = \frac{7.5}{29.8} = 0.25$$

This is how we compute the R^2 value.[12] It also corresponds to the square of the correlation coefficient, r, that is, $R^2 = r^2$.

$$R^2 = 0.25 \qquad\qquad\qquad r = -0.499$$

R^2 is the explained variance

R^2 is always between 0 and 1, inclusive. It tells us the proportion of variation in the y values that is explained by a regression model. The higher the value of R^2, the better the model "explains" the reponse variable.

---

[12] $R^2 = 1 - \frac{s^2_{RES}}{s^2_y}$

⊙ **Guided Practice 8.22**   If a linear model has a very strong negative relationship with a correlation of -0.97, how much of the variation in the response is explained by the explanatory variable?[13]

⊙ **Guided Practice 8.23**   If a linear model has an R^2 or explained variance of 0.94, what is the correlation coefficient?[14]

8.2.7   Calculator: linear correlation and regression

▶ TI-84: finding b_0, b_1, R^2, and r for a linear model

Use STAT, CALC, LinReg(a + bx).

1. Choose STAT.
2. Right arrow to CALC.
3. Down arrow and choose 8:LinReg(a+bx).
   - Caution: choosing 4:LinReg(ax+b) will reverse a and b.
4. Let Xlist be L1 and Ylist be L2 (don't forget to enter the x and y values in L1 and L2 before doing this calculation).
5. Leave FreqList blank.
6. Leave Store RegEQ blank.
7. Choose Calculate and hit ENTER, which returns:
   - a b_0, the y-intercept of the best fit line
   - b b_1, the slope of the best fit line
   - r² R^2, the explained variance
   - r r, the correlation coefficient

TI-83: Do steps 1-3, then enter the x list and y list separated by a comma, e.g. LinReg(a+bx) L1, L2, then hit ENTER.

TIP: What to do if r^2 and r do not show up on a TI-83/84

If r^2 and r do now show up when doing STAT, CALC, LinReg, the *diagnostics* must be turned on. This only needs to be once and the diagnostics will remain on.

1. Hit 2ND 0 (i.e. CATALOG).
2. Scroll down until the arrow points at DiagnosticOn.
3. Hit ENTER and ENTER again. The screen should now say:

   DiagnosticOn
             Done

---

[13]About $R^2 = (-0.97)^2 = 0.94$ or 94% of the variation in aid is explained by the linear model.

[14]We take the square root of R^2 and get 0.97, but we must be careful, because r could be 0.97 *or* -0.97. Without knowing the slope or seeing the scatterplot, we have no way of knowing if r is positive or negative.

TIP: What to do if a TI-83/84 returns: `ERR: DIM MISMATCH`
This error means that the lists, generally L1 and L2, do not have the same length.

1. Choose `1:Quit`.

2. Choose `STAT`, `Edit` and make sure that the lists have the same number of entries.

📹 **Casio fx-9750GII: finding b_0, b_1, R^2, and r for a linear model**

1. Navigate to `STAT` (`MENU` button, then hit the 2 button or select `STAT`).

2. Enter the x and y data into 2 separate lists, e.g. x values in `List 1` and y values in `List 2`. Observation ordering should be the same in the two lists. For example, if $(5, 4)$ is the second observation, then the second value in the x list should be 5 and the second value in the y list should be 4.

3. Navigate to `CALC` (`F2`) and then `SET` (`F6`) to set the regression context.

   - To change the `2Var XList`, navigate to it, select `List` (`F1`), and enter the proper list number. Similarly, set `2Var YList` to the proper list.

4. Hit `EXIT`.

5. Select `REG` (`F3`), `X` (`F1`), and `a+bx` (`F2`), which returns:

a	b_0, the y-intercept of the best fit line
b	b_1, the slope of the best fit line
r	r, the correlation coefficient
r²	R^2, the explained variance
MSe	Mean squared error, which you can ignore

   If you select `ax+b` (`F1`), the a and b meanings will be reversed.

---

	fed_spend	poverty
1	6.07	10.6
2	6.14	12.2
3	8.75	25.0
4	7.12	12.6
5	5.13	13.4
6	8.71	5.6
7	6.70	7.9

Table 8.17: Data for Guided Practice 8.24.

⊙ **Guided Practice 8.24** Table 8.17 contains values of federal spending per capita (rounded to the nearand percent of population in poverty for seven counties. This is a subset of the `countyDF` data set from Chapter 1. Use a calculator to find the equation of the least squares regression line for this partial data set.[15]

---

[15]$a = 5.136$ and $b = 1.056$, therefore $\hat{y} = 5.136 + 1.056x$.

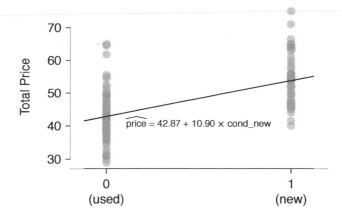

Figure 8.18: Total auction prices for the video game *Mario Kart*, divided into used ($x = 0$) and new ($x = 1$) condition games. The least squares regression line is also shown.

8.2.8 Categorical predictors with two levels (special topic)

Categorical variables are also useful in predicting outcomes. Here we consider a categorical predictor with two levels (recall that a *level* is the same as a *category*). We'll consider Ebay auctions for a video game, *Mario Kart* for the Nintendo Wii, where both the total price of the auction and the condition of the game were recorded.[16] Here we want to predict total price based on game condition, which takes values used and new. A plot of the auction data is shown in Figure 8.18.

To incorporate the game condition variable into a regression equation, we must convert the categories into a numerical form. We will do so using an **indicator variable** called cond_new, which takes value 1 when the game is new and 0 when the game is used. Using this indicator variable, the linear model may be written as

$$\widehat{price} = \beta_0 + \beta_1 \times \texttt{cond\_new}$$

The fitted model is summarized in Table 8.19, and the model with its parameter estimates is given as

$$\widehat{price} = 42.87 + 10.90 \times \texttt{cond\_new}$$

For categorical predictors with just two levels, the linearity assumption will always be satisfied. However, we must evaluate whether the residuals in each group are approximately normal and have approximately equal variance. As can be seen in Figure 8.18, both of these conditions are reasonably satisfied by the auction data.

● **Example 8.25** Interpret the two parameters estimated in the model for the price of *Mario Kart* in eBay auctions.

The intercept is the estimated price when cond_new takes value 0, i.e. when the game is in used condition. That is, the average selling price of a used version of the game is $42.87.

The slope indicates that, on average, new games sell for about $10.90 more than used games.

---

[16]These data were collected in Fall 2009 and may be found at openintro.org/stat.

	Estimate	Std. Error	t value	Pr($>$|t|)
(Intercept)	42.87	0.81	52.67	0.0000
cond_new	10.90	1.26	8.66	0.0000

Table 8.19: Least squares regression summary for the final auction price against the condition of the game.

TIP: Interpreting model estimates for categorical predictors.
The estimated intercept is the value of the response variable for the first category (i.e. the category corresponding to an indicator value of 0). The estimated slope is the average change in the response variable between the two categories.

8.3 Types of outliers in linear regression

In this section, we identify criteria for determining which outliers are important and influential.

Outliers in regression are observations that fall far from the "cloud" of points. These points are especially important because they can have a strong influence on the least squares line.

● **Example 8.26** There are six plots shown in Figure 8.20 along with the least squares line and residual plots. For each scatterplot and residual plot pair, identify any obvious outliers and note how they influence the least squares line. Recall that an outlier is any point that doesn't appear to belong with the vast majority of the other points.

---

(1) There is one outlier far from the other points, though it only appears to slightly influence the line.

(2) There is one outlier on the right, though it is quite close to the least squares line, which suggests it wasn't very influential.

(3) There is one point far away from the cloud, and this outlier appears to pull the least squares line up on the right; examine how the line around the primary cloud doesn't appear to fit very well.

(4) There is a primary cloud and then a small secondary cloud of four outliers. The secondary cloud appears to be influencing the line somewhat strongly, making the least square line fit poorly almost everywhere. There might be an interesting explanation for the dual clouds, which is something that could be investigated.

(5) There is no obvious trend in the main cloud of points and the outlier on the right appears to largely control the slope of the least squares line.

(6) There is one outlier far from the cloud, however, it falls quite close to the least squares line and does not appear to be very influential.

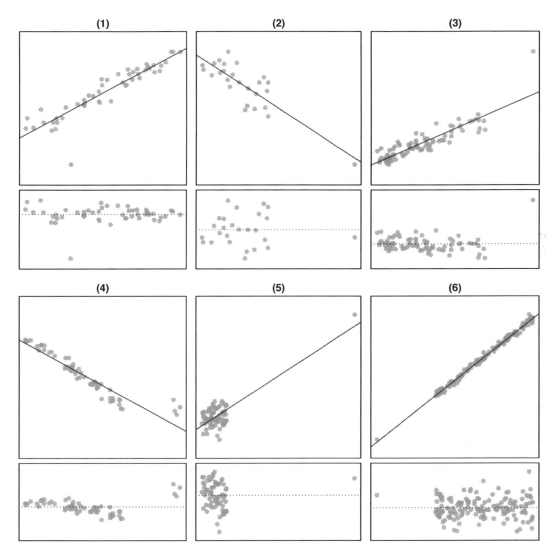

Figure 8.20: Six plots, each with a least squares line and residual plot. All data sets have at least one outlier.

Examine the residual plots in Figure 8.20. You will probably find that there is some trend in the main clouds of (3) and (4). In these cases, the outliers influenced the slope of the least squares lines. In (5), data with no clear trend were assigned a line with a large trend simply due to one outlier (!).

Leverage

Points that fall horizontally away from the center of the cloud tend to pull harder on the line, so we call them points with **high leverage**.

Points that fall horizontally far from the line are points of high leverage; these points can strongly influence the slope of the least squares line. If one of these high leverage points does appear to actually invoke its influence on the slope of the line – as in cases (3), (4), and (5) of Example 8.26 – then we call it an **influential point**. Usually we can say a point is influential if, had we fitted the line without it, the influential point would have been unusually far from the least squares line.

It is tempting to remove outliers. Don't do this without a very good reason. Models that ignore exceptional (and interesting) cases often perform poorly. For instance, if a financial firm ignored the largest market swings – the "outliers" – they would soon go bankrupt by making poorly thought-out investments.

Caution: Don't ignore outliers when fitting a final model
If there are outliers in the data, they should not be removed or ignored without a good reason. Whatever final model is fit to the data would not be very helpful if it ignores the most exceptional cases.

Caution: Outliers for a categorical predictor with two levels
Be cautious about using a categorical predictor when one of the levels has very few observations. When this happens, those few observations become influential points.

8.4   Inference for the slope of a regression line

In this section we discuss uncertainty in the estimates of the slope and y-intercept for a regression line. Just as we identified standard errors for point estimates in previous chapters, we first discuss standard errors for these new estimates. However, in the case of regression, we will identify standard errors using statistical software.

8.4.1   Midterm elections and unemployment

Elections for members of the United States House of Representatives occur every two years, coinciding every four years with the U.S. Presidential election. The set of House elections occurring during the middle of a Presidential term are called midterm elections. In America's two-party system, one political theory suggests the higher the unemployment rate, the worse the President's party will do in the midterm elections.

To assess the validity of this claim, we can compile historical data and look for a connection. We consider every midterm election from 1898 to 2010, with the exception

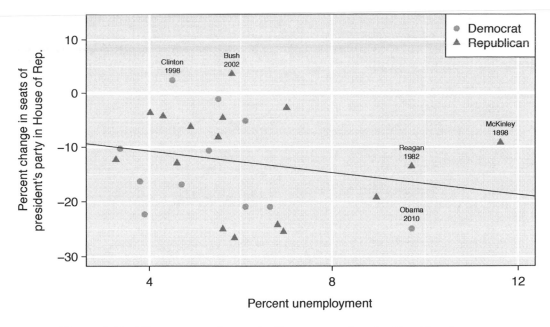

Figure 8.21: The percent change in House seats for the President's party in each election from 1898 to 2010 plotted against the unemployment rate. The two points for the Great Depression have been removed, and a least squares regression line has been fit to the data.

of those elections during the Great Depression. Figure 8.21 shows these data and the least-squares regression line:

$$\% \text{ change in House seats for President's party}$$
$$= -6.71 - 1.00 \times (\text{unemployment rate})$$

We consider the percent change in the number of seats of the President's party (e.g. percent change in the number of seats for Democrats in 2010) against the unemployment rate.

Examining the data, there are no clear deviations from linearity, the constant variance condition, or in the normality of residuals (though we don't examine a normal probability plot here). While the data are collected sequentially, a separate analysis was used to check for any apparent correlation between successive observations; no such correlation was found.

⊙ **Guided Practice 8.27**   The data for the Great Depression (1934 and 1938) were removed because the unemployment rate was 21% and 18%, respectively. Do you agree that they should be removed for this investigation? Why or why not?[17]

---

[17]We will provide two considerations. Each of these points would have very high leverage on any least-squares regression line, and years with such high unemployment may not help us understand what would happen in other years where the unemployment is only modestly high. On the other hand, these are exceptional cases, and we would be discarding important information if we exclude them from a final analysis.

There is a negative slope in the line shown in Figure 8.21. However, this slope (and the y-intercept) are only estimates of the parameter values. We might wonder, is this convincing evidence that the "true" linear model has a negative slope? That is, do the data provide strong evidence that the political theory is accurate? We can frame this investigation into a one-sided statistical hypothesis test:

H_0: $\beta_1 = 0$. The true linear model has slope zero.

H_A: $\beta_1 < 0$. The true linear model has a slope less than zero. The higher the unemployment, the greater the loss for the President's party in the House of Representatives.

We would reject H_0 in favor of H_A if the data provide strong evidence that the true slope parameter is less than zero. To assess the hypotheses, we identify a standard error for the estimate, compute an appropriate test statistic, and identify the p-value.

8.4.2   Understanding regression output from software

Just like other point estimates we have seen before, we can compute a standard error and test statistic for b_1. We will generally label the test statistic using a T, since it follows the t-distribution.

TIP: Hypothesis tests on the slope of the regression line
Use a t-test with $n - 2$ degrees of freedom when performing a hypothesis test on the slope of a regression line.

We will rely on statistical software to compute the standard error and leave the explanation of how this standard error is determined to a second or third statistics course. The table below shows software output for the least squares regression line in Figure 8.21. The row labeled *unemp* represents the information for the slope, which is the coefficient of the unemployment variable.

```
The regression equation is

Change = -6.7142 - 1.0010 unemp

Predictor        Coef    SE Coef        T         P
Constant      -6.7142     5.4567    -1.23    0.2300
unemp         -1.0010     0.8717    -1.15    0.2617

S = 9.624    R-Sq = 0.03%    R-Sq(adj) = -3.7%
```

● **Example 8.28**   What do the first and second columns of numbers in the regression summary represent?

The entries in the first column represent the least squares estimates, b_0 and b_1, and the values in the second column correspond to the standard errors of each estimate.

We previously used a T test statistic for hypothesis testing in the context of numerical data. Regression is very similar. In the hypotheses we consider, the null value for the slope

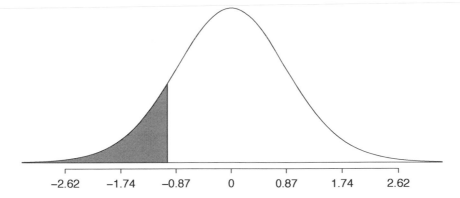

Figure 8.22: The distribution shown here is the sampling distribution for b_1, if the null hypothesis was true. The shaded tail represents the p-value for the hypothesis test evaluating whether there is convincing evidence that higher unemployment corresponds to a greater loss of House seats for the President's party during a midterm election.

is 0, so we can compute the test statistic using the T score formula:

$$T = \frac{\text{estimate} - \text{null value}}{\text{SE}} = \frac{-1.0010 - 0}{0.8717} = -1.15$$

We can look for the one-sided p-value – shown in Figure 8.22 – using the probability table for the t-distribution in Appendix B.3 on page 452.

> ● **Example 8.29** In this example, the sample size $n = 27$. Identify the degrees of freedom and p-value for the hypothesis test.
>
> The degrees of freedom for this test is $n - 2$, or $df = 27 - 2 = 25$. Looking in the 25 degrees of freedom row in Appendix B.3, we see that the absolute value of the test statistic is smaller than any value listed, which means the tail area and therefore also the p-value is larger than 0.100 (one tail!). Because the p-value is so large, we fail to reject the null hypothesis. That is, the data do not provide convincing evidence that a higher unemployment rate has any correspondence with smaller or larger losses for the President's party in the House of Representatives in midterm elections.

We could have identified the T test statistic from the software output of the regression model, shown in the `unemp` row and third column (t value). The entry in the `unemp` row and last column represents the p-value for the two-sided hypothesis test where the null value is zero. The corresponding one-sided test would have a p-value half of the listed value.

> **Inference for regression**
>
> We usually rely on statistical software to identify point estimates and standard errors for parameters of a regression line. After verifying conditions hold for fitting a line, we can use the methods learned in Section 7.1 for the t-distribution to create confidence intervals for regression parameters or to evaluate hypothesis tests.

> **Caution: Don't carelessly use the p-value from regression output**
> The last column in regression output often lists p-values for one particular hypothesis: a two-sided test where the null value is zero. If your test is one-sided and the point estimate is in the direction of H_A, then you can halve the software's p-value to get the one-tail area. If neither of these scenarios match your hypothesis test, be cautious about using the software output to obtain the p-value.

● **Example 8.30** Examine Figure 8.16 on page 390, which relates the Elmhurst College aid and student family income. How sure are you that the slope is statistically significantly different from zero? That is, do you think a formal hypothesis test would reject the claim that the true slope of the line should be zero?

─────────

While the relationship between the variables is not perfect, there is an evident decreasing trend in the data. This suggests the hypothesis test will reject the null claim that the slope is zero.

Recall that $b_1 = r\frac{s_y}{s_x}$. If the slope of the true regression line is zero, the population correlation coefficient must also be zero. The linear regression test for $\beta_1 = 0$ is equivalent, then, to a test for the population correlation coefficient $\rho = 0$.

⊙ **Guided Practice 8.31** The regression summary below shows statistical software output from fitting the least squares regression line shown in Figure 8.16. Use this output to formally evaluate the following hypotheses. H_0: The true slope of the regression line is zero. H_A: The true slope of the regression line is not zero.[18]

```
The regression equation is

aid = 24.31933 - 0.04307 family_income

Predictor        Coef      SE Coef   T         P
Constant         24.31933  1.29145   18.831    < 2e-16
family_income    -0.04307  0.01081   -3.985    0.000229

S = 4.783    R-Sq = 24.86%    R-Sq(adj) = 23.29%
```

> **TIP: Always check assumptions**
> If conditions for fitting the regression line do not hold, then the methods presented here should not be applied. The standard error or distribution assumption of the point estimate – assumed to be normal when applying the T test statistic – may not be valid.

─────────────────────────

[18]We look in the second row corresponding to the family income variable. We see the point estimate of the slope of the line is -0.0431, the standard error of this estimate is 0.0108, and the T test statistic is -3.98. The p-value corresponds exactly to the two-sided test we are interested in: 0.0002. The p-value is so small that we reject the null hypothesis and conclude that family income and financial aid at Elmhurst College for freshman entering in the year 2011 are negatively correlated and the true slope parameter is indeed less than 0, just as we believed in Example 8.30.

8.4.3 Summarizing inference procedures for linear regression

Hypothesis test for the slope of regression line

1. State the name of the test being used.

   - Linear regression t-test

2. Verify conditions.

   - The residual plot has no pattern.

3. Write the hypotheses in plain language. No mathematical notation is needed for this test.

   - H_0: $\beta_1 = 0$, There is no significant linear relationship between [x] and [y].
   - H_A: $\beta_1 \neq$, or $<$, or > 0, There is a significant or significant negative or significant positive linear relationship between [x] and [y].

4. Identify the significance level α.

5. Calculate the test statistic and df: $T = \frac{\text{point estimate} - \text{null value}}{\text{SE of estimate}}$

   - The point estimate is b_1
   - SE can be located on regression summary table next to value of b_1
   - $df = n - 2$

6. Find the p-value, compare it to α, and state whether to reject or not reject the null hypothesis.

7. Write the conclusion in the context of the question.

Constructing a confidence interval for the slope of regression line

1. State the name of the CI being used.

   - t-interval for slope of regression line

2. Verify conditions.

   - The residual plot has no pattern.

3. Plug in the numbers and write the interval in the form

$$\text{point estimate } \pm t^{\star} \times \text{SE of estimate}$$

   - The point estimate is b_1.
   - $df = n - 2$
   - The critical value t^* can be found on the t-table at row $df = n - 2$
   - SE can be located on regression summary table next to value of b_1

4. Evaluate the CI and write in the form (_ , _).

5. Interpret the interval: "We are [XX]% confident that this interval contains the true average increase in [y] for each additional [unit] of [x].

6. State the conclusion to the original question.

8.4.4 Calculator: the linear regression t-test and t-interval

When doing this type of inference, we generally make use of computer output that provides us with the necessary quantities: b and s_b. The calculator functions below require knowing all of the data and are, therefore, rarely used. We describe them here for the sake of completion.

TI-83/84: Linear regression t-test on β_1

Use STAT, TESTS, LinRegTTest.

1. Choose STAT.

2. Right arrow to TESTS.

3. Down arrow and choose F:LinRegTest. (On TI-83 it is E:LinRegTTest).

4. Let Xlist be L1 and Ylist be L2. (Don't forget to enter the x and y values in L1 and L2 before doing this test.)

5. Let Freq be 1.

6. Choose $\neq$, $<$, or $>$ to correspond to H_A.

7. Leave RegEQ blank.

8. Choose Calculate and hit ENTER, which returns:

t	t statistic	b	b_1, slope of the line
p	p-value	s	st. dev. of the residuals
df	degrees of freedom for the test	r²	R^2, explained variance
a	b_0, y-intercept of the line	r	r, correlation coefficient

Casio fx-9750GII: Linear regression t-test on β_1

1. Navigate to STAT (MENU button, then hit the 2 button or select STAT).

2. Enter your data into 2 lists.

3. Select TEST (F3), t (F2), and REG (F3).

4. If needed, update the sidedness of the test and the XList and YList lists. The Freq should be set to 1.

5. Hit EXE, which returns:

t	t statistic	b	b_1, slope of the line
p	p-value	s	st. dev. of the residuals
df	degrees of freedom for the test	r	r, correlation coefficient
a	b_0, y-intercept of the line	r²	R^2, explained variance

▶ TI-84: t-interval for β_1

Use STAT, TESTS, LinRegTInt.

1. Choose STAT.

2. Right arrow to TESTS.

3. Down arrow and choose G: LinRegTest.

    - This test is not built into the TI-83.

4. Let Xlist be L1 and Ylist be L2. (Don't forget to enter the x and y values in L1 and L2 before doing this interval.)

5. Let Freq be 1.

6. Enter the desired confidence level.

7. Leave RegEQ blank.

8. Choose Calculate and hit ENTER, which returns:

(__,__)	the confidence interval
b	b_1, the slope of best fit line of the sample data
df	degrees of freedom associated with this confidence interval
s	standard deviation of the residuals
a	b_0, the y-intercept of the best fit line of the sample data
r²	R^2, the explained variance
r	r, the correlation coefficient

8.5   Transformations for nonlinear data

8.5.1   Untransformed

● **Example 8.32**   Consider the scatterplot and residual plot in Figure 8.23. The regression output is also provided. Would a linear model be a good model for the data shown?

First, we can note the R^2 value is fairly large. However, this alone does not mean that the model is good. Another model might be much better. When assessing the appropriateness of a linear model, we should look at the residual plot. The U pattern in the residual plot tells us the original data is curved. If we inspect the two plots, we can see that for small and large values of x we systematically underestimate y, whereas for middle values of x, we systematically overestimate y. Because of this, the model is not appropriate, and we should not carry out a linear regression t-test because the conditions for inference are not met. However, we might be able to use a transformation to linearize the data.

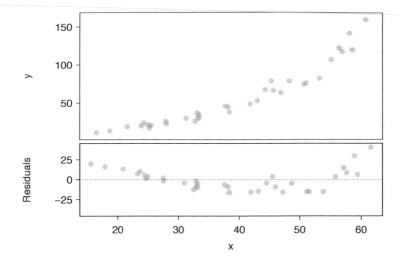

Figure 8.23: Variable y is plotted against x. A nonlinear relationship is evident by the "U" shown in the residual plot. The curvature is also visible in the original plot.

```
The regression equation is

y = -52.3564 + 2.7842 x

Predictor         Coef    SE Coef          T          P
Constant      -52.3564     7.2757     -7.196       3e-08
x               2.7842     0.1768     15.752     < 2e-16

S = 13.76    R-Sq = 88.26%    R-Sq(adj) = 87.91%
```

8.5.2   Transformed

Regression analysis is easier to perform on linear data. When data are nonlinear, we sometimes **transform** the data in a way that makes the resulting relationship linear. The most common **transformation** is log (or ln) of the y values. Sometimes we also apply a transformation to the x values. We generally use the residuals as a way to evaluate whether the transformed data are more linear. If so, we can say that a better model has been found.

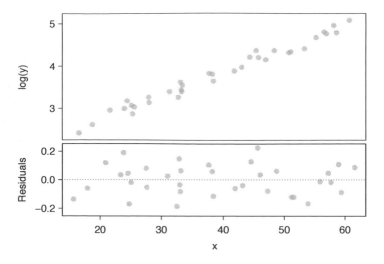

Figure 8.24: A plot of $\log(y)$ against x. The residuals don't show any evident patterns, which suggests the transformed data is well-fit by a linear model.

● **Example 8.33** Using the regression output for the transformed data, write the new linear regression equation

$$\widehat{\log(y)} = 1.723 + 0.053x$$

```
The regression equation is

log(y) = 1.722540 + 0.052985 x

Predictor        Coef      SE Coef        T          P
Constant      1.722540    0.056731      30.36    < 2e-16
x             0.052985    0.001378      38.45    < 2e-16


S = 0.1073    R-Sq = 97.82%    R-Sq(adj) = 97.75%
```

⊙ **Guided Practice 8.34** Which of the following statements are true? There may be more than one.[19]

(a) There is an apparent linear relationship between x and y.

(b) There is an apparent linear relationship between x and $\widehat{\log(y)}$.

(c) The model provided by Regression I ($\hat{y} = -52.3564 + 2.7842x$) yields a better fit.

(d) The model provided by Regression II ($\widehat{\log(y)} = 1.723 + 0.053x$) yields a better fit.

---

[19]Part (a) is *false* since there is a nonlinear (curved) trend in the data. Part (b) is true. Since the transformed data shows a stronger linear trend, it is a better fit, i.e. Part (c) is *false*, and Part (d) is true.

8.6 Exercises

8.6.1 Line fitting, residuals, and correlation

8.1 Visualize the residuals. The scatterplots shown below each have a superimposed regression line. If we were to construct a residual plot (residuals versus x) for each, describe what those plots would look like.

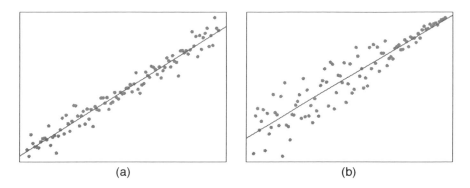

(a)   (b)

8.2 Trends in the residuals. Shown below are two plots of residuals remaining after fitting a linear model to two different sets of data. Describe important features and determine if a linear model would be appropriate for these data. Explain your reasoning.

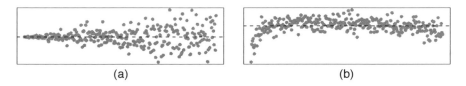

(a)   (b)

8.3 Identify relationships, Part I. For each of the six plots, identify the strength of the relationship (e.g. weak, moderate, or strong) in the data and whether fitting a linear model would be reasonable.

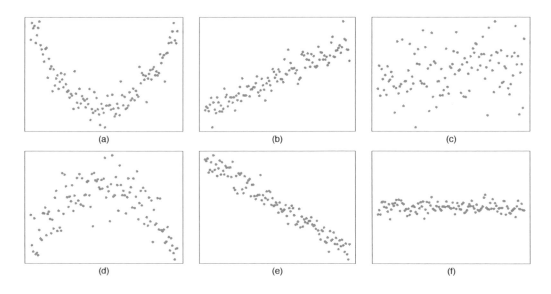

(a)   (b)   (c)

(d)   (e)   (f)

8.4 Identify relationships, Part II. For each of the six plots, identify the strength of the relationship (e.g. weak, moderate, or strong) in the data and whether fitting a linear model would be reasonable.

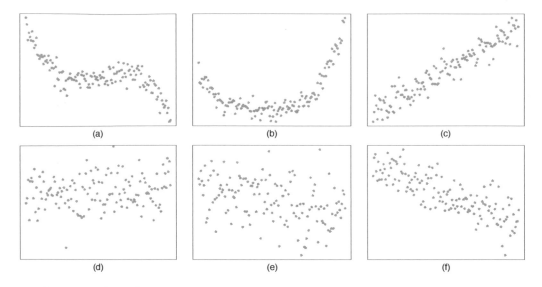

8.5 Exams and grades. The two scatterplots below show the relationship between final and mid-semester exam grades recorded during several years for a Statistics course at a university.

(a) Based on these graphs, which of the two exams has the strongest correlation with the final exam grade? Explain.

(b) Can you think of a reason why the correlation between the exam you chose in part (a) and the final exam is higher?

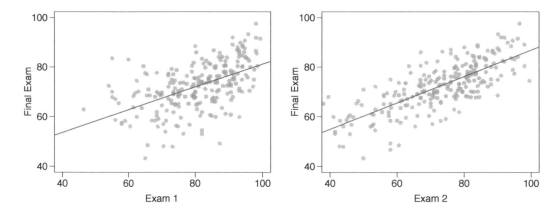

8.6 Husbands and wives, Part I. The Great Britain Office of Population Census and Surveys once collected data on a random sample of 170 married couples in Britain, recording the age (in years) and heights (converted here to inches) of the husbands and wives.[20] The scatterplot on the left shows the wife's age plotted against her husband's age, and the plot on the right shows wife's height plotted against husband's height.

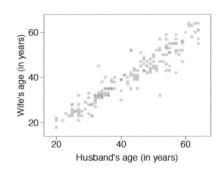
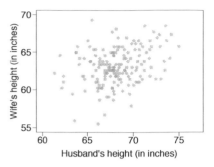

(a) Describe the relationship between husbands' and wives' ages.

(b) Describe the relationship between husbands' and wives' heights.

(c) Which plot shows a stronger correlation? Explain your reasoning.

(d) Data on heights were originally collected in centimeters, and then converted to inches. Does this conversion affect the correlation between husbands' and wives' heights?

8.7 Match the correlation, Part I.
Match the calculated correlations to the corresponding scatterplot.

(a) $r = -0.7$

(b) $r = 0.45$

(c) $r = 0.06$

(d) $r = 0.92$

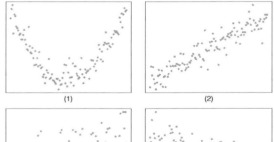

8.8 Match the correlation, Part II.
Match the calculated correlations to the corresponding scatterplot.

(a) $r = 0.49$

(b) $r = -0.48$

(c) $r = -0.03$

(d) $r = -0.85$

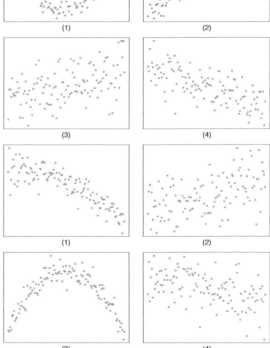

[20]D.J. Hand. *A handbook of small data sets.* Chapman & Hall/CRC, 1994.

8.9 True / False. Determine if the following statements are true or false. If false, explain why.

(a) A correlation coefficient of -0.90 indicates a stronger linear relationship than a correlation coefficient of 0.5.

(b) Correlation is a measure of the association between any two variables.

8.10 Guess the correlation. Eduardo and Rosie are both collecting data on number of rainy days in a year and the total rainfall for the year. Eduardo records rainfall in inches and Rosie in centimeters. How will their correlation coefficients compare?

8.11 Speed and height. 1,302 UCLA students were asked to fill out a survey where they were asked about their height, fastest speed they have ever driven, and gender. The scatterplot on the left displays the relationship between height and fastest speed, and the scatterplot on the right displays the breakdown by gender in this relationship.

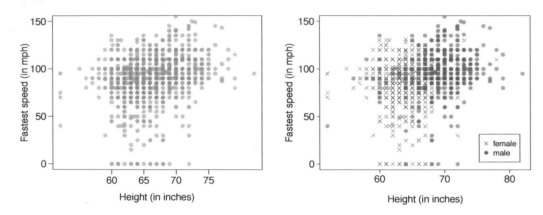

(a) Describe the relationship between height and fastest speed.

(b) Why do you think these variables are positively associated?

(c) What role does gender play in the relationship between height and fastest driving speed?

8.12 Trees. The scatterplots below show the relationship between height, diameter, and volume of timber in 31 felled black cherry trees. The diameter of the tree is measured 4.5 feet above the ground.[21]

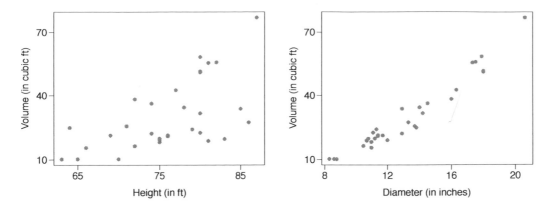

(a) Describe the relationship between volume and height of these trees.

(b) Describe the relationship between volume and diameter of these trees.

(c) Suppose you have height and diameter measurements for another black cherry tree. Which of these variables would be preferable to use to predict the volume of timber in this tree using a simple linear regression model? Explain your reasoning.

8.13 The Coast Starlight, Part I. The Coast Starlight Amtrak train runs from Seattle to Los Angeles. The scatterplot below displays the distance between each stop (in miles) and the amount of time it takes to travel from one stop to another (in minutes).

(a) Describe the relationship between distance and travel time.

(b) How would the relationship change if travel time was instead measured in hours, and distance was instead measured in kilometers?

(c) Correlation between travel time (in miles) and distance (in minutes) is $r = 0.636$. What is the correlation between travel time (in kilometers) and distance (in hours)?

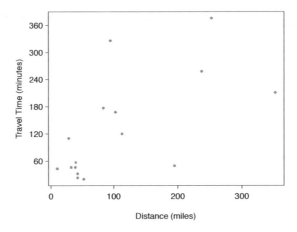

---

[21]Source: R Dataset, stat.ethz.ch/R-manual/R-patched/library/datasets/html/trees.html.

8.14   Crawling babies, Part I. A study conducted at the University of Denver investigated whether babies take longer to learn to crawl in cold months, when they are often bundled in clothes that restrict their movement, than in warmer months.[22] Infants born during the study year were split into twelve groups, one for each birth month. We consider the average crawling age of babies in each group against the average temperature when the babies are six months old (that's when babies often begin trying to crawl). Temperature is measured in degrees Fahrenheit (°F) and age is measured in weeks.

(a) Describe the relationship between temperature and crawling age.

(b) How would the relationship change if temperature was measured in degrees Celsius (°C) and age was measured in months?

(c) The correlation between temperature in °F and age in weeks was $r = -0.70$. If we converted the temperature to °C and age to months, what would the correlation be?

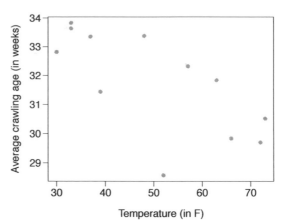

8.15   Body measurements, Part I. Researchers studying anthropometry collected body girth measurements and skeletal diameter measurements, as well as age, weight, height and gender for 507 physically active individuals.[23] The scatterplot below shows the relationship between height and shoulder girth (over deltoid muscles), both measured in centimeters.

(a) Describe the relationship between shoulder girth and height.

(b) How would the relationship change if shoulder girth was measured in inches while the units of height remained in centimeters?

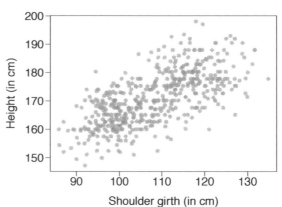

[22] J.B. Benson. "Season of birth and onset of locomotion: Theoretical and methodological implications". In: *Infant behavior and development* 16.1 (1993), pp. 69–81. ISSN: 0163-6383.

[23] G. Heinz et al. "Exploring relationships in body dimensions". In: *Journal of Statistics Education* 11.2 (2003).

8.16 Body measurements, Part II. The scatterplot below shows the relationship between weight measured in kilograms and hip girth measured in centimeters from the data described in Exercise 8.15.

(a) Describe the relationship between hip girth and weight.

(b) How would the relationship change if weight was measured in pounds while the units for hip girth remained in centimeters?

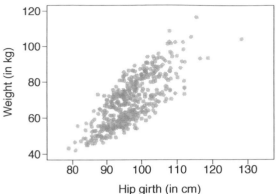

8.17 Correlation, Part I. What would be the correlation between the ages of husbands and wives if men always married woman who were

(a) 3 years younger than themselves?

(b) 2 years older than themselves?

(c) half as old as themselves?

8.18 Correlation, Part II. What would be the correlation between the annual salaries of males and females at a company if for a certain type of position men always made

(a) $5,000 more than women?

(b) 25% more than women?

(c) 15% less than women?

8.6.2 Fitting a line by least squares regression

8.19 Units of regression. Consider a regression predicting weight (kg) from height (cm) for a sample of adult males. What are the units of the correlation coefficient, the intercept, and the slope?

8.20 Which is higher? Determine if I or II is higher or if they are equal. Explain your reasoning.

For a regression line, the uncertainty associated with the slope estimate, b_1, is higher when

   I. there is a lot of scatter around the regression line or

   II. there is very little scatter around the regression line

8.21 Over-under, Part I. Suppose we fit a regression line to predict the shelf life of an apple based on its weight. For a particular apple, we predict the shelf life to be 4.6 days. The apple's residual is -0.6 days. Did we over or under estimate the shelf-life of the apple? Explain your reasoning.

8.22 Over-under, Part II. Suppose we fit a regression line to predict the number of incidents of skin cancer per 1,000 people from the number of sunny days in a year. For a particular year, we predict the incidence of skin cancer to be 1.5 per 1,000 people, and the residual for this year is 0.5. Did we over or under estimate the incidence of skin cancer? Explain your reasoning.

8.23   Tourism spending. The Association of Turkish Travel Agencies reports the number of foreign tourists visiting Turkey and tourist spending by year.[24] Three plots are provided: scatterplot showing the relationship between these two variables along with the least squares fit, residuals plot, and histogram of residuals.

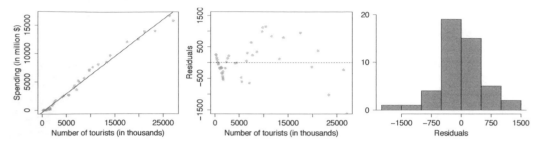

(a) Describe the relationship between number of tourists and spending.

(b) What are the explanatory and response variables?

(c) Why might we want to fit a regression line to these data?

(d) Do the data meet the conditions required for fitting a least squares line? In addition to the scatterplot, use the residual plot and histogram to answer this question.

8.24   Nutrition at Starbucks, Part I. The scatterplot below shows the relationship between the number of calories and amount of carbohydrates (in grams) Starbucks food menu items contain.[25] Since Starbucks only lists the number of calories on the display items, we are interested in predicting the amount of carbs a menu item has based on its calorie content.

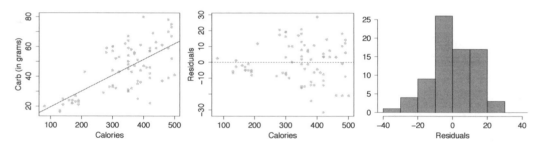

(a) Describe the relationship between number of calories and amount of carbohydrates (in grams) that Starbucks food menu items contain.

(b) In this scenario, what are the explanatory and response variables?

(c) Why might we want to fit a regression line to these data?

(d) Do these data meet the conditions required for fitting a least squares line?

---

[24]Association of Turkish Travel Agencies, Foreign Visitors Figure & Tourist Spendings By Years.
[25]Source: Starbucks.com, collected on March 10, 2011, www.starbucks.com/menu/nutrition.

8.25 The Coast Starlight, Part II. Exercise 8.13 introduces data on the Coast Starlight Amtrak train that runs from Seattle to Los Angeles. The mean travel time from one stop to the next on the Coast Starlight is 129 mins, with a standard deviation of 113 minutes. The mean distance traveled from one stop to the next is 108 miles with a standard deviation of 99 miles. The correlation between travel time and distance is 0.636.

(a) Write the equation of the regression line for predicting travel time.

(b) Interpret the slope and the intercept in this context.

(c) Calculate R^2 of the regression line for predicting travel time from distance traveled for the Coast Starlight, and interpret R^2 in the context of the application.

(d) The distance between Santa Barbara and Los Angeles is 103 miles. Use the model to estimate the time it takes for the Starlight to travel between these two cities.

(e) It actually takes the Coast Starlight about 168 mins to travel from Santa Barbara to Los Angeles. Calculate the residual and explain the meaning of this residual value.

(f) Suppose Amtrak is considering adding a stop to the Coast Starlight 500 miles away from Los Angeles. Would it be appropriate to use this linear model to predict the travel time from Los Angeles to this point?

8.26 Body measurements, Part III. Exercise 8.15 introduces data on shoulder girth and height of a group of individuals. The mean shoulder girth is 107.20 cm with a standard deviation of 10.37 cm. The mean height is 171.14 cm with a standard deviation of 9.41 cm. The correlation between height and shoulder girth is 0.67.

(a) Write the equation of the regression line for predicting height.

(b) Interpret the slope and the intercept in this context.

(c) Calculate R^2 of the regression line for predicting height from shoulder girth, and interpret it in the context of the application.

(d) A randomly selected student from your class has a shoulder girth of 100 cm. Predict the height of this student using the model.

(e) The student from part (d) is 160 cm tall. Calculate the residual, and explain what this residual means.

(f) A one year old has a shoulder girth of 56 cm. Would it be appropriate to use this linear model to predict the height of this child?

8.27 Nutrition at Starbucks, Part II. Exercise 8.24 introduced a data set on nutrition information on Starbucks food menu items. Based on the scatterplot and the residual plot provided, describe the relationship between the protein content and calories of these menu items, and determine if a simple linear model is appropriate to predict amount of protein from the number of calories.

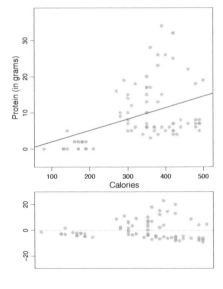

8.28 Helmets and lunches. The scatterplot shows the relationship between socioeconomic status measured as the percentage of children in a neighborhood receiving reduced-fee lunches at school (`lunch`) and the percentage of bike riders in the neighborhood wearing helmets (`helmet`). The average percentage of children receiving reduced-fee lunches is 30.8% with a standard deviation of 26.7% and the average percentage of bike riders wearing helmets is 38.8% with a standard deviation of 16.9%.

(a) If the R^2 for the least-squares regression line for these data is 72%, what is the correlation between `lunch` and `helmet`?

(b) Calculate the slope and intercept for the least-squares regression line for these data.

(c) Interpret the intercept of the least-squares regression line in the context of the application.

(d) Interpret the slope of the least-squares regression line in the context of the application.

(e) What would the value of the residual be for a neighborhood where 40% of the children receive reduced-fee lunches and 40% of the bike riders wear helmets? Interpret the meaning of this residual in the context of the application.

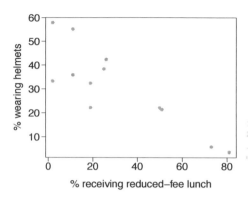

8.29 Murders and poverty, Part I. The following regression output is for predicting annual murders per million from percentage living in poverty in a random sample of 20 metropolitan areas.

	Estimate	Std. Error	t value	Pr(>|t|)
(Intercept)	-29.901	7.789	-3.839	0.001
poverty%	2.559	0.390	6.562	0.000

$s = 5.512$ $R^2 = 70.52\%$ $R^2_{adj} = 68.89\%$

(a) Write out the linear model.

(b) Interpret the intercept.

(c) Interpret the slope.

(d) Interpret R^2.

(e) Calculate the correlation coefficient.

8.30 Cats, Part I. The following regression output is for predicting the heart weight (in g) of cats from their body weight (in kg). The coefficients are estimated using a dataset of 144 domestic cats.

	Estimate	Std. Error	t value	Pr(>|t|)
(Intercept)	-0.357	0.692	-0.515	0.607
body wt	4.034	0.250	16.119	0.000

$s = 1.452$ $R^2 = 64.66\%$ $R^2_{adj} = 64.41\%$

(a) Write out the linear model.

(b) Interpret the intercept.

(c) Interpret the slope.

(d) Interpret R^2.

(e) Calculate the correlation coefficient.

8.6.3 Types of outliers in linear regression

8.31 Outliers, Part I. Identify the outliers in the scatterplots shown below, and determine what type of outliers they are. Explain your reasoning.

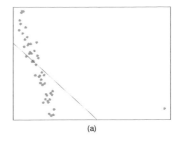
(a)

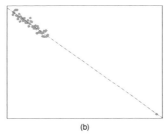
(b)

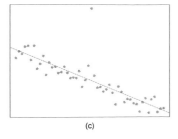
(c)

8.32 Outliers, Part II. Identify the outliers in the scatterplots shown below and determine what type of outliers they are. Explain your reasoning.

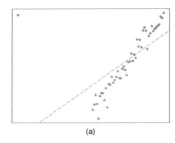
(a)

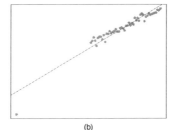
(b)

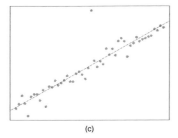
(c)

8.33 Urban homeowners, Part I. The scatterplot below shows the percent of families who own their home vs. the percent of the population living in urban areas in 2010.[26] There are 52 observations, each corresponding to a state in the US. Puerto Rico and District of Columbia are also included.

(a) Describe the relationship between the percent of families who own their home and the percent of the population living in urban areas in 2010.

(b) The outlier at the bottom right corner is District of Columbia, where 100% of the population is considered urban. What type of outlier is this observation?

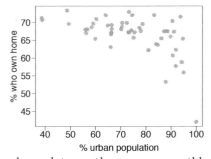

8.34 Crawling babies, Part II. Exercise 8.14 introduces data on the average monthly temperature during the month babies first try to crawl (about 6 months after birth) and the average first crawling age for babies born in a given month. A scatterplot of these two variables reveals a potential outlying month when the average temperature is about 53°F and average crawling age is about 28.5 weeks. Does this point have high leverage? Is it an influential point?

---

[26]United States Census Bureau, 2010 Census Urban and Rural Classification and Urban Area Criteria and Housing Characteristics: 2010.

8.6.4   Inference for the slope of a regression line

In the following exercises, visually check the conditions for fitting a least squares regression line, but you do not need to report these conditions in your solutions.

8.35 Body measurements, Part IV. The scatterplot and least squares summary below show the relationship between weight measured in kilograms and height measured in centimeters of 507 physically active individuals.

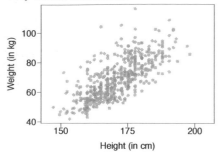

	Estimate	Std. Error	t value	Pr(>|t|)
(Intercept)	-105.0113	7.5394	-13.93	0.0000
height	1.0176	0.0440	23.13	0.0000

(a) Describe the relationship between height and weight.

(b) Write the equation of the regression line. Interpret the slope and intercept in context.

(c) Do the data provide strong evidence that an increase in height is associated with an increase in weight? State the null and alternative hypotheses, report the p-value, and state your conclusion.

(d) The correlation coefficient for height and weight is 0.72. Calculate R^2 and interpret it in context.

8.36 **Beer and blood alcohol content.** Many people believe that gender, weight, drinking habits, and many other factors are much more important in predicting blood alcohol content (BAC) than simply considering the number of drinks a person consumed. Here we examine data from sixteen student volunteers at Ohio State University who each drank a randomly assigned number of cans of beer. These students were evenly divided between men and women, and they differed in weight and drinking habits. Thirty minutes later, a police officer measured their blood alcohol content (BAC) in grams of alcohol per deciliter of blood.[27] The scatterplot and regression table summarize the findings.

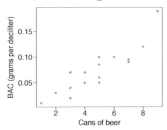

	Estimate	Std. Error	t value	Pr(>|t|)
(Intercept)	-0.0127	0.0126	-1.00	0.3320
beers	0.0180	0.0024	7.48	0.0000

(a) Describe the relationship between the number of cans of beer and BAC.

(b) Write the equation of the regression line. Interpret the slope and intercept in context.

(c) Do the data provide strong evidence that drinking more cans of beer is associated with an increase in blood alcohol? State the null and alternative hypotheses, report the p-value, and state your conclusion.

(d) The correlation coefficient for number of cans of beer and BAC is 0.89. Calculate R^2 and interpret it in context.

(e) Suppose we visit a bar, ask people how many drinks they have had, and also take their BAC. Do you think the relationship between number of drinks and BAC would be as strong as the relationship found in the Ohio State study?

8.37 **Husbands and wives, Part II.** The scatterplot below summarizes husbands' and wives' heights in a random sample of 170 married couples in Britain, where both partners' ages are below 65 years. Summary output of the least squares fit for predicting wife's height from husband's height is also provided in the table.

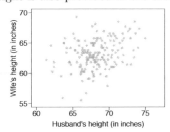

	Estimate	Std. Error	t value	Pr(>|t|)
(Intercept)	43.5755	4.6842	9.30	0.0000
height_husband	0.2863	0.0686	4.17	0.0000

(a) Is there strong evidence that taller men marry taller women? State the hypotheses and include any information used to conduct the test.

(b) Write the equation of the regression line for predicting wife's height from husband's height.

(c) Interpret the slope and intercept in the context of the application.

(d) Given that $R^2 = 0.09$, what is the correlation of heights in this data set?

(e) You meet a married man from Britain who is 5'9" (69 inches). What would you predict his wife's height to be? How reliable is this prediction?

(f) You meet another married man from Britain who is 6'7" (79 inches). Would it be wise to use the same linear model to predict his wife's height? Why or why not?

---

[27]J. Malkevitch and L.M. Lesser. *For All Practical Purposes: Mathematical Literacy in Today's World.* WH Freeman & Co, 2008.

8.38 Husbands and wives, Part III. Exercise 8.6 presents a scatterplot displaying the relationship between husbands' and wives' ages in a random sample of 170 married couples in Britain, where both partners' ages are below 65 years. Given below is summary output of the least squares fit for predicting wife's age from husband's age.

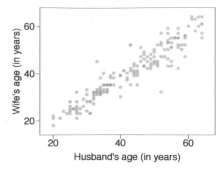

	Estimate	Std. Error	t value	Pr(>|t|)
(Intercept)	1.5740	1.1501	1.37	0.1730
age_husband	0.9112	0.0259	35.25	0.0000
				$df = 168$

(a) We might wonder, is the age difference between husbands and wives consistent across ages? If this were the case, then the slope parameter would be $\beta_1 = 1$. Use the information above to evaluate if there is strong evidence that the difference in husband and wife ages differs for different ages.

(b) Write the equation of the regression line for predicting wife's age from husband's age.

(c) Interpret the slope and intercept in context.

(d) Given that $R^2 = 0.88$, what is the correlation of ages in this data set?

(e) You meet a married man from Britain who is 55 years old. What would you predict his wife's age to be? How reliable is this prediction?

(f) You meet another married man from Britain who is 85 years old. Would it be wise to use the same linear model to predict his wife's age? Explain.

8.39 Urban homeowners, Part II. Exercise 8.33 gives a scatterplot displaying the relationship between the percent of families that own their home and the percent of the population living in urban areas. Below is a similar scatterplot, excluding District of Columbia, as well as the residuals plot. There were 51 cases.

(a) For these data, $R^2 = 0.28$. What is the correlation? How can you tell if it is positive or negative?

(b) Examine the residual plot. What do you observe? Is a simple least squares fit appropriate for these data?

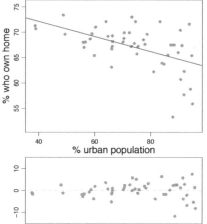

8.40 Rate my professor. Many college courses conclude by giving students the opportunity to evaluate the course and the instructor anonymously. However, the use of these student evaluations as an indicator of course quality and teaching effectiveness is often criticized because these measures may reflect the influence of non-teaching related characteristics, such as the physical appearance of the instructor. Researchers at University of Texas, Austin collected data on teaching evaluation score (higher score means better) and standardized beauty score (a score of 0 means average, negative score means below average, and a positive score means above average) for a sample of 463 professors.[28] The scatterplot below shows the relationship between these variables,

---

[28]Daniel S Hamermesh and Amy Parker. "Beauty in the classroom: Instructors pulchritude and putative pedagogical productivity". In: *Economics of Education Review* 24.4 (2005), pp. 369–376.

and also provided is a regression output for predicting teaching evaluation score from beauty score.

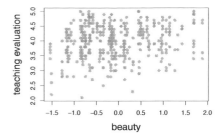

	Estimate	Std. Error	t value	Pr(>|t|)
(Intercept)	4.010	0.0255	157.21	0.0000
beauty		0.0322	4.13	0.0000

(a) Given that the average standardized beauty score is -0.0883 and average teaching evaluation score is 3.9983, calculate the slope. Alternatively, the slope may be computed using just the information provided in the model summary table.

(b) Do these data provide convincing evidence that the slope of the relationship between teaching evaluation and beauty is positive? Explain your reasoning.

(c) List the conditions required for linear regression and check if each one is satisfied for this model based on the following diagnostic plots.

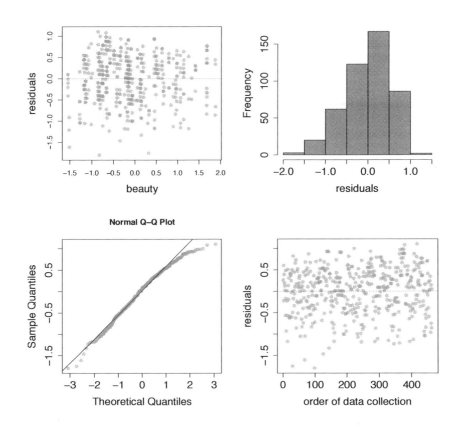

8.41 Murders and poverty, Part II. Exercise 8.29 presents regression output from a model for predicting annual murders per million from percentage living in poverty based on a random sample of 20 metropolitan areas. The model output is also provided below.

	Estimate	Std. Error	t value	Pr(>|t|)
(Intercept)	-29.901	7.789	-3.839	0.001
poverty%	2.559	0.390	6.562	0.000

$$s = 5.512 \qquad R^2 = 70.52\% \qquad R^2_{adj} = 68.89\%$$

(a) What are the hypotheses for evaluating whether poverty percentage is a significant predictor of murder rate?

(b) State the conclusion of the hypothesis test from part (a) in context of the data.

(c) Calculate a 95% confidence interval for the slope of poverty percentage, and interpret it in context of the data.

(d) Do your results from the hypothesis test and the confidence interval agree? Explain.

8.42 Babies. Is the gestational age (time between conception and birth) of a low birth-weight baby useful in predicting head circumference at birth? Twenty-five low birth-weight babies were studied at a Harvard teaching hospital; the investigators calculated the regression of head circumference (measured in centimeters) against gestational age (measured in weeks). The estimated regression line is

$$\widehat{head\_circumference} = 3.91 + 0.78 \times gestational\_age$$

(a) What is the predicted head circumference for a baby whose gestational age is 28 weeks?

(b) The standard error for the coefficient of gestational age is 0.35, which is associated with $df = 23$. Does the model provide strong evidence that gestational age is significantly associated with head circumference?

8.43 Murders and poverty, Part III. In Exercises 8.41 you evaluated whether poverty percentage is a significant predictor of murder rate. How, if at all, would your answer change if we wanted to find out whether poverty percentage is positively associated with murder rate. Make sure to include the appropriate p-value for this hypothesis test in your answer.

8.44 Cats, Part II. Exercise 8.30 presents regression output from a model for predicting the heart weight (in g) of cats from their body weight (in kg). The coefficients are estimated using a dataset of 144 domestic cat. The model output is also provided below.

	Estimate	Std. Error	t value	Pr(>|t|)
(Intercept)	-0.357	0.692	-0.515	0.607
body wt	4.034	0.250	16.119	0.000
$s = 1.452$	$R^2 = 64.66\%$		$R^2_{adj} = 64.41\%$	

(a) What are the hypotheses for evaluating whether body weight is positively associated with heart weight in cats?

(b) State the conclusion of the hypothesis test from part (a) in context of the data.

(c) Calculate a 95% confidence interval for the slope of body weight, and interpret it in context of the data.

(d) Do your results from the hypothesis test and the confidence interval agree? Explain.

8.6.5 Transformations for nonlinear data

8.45 Used trucks. The scatterplot below shows the relationship between year and price (in thousands of $) of a random sample of 42 pickup trucks. Also shown is a residuals plot for the linear model for predicting price from year.

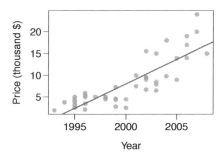
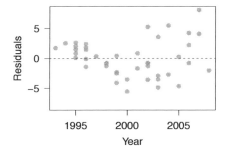

(a) Describe the relationship between these two variables and comment on whether a linear model is appropriate for modeling the relationship between year and price.

(b) The scatterplot below shows the relationship between logged (natural log) price and year of these trucks, as well as the residuals plot for modeling these data. Comment on which model (linear model from earlier or logged model presented here) is a better fit for these data.

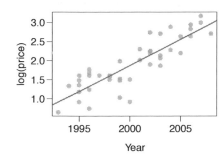
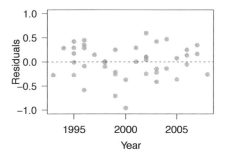

(c) The output for the logged model is given below. Interpret the slope in context of the data.

	Estimate	Std. Error	t value	Pr(>|t|)
(Intercept)	-271.981	25.042	-10.861	0.000
Year	0.137	0.013	10.937	0.000

8.46 Income and hours worked. The scatterplot below shows the relationship between income and years worked for a random sample of 787 Americans. Also shown is a residuals plot for the linear model for predicting income from hours worked. The data come from the 2012 American Community Survey.[29]

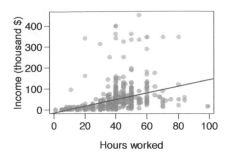 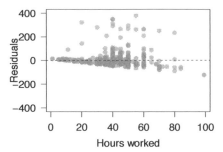

(a) Describe the relationship between these two variables and comment on whether a linear model is appropriate for modeling the relationship between year and price.

(b) The scatterplot below shows the relationship between logged (natural log) income and hours worked, as well as the residuals plot for modeling these data. Comment on which model (linear model from earlier or logged model presented here) is a better fit for these data.

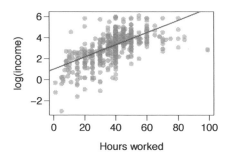 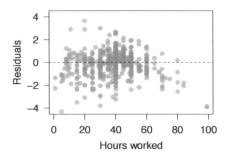

(c) The output for the logged model is given below. Interpret the slope in context of the data.

	Estimate	Std. Error	t value	Pr(>|t|)
(Intercept)	1.017	0.113	9.000	0.000
hrs_work	0.058	0.003	21.086	0.000

---

[29]United States Census Bureau. Summary File. 2012 American Community Survey. U.S. Census Bureaus American Community Survey Office, 2013. Web.

Appendix A

End of chapter exercise solutions

1 Data collection

1.1 (a) Treatment: $10/43 = 0.23 \to 23\%$. Control: $2/46 = 0.04 \to 4\%$. (b) There is a 19% difference between the pain reduction rates in the two groups. At first glance, it appears patients in the treatment group are more likely to experience pain reduction from the acupuncture treatment. (c) Answers may vary but should be sensible. Two possible answers: [1]Though the groups' difference is big, I'm skeptical the results show a real difference and think this might be due to chance. [2]The difference in these rates looks pretty big, so I suspect acupuncture is having a positive impact on pain.

1.3 (a-i) 143,196 eligible study subjects born in Southern California between 1989 and 1993. (a-ii) Measurements of carbon monoxide, nitrogen dioxide, ozone, and particulate matter less than $10\mu g/m^3$ (PM_{10}) collected at air-quality-monitoring stations as well as length of gestation. These are continuous numerical variables. (a-iii) The research question: "Is there an association between air pollution exposure and preterm births?" (b-i) 600 adult patients aged 18-69 years diagnosed and currently treated for asthma. (b-ii) The variables were whether or not the patient practiced the Buteyko method (categorical) and measures of quality of life, activity, asthma symptoms and medication reduction of the patients (categorical, ordinal). It may also be reasonable to treat the ratings on a scale of 1 to 10 as discrete numerical variables. (b-iii) The research question: "Do asthmatic patients who practice the Buteyko method experience improvement in their condition?"

1.5 (a) $50 \times 3 = 150$. (b) Four continuous numerical variables: sepal length, sepal width, petal length, and petal width. (c) One categorical variable, species, with three levels: *setosa*, *versicolor*, and *virginica*.

1.7 (a) Population of interest: all births in Southern California. Sample: 143,196 births between 1989 and 1993 in Southern California. If births in this time span can be considered to be representative of all births, then the results are generalizable to the population of Southern California. However, since the study is observational, the findings do not imply causal relationships. (b) Population: all 18-69 year olds diagnosed and currently treated for asthma. Sample: 600 adult patients aged 18-69 years diagnosed and currently treated for asthma. Since the sample consists of voluntary patients, the results cannot necessarily be generalized to the population at large. However, since the study is an experiment, the findings can be used to establish causal relationships.

1.9 (a) Observation (b) Variable (c) Sample statistic (d) Population parameter

1.11 (a) Explanatory: number of study hours per week. Response: GPA. (b) There is a slight positive relationship between the two variables. One respondent reported a GPA above 4.0, which is a data error. There are also a few respondents who reported unusually high study hours (60 and 70 hours/week). The variability in GPA also appears to be larger for students who study less than those who study more. Since the data become sparse as the number of study hours increases, it is somewhat difficult to evaluate the strength of the relationship and also the variability across different numbers of study hours. (c) Observational. (d) Since this is an observational study, a causal relationship is not implied.

1.13 (a) Observational. (b) The professor suspects students in a given section may have similar feelings about the course. To ensure each section is reasonably represented, she may choose to randomly select a fixed number of students, say 10, from each section for a total sample size of 40 students. Since a random sample of fixed size was taken within each section in this scenario, this represents stratified sampling.

1.15 (a) The relationship between life expectancy and percentage of internet users is positive, non-linear, and somewhat strong. (b) This is an observational study. (c) Countries in which a higher percentage of the population have access to the Internet are most probably developed countries which also tend to have a higher quality of life in general and also better health care. Whether or not the country is developed is a lurking variable here, since level of Internet access varies for underdeveloped, developing, and developed countries. (Note: Answers may vary.)

1.17 Sampling from the phone book would miss unlisted phone numbers, so this would result in bias. People who do not have their numbers listed may share certain characteristics, e.g. consider that cell phones are not listed in phone books, so a sample from the phone book would not necessarily be a representative of the population.

1.19 The estimate will be biased, and it will tend to overestimate the true family size. For example, suppose we had just two families: the first with 2 parents and 5 children, and the second with 2 parents and 1 child. Then if we draw one of the six children at random, 5 times out of 6 we would sample the larger family

1.21 (a) Simple random sampling. This is usu-ally an effective method as it assigns equal probability to each household to be picked. (b) Stratified sampling. This is an effective method in this setting since neighborhoods are unique and this method allows us to sample from each neighborhood. (c) Cluster sampling. This is not an effective method in this setting since the resulting sample will not contain households from certain neighborhoods and we are told that some neighborhoods are very different from others. (d) Multi-stage sampling. This method will suffer from the same issue discussed in part (d). (e) Convenience sampling. This is not an effective method since it will result in a biased sample for households that are similar to each other (in the same neighborhood) and the sample will not contain any houses from neighborhoods far from the city council offices.

1.23 (a) No, this is an observational study. (b) This statement is not justified; it implies a causal association between sleep disorders and bullying. However, this was an observational study. A better conclusion would be "School children identified as bullies are more likely to suffer from sleep disorders than non-bullies."

1.25 (a) Experiment (b) Yes

1.27 (a) Experiment, as the treatment was assigned to each patient. (b) Response: Duration of the cold. Explanatory: Treatment, with 4 levels: *placebo, 1g, 3g, 3g with additives.* (c) Patients were blinded. (d) Double-blind with respect to the researchers evaluating the patients, but the nurses who briefly interacted with patients during the distribution of the medication were not blinded. We could say the study was partly double-blind. (e) No. The patients were randomly assigned to treatment groups and were blinded, so we would expect about an equal number of patients in each group to not adhere to the treatment.

1.29 Recruit 30 friends and randomly assign them to three groups: no music, instrumental music, and music with lyrics. Have each participant read a passage to learn about a new concept, and then give them a short quiz assessing what they have learned. Compare the number of questions participants got correct on average across the three groups.

1.31 (a) Experiment. (b) Treatment is exercise twice a week. Control is no exercise. (c) Yes, the blocking variable is age. (d) No. (e) This is an experiment, so a causal conclusion is reasonable. Since the sample is random, the conclusion can be generalized to the population at large. However, we must consider that a placebo effect is possible. (f) Yes. Randomly sampled people should not be required to participate in a clinical trial, and there are also ethical concerns about the plan to instruct one group not to participate in a healthy behavior, which in this case is exercise.

1.33 (a) (i) Observational study. (ii) There is random sampling. (iii) There is no random assignment. (iv) Since only random sampling from a known population is performed, one may infer the characteristics of the sample tend to mirror corresponding characteristics of the population. Therefore, we can say that among high school students those who do not watch TV while doing homework tend to do better on average. However we cannot infer causation based on this study. (v) Neither stratifying nor blocking was used in this study. (b) (i) Observational study. (ii) There is random sampling. (iii) There is no random assignment. (iv) Since only random sampling from a known population is performed, one may infer the characteristics of the sample tend to mirror corresponding characteristics of the population. Therefore, we can say that among high school males and females those who do not watch TV while doing homework tend to do better on average. However we cannot infer causation based on this study. (v) Stratifying (separating into groups during the process of sampling) is used but no blocking. (c) (i) Experiment. (ii) There is random sampling.

(iii) There is no random assignment. (iv) Since both random sampling and random assignment to groups are performed, one may draw cause-effect inferences about the sample results, as well as generalize to the larger population from which the sample was drawn. Therefore, we can say that there appears to be a cause-and-effect relationship between watching TV while doing homework and low grade point averages. (v) Neither stratifying nor blocking was used in this study.

1.35 A sample of everyone in your homeroom class is not a random sample of all high schoolers, therefore findings from this study should not be generalized to all high schoolers. Also, this is an observational study with no random assignment, therefore causal inferences cannot be drawn.

1.37 Match subjects on attributes that might be associated with running faster: age, height, weight, fitness level, running experience, etc. Then, randomly assign one student from each pair to consume a sports drink that replenishes electrolytes and the other to consume a placebo drink (a drink that looks and tastes the same but does not replenish electrolytes). Then, time the students running the same pre-specified distance (e.g. 1 mile) and compare the average finishing times for the two groups.

1.39 Design B is the most appropriate since there are an equal number of cups from each group at each of the possible distances from the window (2 cups from each group are next to the window, 2 cups from each group are middle distance from the window, 2 cups from each group are far from the window). This is important since the amount of sunlight is likely an important factor in the plants' growth.

2 Summarizing data

2.1 (a) There is a weak and positive relationship between age and income. With so few points it is difficult to tell the form of the relationship (linear or not) however the relationship does look somewhat curved.

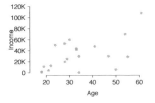

(b)

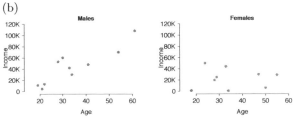

(c) For males as age increases so does income, however this pattern is not apparent for females.

2.3 (a)

```
0 | 000003333333
0 | 7779
1 | 0011
```

Legend: 1 | 0 = 10%

(b)

(c)

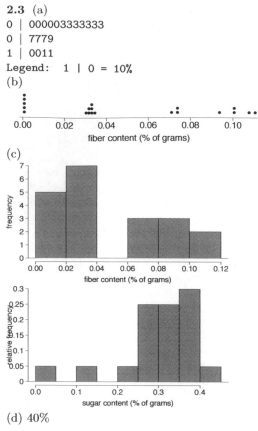

(d) 40%

2.5 (a) Positive association: mammals with longer gestation periods tend to live longer as well. (b) Association would still be positive. (c) No, they are not independent. See part (a).

2.7

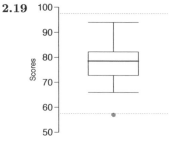

2.9 Population mean = 5.5. Sample mean = 6.25.

2.11 (a) Decrease: the new score is smaller than the mean of the 24 previous scores. (b) Calculate a weighted mean. Use a weight of 24 for the old mean and 1 for the new mean: $(24 \times 74 + 1 \times 64)/(24 + 1) = 73.6$. There are other ways to solve this exercise that do not use a weighted mean. (c) The new score is more than 1 standard deviation away from the previous mean, so increase.

2.13 Both distributions are right skewed and bimodal with modes at 10 and 20 cigarettes;

note that people may be rounding their answers to half a pack or a whole pack. The median of each distribution is between 10 and 15 cigarettes. The middle 50% of the data (the IQR) appears to be spread equally in each group and have a width of about 10 to 15. There are potential outliers above 40 cigarettes per day. It appears that respondents who smoke only a few cigarettes (0 to 5) smoke more on the weekdays than on weekends.

2.15 (a) $\bar{x}_{amtWeekends} = 20$, $\bar{x}_{amtWeekdays} = 16$. (b) $s_{amtWeekends} = 0$, $s_{amtWeekdays} = 4.18$. In this very small sample, higher on weekdays.

2.17 (a) Both distributions have the same median and IQR. (b) Second distribution has a higher median and higher IQR. (c) Second distribution has higher median. IQRs are equal. (d) Second distribution has higher median and larger IQR.

2.19

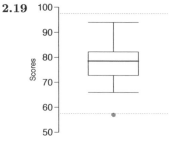

2.21 Descriptions will vary a little. (a) 2. Unimodal, symmetric, centered at 60, standard deviation of roughly 3. (b) 3. Symmetric and approximately evenly distributed from 0 to 100. (c) 1. Right skewed, unimodal, centered at about 1.5, with most observations falling between 0 and 3. A very small fraction of observations exceed a value of 5.

2.23 The histogram shows that the distribution is bimodal, which is not apparent in the box plot. The box plot makes it easy to identify more precise values of observations outside of the whiskers.

2.25 (a) Both distributions are right skewed however the distribution of incomes of males has a much higher median (around \$40K) compared to females (around \$20K). (b) We could also use side-by-side box plots for displaying and easily comparing the distributions of incomes of males and females.

2.27 (a) The distribution of number of pets per household is likely right skewed as there is a natural boundary at 0 and only a few people have many pets. Therefore the center would be best described by the median, and variability would be best described by the IQR. (b) The distribution of number of distance to work is likely right skewed as there is a natural boundary at 0 and only a few people live a very long distance from work. Therefore the center would be best described by the median, and variability would be best described by the IQR. (c) The distribution of heights of males is likely symmetric. Therefore the center would be best described by the mean, and variability would be best described by the standard deviation.

2.29 No, we would expect this distribution to be right skewed. There are two reasons for this: (1) there is a natural boundary at 0 (it is not possible to watch less than 0 hours of TV), (2) the standard deviation of the distribution is very large compared to the mean.

2.31 The statement "50% of Facebook users have over 100 friends" means that the median number of friends is 100, which is lower than the mean number of friends (190), which suggests a right skewed distribution for the number of friends of Facebook users.

2.33 (a) The median is better; the mean is substantially affected by the two extreme observations. (b) The IQR is better; the standard deviation, like the mean, is substantially affected by the two high salaries.

2.35 The distribution is unimodal and symmetric with a mean of about 25 minutes and a standard deviation of about 5 minutes. There does not appear to be any counties with unusually high or low mean travel times. Since the distribution is already unimodal and symmetric, a log transformation is not necessary.

2.37 mean $= 65,090 \times 1.05 = 68,344.50$; median $= 65,240 \times 1.05 = 68,502$; mean $= 2,122 \times 1.05 = 2,228.10$

2.39 mean $= 16 - 3 = 13$; sd $= 4.18$

2.41 Answers will vary. There are pockets of longer travel time around DC, Southeastern NY, Chicago, Minneapolis, Los Angeles, and many other big cities. There is also a large section of shorter average commute times that overlap with farmland in the Midwest. Many farmers' homes are adjacent to their farmland, so their commute would be 0 minutes, which may explain why the average commute time for these counties is relatively low.

2.43 (a) We see the order of the categories and the relative frequencies in the bar plot. (b) There are no features that are apparent in the pie chart but not in the bar plot. (c) We usually prefer to use a bar plot as we can also see the relative frequencies of the categories in this graph.

2.45 (a) False. Instead of comparing counts, we should compare percentages. (b) True. (c) False. We cannot infer a causal relationship from an association in an observational study. However, we can say the drug a person is on affects his risk in this case, as he chose that drug and his choice may be associated with other variables, which is why part (b) is true. The difference in these statements is subtle but important. (d) True.

2.47 (a) Proportion who had heart attack: $\frac{7,979}{227,571} \approx 0.035$ (b) Expected number of cardiovascular problems in the rosiglitazone group if having cardiovascular problems and treatment were independent can be calculated as the number of patients in that group multiplied by the overall rate of cardiovascular problems in the study: $67,593 \times \frac{7,979}{227,571} \approx 2370$. (c-i) H_0: Independence model. The treatment and cardiovascular problems are independent. They have no relationship, and the difference in incidence rates between the rosiglitazone and pioglitazone groups is due to chance. H_A: Alternate model. The treatment and cardiovascular problems are not independent. The difference in the incidence rates between the rosiglitazone and pioglitazone groups is not due to chance, and rosiglitazone is associated with an increased risk of serious cardiovascular problems. (c-ii) A higher number of patients with cardiovascular problems in the rosiglitazone group than expected under the assumption of independence would provide support for the alternative hypothesis. This would suggest that rosiglitazone increases the risk of such problems. (c-iii) In the actual study, we observed 2,593 cardiovascular events in the rosiglitazone group. In the 1,000 simulations under the independence model, we observed somewhat less than 2,593 in all simulations, which suggests that the actual results did not come from the independence model. That is, the analysis provides strong evidence that the variables are not independent, and we reject the independence model in favor of the alternative. The study's results provide strong evidence that rosiglitazone is associated with an increased risk of cardiovascular problems.

3 Probability

3.1 (a) False. These are independent trials. (b) False. There are red face cards. (c) True. A card cannot be both a face card and an ace.

3.3 (a) 10 tosses. Fewer tosses mean more variability in the sample fraction of heads, meaning there's a better chance of getting at least 60% heads. (b) 100 tosses. More flips means the observed proportion of heads would often be closer to the average, 0.50, and therefore also above 0.40. (c) 100 tosses. With more flips, the observed proportion of heads would often be closer to the average, 0.50. (d) 10 tosses. Fewer flips would increase variability in the fraction of tosses that are heads.

3.5 (a) $0.5^{10} = 0.00098$. (b) $0.5^{10} = 0.00098$. (c) $P(\text{at least one tails}) = 1 - P(\text{no tails}) = 1 - (0.5^{10}) \approx 1 - 0.001 = 0.999$.

3.7 (a) No, there are voters who are both politically Independent and also swing voters. (b) Venn diagram below:

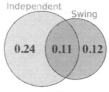

Independent   Swing

0.24   0.11   0.12

0.53

(c) 24%. (d) Add up the corresponding disjoint sections in the Venn diagram: $0.24 + 0.11 + 0.12 = 0.47$. Alternatively, use the General Addition Rule: $0.35 + 0.23 - 0.11 = 0.47$. (e) $1 - 0.47 = 0.53$. (f) $P(\text{Independent}) \times P(\text{swing}) = 0.35 \times 0.23 = 0.08$, which does not equal P(Independent and swing) $= 0.11$, so the events are dependent. If you stated that this difference might be due to sampling variability in the survey, that answer would also be reasonable (we'll dive into this topic more in later chapters).

3.9 (a) If the class is not graded on a curve, they are independent. If graded on a curve, then neither independent nor disjoint (unless the instructor will only give one A, which is a situation we will ignore in parts (b) and (c)). (b) They are probably not independent: if you study together, your study habits would be related, which suggests your course performances are also related. (c) No. See the answer to part (a) when the course is not graded on a curve. More generally: if two things are un-related (independent), then one occurring does not preclude the other from occurring.

3.11 (a) $0.16 + 0.09 = 0.25$. (b) $0.17 + 0.09 = 0.26$. (c) Assuming that the education level of the husband and wife are independent: $0.25 \times 0.26 = 0.065$. You might also notice we actually made a second assumption: that the decision to get married is unrelated to education level. (d) The husband/wife independence assumption is probably not reasonable, because people often marry another person with a comparable level of education. We will leave it to you to think about whether the second assumption noted in part (c) is reasoanble.

3.13 (a) Invalid. Sum is greater than 1. (b) Valid. Probabilities are between 0 and 1, and they sum to 1. In this class, every student gets a C. (c) Invalid. Sum is less than 1. (d) Invalid. There is a negative probability. (e) Valid. Probabilities are between 0 and 1, and they sum to 1. (f) Invalid. There is a negative probability.

3.15 (a) No, but we could if A and B are independent. (b-i) 0.21. (b-ii) $0.3+0.7-0.21 = 0.79$. (b-iii) Same as $P(A)$: 0.3. (c) No, because $0.1 \neq 0.21$, where 0.21 was the value computed under independence from part (a). (d) $P(A|B) = 0.1/0.7 = 0.143$.

3.17 (a) No, these events are not mutually exclusive, there are people who believe the earth is warming and are liberal Democrats. (a) $0.60 + 0.20 - 0.18 = 0.62$. (b) $0.18/0.20 = 0.90$. (c) $0.11/0.33 \approx 0.33$. (d) No, otherwise the final answers of parts (b) and (c) would have been equal. (e) $0.06/0.34 \approx 0.18$.

3.19 (a) No, these events are not mutually exclusive, there are females who like Five Guys Burgers. (a) $162/248 = 0.65$. (b) $181/252 = 0.72$ (c) Under the assumption of a dating choices being independent of hamburger preference, which on the surface seems reasonable: $0.65 \times 0.72 = 0.468$. (d) $(252+6-1)/500 = 0.514$

3.21 (a) 0.3. (b) 0.3. (c) 0.3. (d) $0.3 \times 0.3 = 0.09$. (e) Yes, the population that is being sampled from is identical in each draw.

3.23 (a) 2/9. (b) 3/9 = 1/3. (c) (3/10) × (2/9) ≈ 0.067. (d) No. In this small population of marbles, removing one marble meaningfully changes the probability of what might be drawn next.

3.25 For 1 leggings (L) and 2 jeans (J), there are three possible orderings: LJJ, JLJ, and JJL. The probability for LJJ is (5/24) × (7/23) × (6/22) = 0.0173. The other two orderings have the same probability, and these three possible orderings are disjoint events. Final answer: 0.0519.

3.27 (a) The tree diagram:

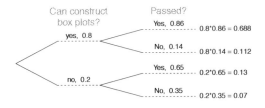

(b) $P(\text{can construct}|\text{pass}) = \frac{P(\text{can construct and pass})}{P(\text{pass})}$
$= \frac{0.8 \times 0.86}{0.8 \times 0.86 + 0.2 \times 0.65} = \frac{0.688}{0.818} \approx 0.84$.

3.29 First draw a tree diagram:

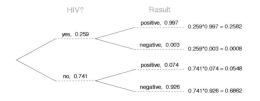

Then compute the probability: $P(HIV|+) = \frac{P(HIV \text{ and } +)}{P(+)} = \frac{0.259 \times 0.997}{0.259 \times 0.997 + 0.741 \times 0.074} = \frac{0.2582}{0.3131} = 0.8247$.

3.31 A tree diagram of the situation:

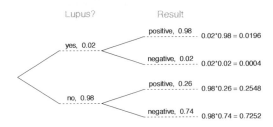

$P(\text{lupus}|\text{positive}) = \frac{P(\text{lupus and positive})}{P(\text{positive})} = \frac{0.0196}{0.0196 + 0.2548} = 0.0714$. Even when a patient tests positive for lupus, there is only a 7.14% chance that he actually has lupus. While House is not exactly right – it is possible that the patient has lupus – his implied skepticism is warranted.

3.33 (a) $\underset{Anna}{1/5} \times \underset{Ben}{1/4} \times \underset{Carl}{1/3} \times \underset{Damian}{1/2} \times \underset{Eddy}{1/1} = 1/5! = 1/120$. (b) Since the probabilities must add to 1, there must be 5! = 120 possible orderings. (c) 8! = 40,320.

3.35 (a) Yes. The conditions are satisfied: independence, fixed number of trials, either success or failure for each trial, and probability of success being constant across trials. (b) 0.200. (c) 0.200. (d) 0.0024 + 0.0284 + 0.1323 = 0.1631. (e) 1 − 0.0024 = 0.9976.

3.37 (a) P(pass) = 0.5 (b) P(pass) = 0.2 (c) P(pass) = 0.17

3.39 (a) Starting at row 3 of the random number table, we will read across the table two digits at a time. If the random number is between 00-15, the car will fail the pollution test. If the number is between 16-99, the car will pass the test. (Answers may vary.) (b) Fleet 1: 18-52-97-32-85-95-29 → P-P-P-P-P-P-P → fleet passes
Fleet 2: 14-96-06-67-17-49-59 → F-P-F-P-P-P-P → fleet fails
Fleet 3: 05-33-67-97-58-11-81 → F-P-P-P-P-F-P → fleet fails
Fleet 4: 23-81-83-21-71-08-50 → P-P-P-P-P-F-P → fleet fails
Fleet 5: 82-84-39-31-83-14-34 → P-P-P-P-P-F-P → fleet fails (c) 4 / 5 = 0.80

3.41 (a) If 13% of the students smoke, then we expect about 0.13 × 100 = 13 to smoke in the sample. (b) No, since this is not a random sample of students, so it may not be representative with respect to smoking behavior.

3.43 (a) The table below summarizes the probability model:

Event	X	P(X)	X · P(X)	$(X - E(X))^2$	$(X - E(X))^2 \cdot P(X)$
3 hearts	50	$\frac{13}{52} \times \frac{12}{51} \times \frac{11}{50} = 0.0129$	0.65	$(50 - 3.59)^2 = 2154.1$	$2154.1 \times 0.0129 = 27.9$
3 blacks	25	$\frac{26}{52} \times \frac{25}{51} \times \frac{24}{50} = 0.1176$	2.94	$(25 - 3.59)^2 = 458.5$	$458.5 \times 0.1176 = 53.9$
Else	0	$1 - (0.0129 + 0.1176) = 0.8695$	0	$(0 - 3.59)^2 = 12.9$	$12.9 \times 0.8695 = 11.2$
			$E(X) = \$3.59$		$V(X) = 93.0$ $SD(X) = \sqrt{V(X)} = 9.64$

(b) $E(X-5) = E(X) - 5 = 3.59 - 5 = -\$1.41$. The standard deviation is the same as the standard deviation of X: \$9.64.

(c) No. The expected earnings is negative, so on average you would lose money playing the game.

3.45

Event	X	$P(X)$	$X \cdot P(X)$
Boom	0.18	$\frac{1}{3}$	$0.18 \times \frac{1}{3} = 0.06$
Normal	0.09	$\frac{1}{3}$	$0.09 \times \frac{1}{3} = 0.03$
Recession	-0.12	$\frac{1}{3}$	$-0.12 \times \frac{1}{3} = -0.04$
			$E(X) = 0.05$

The expected return is a 5% increase in value for a single year.

3.47 (a) Expected: -\$0.16. Variance: 8.95. SD: \$2.99. (b) Expected: -\$0.16. SD: \$1.73. (c) Expected values are the same, but the SDs differ. The SD from the game with tripled winnings/losses is larger, since the three independent games might go in different directions (e.g. could win one game and lose two games). So the three independent games is lower risk, but in this context it just means we are likely to lose a more stable amount since the expected value is still negative.

3.49 A fair game has an expected value of zero. From the friend's perspective: $-\$5 \times 0.54 + x \times 0.46 = 0$. Solving for x: \$5.87. You would bet \$5.87 for the Padres to make the game fair.

3.51 (a) Expected: \$3.90. SD: \$0.34. (b) Expected: \$27.30. SD: \$0.89. If you computed part (b) using part (a), you should have obtained an SD of \$0.90.

3.53 Approximate answers are OK. Answers are only estimates based on the sample. (a) $(29 + 32)/144 = 0.42$. (b) $21/144 = 0.15$. (c) $(26 + 12 + 15)/144 = 0.37$.

4 Distributions of random variables

4.1 (a) 8.85%. (b) 6.94%. (c) 58.86%. (d) 4.56%.

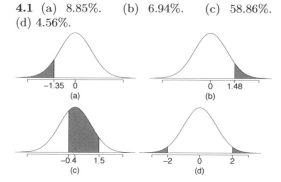

4.3 (a) Verbal: $N(\mu = 151, \sigma = 7)$, Quant: $N(\mu = 153, \sigma = 7.67)$. (b) $Z_{VR} = 1.29$, $Z_{QR} = 0.52$.

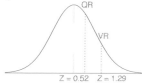

(c) She scored 1.29 standard deviations above the mean on the Verbal Reasoning section and 0.52 standard deviations above the mean on the Quantitative Reasoning section. (d) She did better on the Verbal Reasoning section since her Z-score on that section was higher. (e) $Perc_{VR} = 0.9007 \approx 90\%$, $Perc_{QR} = 0.6990 \approx 70\%$. (f) $100\% - 90\% = 10\%$ did better than her on VR, and $100\% - 70\% = 30\%$ did better than her on QR. (g) We cannot compare the raw scores since they are on different scales. Comparing her percentile scores is more appropriate when comparing her performance to others. (h) Answer to part (b) would not change as Z-scores can be calculated for distributions that are not normal. However, we could not answer parts (d)-(f) since we cannot use the normal probability table to calculate probabilities and percentiles without a normal model.

4.5 (a) $Z = 0.84$, which corresponds to approximately 160 on QR. (b) $Z = -0.52$, which corresponds to approximately 147 on VR.

4.7 (a) $Z = 1.2 \rightarrow 0.1151$. (b) $Z = -1.28 \rightarrow 70.6°F$ or colder.

4.9 (a) $N(25, 2.78)$. (b) $Z = 1.08 \rightarrow 0.1401$. (c) The answers are very close because only the units were changed. (The only reason why they are a little different is because $28°C$ is $82.4°F$, not precisely $83°F$.) (c) Since $IQR = Q3 - Q1$, we first need to find $Q3$ and $Q1$ and take the difference between the two. Remember that $Q3$ is the 75th and $Q1$ is the 25th percentile of a distribution. Q1 = 23.13, Q3 = 26.86, IQR = 26.86 - 23.13 = 3.73.

4.11 (a) $Z = 0.67$. (b) $\mu = \$1650$, $x = \$1800$. (c) $0.67 = \frac{1800 - 1650}{\sigma} \rightarrow \sigma = \$223.88$.

4.13 $Z = 1.56 \rightarrow 0.0594$, i.e. 6%.

4.15 (a) $Z = 0.73 \rightarrow 0.2327$. (b) If you are bidding on only one auction and set a low maximum bid price, someone will probably outbid you. If you set a high maximum bid price, you may win the auction but pay more than is necessary. If bidding on more than one auction, and you set your maximum bid price very low, you probably won't win any of the auctions. However, if the maximum bid price is even modestly high, you are likely to win multiple auctions. (c) An answer roughly equal to the 10th percentile would be reasonable. Regrettably, no percentile cut-off point guarantees beyond any possible event that you win at least one auction. However, you may pick a higher percentile if you want to be more sure of winning an auction. (d) Answers will vary a little but should correspond to the answer in part (c). We use the 10th percentile: $Z = -1.28 \rightarrow \$69.80$.

4.17 $14/20 = 70\%$ are within 1 SD. Within 2 SD: $19/20 = 95\%$. Within 3 SD: $20/20 = 100\%$. They follow this rule closely.

4.19 The distribution is unimodal and symmetric. The superimposed normal curve approximates the distribution pretty well. The points on the normal probability plot also follow a relatively straight line. There is one slightly distant observation on the lower end, but it is not extreme. The data appear to be reasonably approximated by the normal distribution.

4.21 (a) Let X represent the amount of lemonade in the pitcher, Y represent the amount of lemonade in a glass, and W represent the amount left over after. Then, $\mu_W = E(X-Y) = 64 - 12 = 62$ (b) $\sigma_W = \sqrt{SD(X)^2 + SD(Y)^2} = \sqrt{1.732^2 + 1^2} \approx \sqrt{4} = 2$ (c) $P(W > 60) = P\left(Z > \frac{60 - 62}{2}\right) = P(Z > -1) = 1 - 0.1587 = 0.8413$

4.23 The combined scores follow a normal distribution with $\mu_{combined} = 304$ and $\sigma_{combined} = 10.38$. Then, P(combined score > 320) is approximately 0.06.

4.25 (a) The distribution is unimodal and strongly right skewed with a median between 5 and 10 years old. Ages range from 0 to slightly over 50 years old, and the middle 50% of the distribution is roughly between 5 and 15 years old. There are potential outliers on the higher end. (b) When the sample size is small, the sampling distribution is right skewed, just like the population distribution. As the sample size increases, the sampling distribution gets more unimodal, symmetric, and approaches normality. The variability also decreases. This is consistent with the Central Limit Theorem.

4.27 (a) Right skewed. There is a long tail on the higher end of the distribution but a much shorter tail on the lower end. (b) Less than, as the median would be less than the mean in a right skewed distribution. (c) We should not. (d) Even though the population distribution is not normal, the conditions for inference are reasonably satisfied, with the possible exception of skew. If the skew isn't very strong (we should ask to see the data), then we can use the Central Limit Theorem to estimate this probability. For now, we'll assume the skew isn't very strong, though the description suggests it is at least moderate to strong. Use $N(1.3, SD_{\bar{x}} = 0.3/\sqrt{60})$: $Z = 2.58 \rightarrow 0.0049$. (e) It would decrease it by a factor of $1/\sqrt{2}$.

4.29 The centers are the same in each plot, and each data set is from a nearly normal distribution (see Section 7.1.1), though the histograms may not look very normal since each represents only 100 data points. The only way to tell which plot corresponds to which scenario is to examine the variability of each distribution. Plot B is the most variable, followed by Plot A, then Plot C. This means Plot B will correspond to the original data, Plot A to the sample means with size 5, and Plot C to the sample means with size 25.

4.31 (a) $Z = -3.33 \rightarrow 0.0004$. (b) The population SD is known and the data are nearly normal, so the sample mean will be nearly normal with distribution $N(\mu, \sigma/\sqrt{n})$, i.e. $N(2.5, 0.0095)$. (c) $Z = -10.54 \rightarrow \approx 0$. (d) See below:

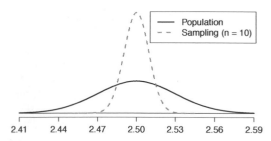

(e) We could not estimate (a) without a nearly normal population distribution. We also could not estimate (c) since the sample size is not sufficient to yield a nearly normal sampling distribution if the population distribution is not nearly normal.

4.33 (a) We cannot use the normal model for this calculation, but we can use the histogram. About 500 songs are shown to be longer than 5 minutes, so the probability is about $500/3000 = 0.167$. (b) Two different answers are reasonable. $^{Option\ 1}$Since the population distribution is only slightly skewed to the right, even a small sample size will yield a nearly normal sampling distribution. We also know that the songs are sampled randomly and the sample size is less than 10% of the population, so the length of one song in the sample is independent of another. We are looking for the probability that the total length of 15 songs is more than 60 minutes, which means that the average song should last at least $60/15 = 4$ minutes. Using $SD_{\bar{x}} = 1.63/\sqrt{15}$, $Z = 1.31 \rightarrow 0.0951$. $^{Option\ 2}$Since the population distribution is not normal, a small sample size may not be sufficient to yield a nearly normal sampling distribution. Therefore, we cannot estimate the probability using the tools we have learned so far. (c) We can now be confident that the conditions are satisfied. $Z = 0.92 \rightarrow 0.1788$.

4.35 (a) $SD_{\bar{x}} = \frac{25}{\sqrt{75}} = 2.89$. (b) $Z = 1.73$, which indicates that the two values are not unusually distant from each other when accounting for the uncertainty in John's point estimate.

4.37 This is the same as checking that the average bag weight of the 10 bags is greater than 46 lbs. $SD_{\bar{x}} = \frac{3.2}{\sqrt{10}} = 1.012$; $z = \frac{46-45}{1.012} = 0.988$;

$P(z > 0.988) = 0.162 = 16.2\%$.

4.39 (a) No. The cards are not independent. For example, if the first card is an ace of clubs, that implies the second card cannot be an ace of clubs. Additionally, there are many possible categories, which would need to be simplified. (b) No. There are six events under consideration. The Bernoulli distribution allows for only two events or categories. Note that rolling a die could be a Bernoulli trial if we simply to two events, e.g. rolling a 6 and not rolling a 6, though specifying such details would be necessary.

4.41 (a) $(1 - 0.471)^2 \times 0.471 = 0.1318$. (b) $0.471^3 = 0.1045$. (c) $\mu = 1/0.471 = 2.12$, $\sigma = \sqrt{2.38} = 1.54$. (d) $\mu = 1/0.30 = 3.33$, $\sigma = 2.79$. (e) When p is smaller, the event is rarer, meaning the expected number of trials before a success and the standard deviation of the waiting time are higher.

4.43 (a) $0.875^2 \times 0.125 = 0.096$. (b) $\mu = 8$, $\sigma = 7.48$.

4.45 (a) $\mu = 35$, $\sigma = 3.24$. (b) Yes. $Z = 3.09$. Since 45 is more than 2 standard deviations from the mean, it would be considered unusual. Note that the normal model is not required to apply this rule of thumb. (c) Using a normal model: 0.0010. This does indeed appear to be an unusual observation. If using a normal model with a 0.5 correction, the probability would be calculated as 0.0017.

4.47 Want to find the probability that there will be 1,786 or more enrollees. Using the normal: 0.0582. With a 0.5 correction: 0.0559.

4.49 (a) $1 - 0.75^3 = 0.5781$. (b) 0.1406. (c) 0.4219. (d) $1 - 0.25^3 = 0.9844$.

4.51 (a) Geometric distribution: 0.109. (b) Binomial: 0.219. (c) Binomial: 0.137. (d) $1 - 0.875^6 = 0.551$. (e) Geometric: 0.084. (f) Using a binomial distribution with $n = 6$ and $p = 0.75$, we see that $\mu = 4.5$, $\sigma = 1.06$, and $Z = ?2.36$. Since this is not within 2 SD, it may be considered unusual.

4.53 0 wins (-\$3): 0.1458. 1 win (-\$1): 0.3936. 2 wins (+\$1): 0.3543. 3 wins (+\$3): 0.1063.

4.55 (a) Each observation in each of the distributions represents the sample proportion ($\hat{p}$) from samples of size $n = 20$, $n = 100$, and $n = 500$, respectively. (b) The centers for all three distributions are at 0.95, the true population parameter. When n is small, the distribution is skewed to the left and not smooth. As n increases, the variability of the distribution (standard deviation) decreases, and the shape of the distribution becomes more unimodal and symmetric.

4.57 (a) $SD_{\hat{p}} = \sqrt{p(1-p)/n} = 0.0707$. This describes the typical distance that the sample proportion will deviate from the true proportion, $p = 0.5$. (b) $\hat{p}$ approximately follows $N(0.5, 0.0707)$. $Z = (0.55 - 0.50)/0.0707 \approx 0.71$. This corresponds to an upper tail of about 0.2389. That is, $P(\hat{p} > 0.55) \approx 0.24$.

4.59 (a) First we need to check that the necessary conditions are met. There are $200 \times 0.08 = 16$ expected successes and $200 \times (1-0.08) = 184$ expected failures, therefore the success-failure condition is met. Then the binomial distribution can be approximated by $N(\mu = 3.84, \sigma = 3.84)$. $P(X < 12) = P(Z < -1.04) = 0.1492$. (b) Since the success-failure condition is met the sampling distribution of $\hat{p} \sim N(\mu = 0.08, \sigma = 0.0192)$. $P(\hat{p} < 0.06) = P(Z < -1.04) = 0.1492$. (c) As expected, the two answers are the same.

4.61 First we need to check that the necessary conditions are met. There are $100 \times 0.389 = 38.9$ expected successes and $100 \times (1 - 0.389) = 61.1$ expected failures, therefore the success-failure condition is met. Calculate using either (1) the normal approximation to the binomial distribution or (2) the sampling distribution of $\hat{p}$. (1) The binomial distribution can be approximated by $N(\mu = 0.389, \sigma = 4.88)$. $P(X \geq 35) = P(Z > -0.80) = 1 - 0.2119 = 0.7881$. (2) The sampling distribution of $\hat{p} \sim N(\mu = 0.389, \sigma = 0.0488)$. $P(\hat{p} > 0.35) = P(Z > -0.8) = 0.7881$.

5 Foundation for inference

5.1 (a) Mean. Each student reports a numerical value: a number of hours. (b) Mean. Each student reports a number, which is a percentage, and we can average over these percentages. (c) Proportion. Each student reports Yes or No, so this is a categorical variable and we use a proportion. (d) Mean. Each student reports a number, which is a percentage like in part (b). (e) Proportion. Each student reports whether or not s/he expects to get a job, so this is a categorical variable and we use a proportion.

5.3 (a) Mean: 13.65. Median: 14. (b) SD: 1.91. IQR: $15 - 13 = 2$. (c) $Z_{16} = 1.23$, which is not unusual since it is within 2 SD of the mean. $Z_{18} = 2.23$, which is generally considered unusual. (d) No. Point estimates that are based on samples only approximate the population parameter, and they vary from one sample to another. (e) We use the SE, which is $1.91/\sqrt{100} = 0.191$ for this sample's mean.

5.5 (a) We are building a distribution of sample statistics, in this case the sample mean. Such a distribution is called a sampling distribution. (b) Because we are dealing with the distribution of sample means, we need to check to see if the Central Limit Theorem applies. Our sample size is greater than 30, and we are told that random sampling is employed. With these conditions met, we expect that the distribution of the sample mean will be nearly normal and therefore symmetric. (c) Because we are dealing with a sampling distribution, we measure its variability with the standard error. $SE = 18.2/\sqrt{45} = 2.713$. (d) The sample means will be more variable with the smaller sample size.

5.7 Recall that the general formula is

$$\text{point estimate} \pm z^{\star} \times SE$$

First, identify the three different values. The point estimate is 45%, $z^{\star} = 1.96$ for a 95% confidence level, and $SE = 1.2\%$. Then, plug the values into the formula:

$$45\% \pm 1.96 \times 1.2\% \quad \rightarrow \quad (42.6\%, 47.4\%)$$

We are 95% confident that the proportion of US adults who live with one or more chronic conditions is between 42.6% and 47.4%.

5.9 (a) False. Confidence intervals provide a range of plausible values, and sometimes the truth is missed. A 95% confidence interval "misses" about 5% of the time. (b) True. Notice that the description focuses on the true population value. (c) True. If we examine the 95% confidence interval computed in Exercise 5.7, we can see that 50% is not included in this interval. This means that in a hypothesis test, we would reject the null hypothesis that the proportion is 0.5. (d) False. The standard error describes the uncertainty in the overall estimate from natural fluctuations due to randomness, not the uncertainty corresponding to individuals' responses.

5.11 (a) We are 95% confident that Americans spend an average of 1.38 to 1.92 hours per day relaxing or pursuing activities they enjoy. (b) Their confidence level must be higher as the width of the confidence interval increases as the confidence level increases. (c) The new margin of error will be smaller since as the sample size increases the standard error decreases, which will decrease the margin of error.

5.13 (a) (40.7%, 45.3%). We are 95% confident that 40.7% to 45.3% of Americans believe women are held to higher standards than men when being considered for top executive business positions. (b) Narrower, since as the confidence level decreases the margin of error of the confidence interval decreases as well. (c) (41.1%, 44.9%). We are 90% confident that 41.1% to 44.9% of Americans believe women are held to higher standards than men when being considered for top executive business positions.

5.15 The subscript $_{pr}$ corresponds to provocative and $_{con}$ to conservative. (a) $H_0 : p_{pr} = p_{con}$. $H_A : p_{pr} \neq p_{con}$. (b) -0.35. (c) The left tail for the p-value is calculated by adding up the two left bins: $0.005 + 0.015 = 0.02$. Doubling the one tail, the p-value is 0.04. (Students may have approximate results, and a small number of students may have a p-value of about 0.05.) Since the p-value is low, we reject H_0. The data provide strong evidence that people react differently under the two scenarios.

5.17 The primary concern is confirmation bias. If researchers look only for what they suspect to be true using a one-sided test, then they are formally excluding from consideration the possibility that the opposite result is true. Additionally, if other researchers believe the opposite possibility might be true, they would be very skeptical of the one-sided test.

5.19 (a) $H_0 : p = 0.69$. $H_A : p \neq 0.69$. (b) $\hat{p} = \frac{17}{30} = 0.57$. (c) The success-failure condition is not satisfied; note that it is appropriate to use the null value ($p_0 = 0.69$) to compute the expected number of successes and failures. (d) Answers may vary. Each student can be represented with a card. Take 100 cards, 69 black cards representing those who follow the news about Egypt and 31 red cards representing those who do not. Shuffle the cards and draw with replacement (shuffling each time in between draws) 30 cards representing the 30 high school students. Calculate the proportion of black cards in this sample, $\hat{p}_{sim}$, i.e. the proportion of those who follow the news in the simulation. Repeat this many times (e.g. 10,000 times) and plot the resulting sample proportions. The p-value will be two times the proportion of simulations where $\hat{p}_{sim} \leq 0.57$. (Note: we would generally use a computer to perform these simulations.) (e) The p-value is about $0.001 + 0.005 + 0.020 + 0.035 + 0.075 = 0.136$, meaning the two-sided p-value is about 0.272. Your p-value may vary slightly since it is based on a visual estimate. Since the p-value is greater than 0.05, we fail to reject H_0. The data do not provide strong evidence that the proportion of high school students who followed the news about Egypt is different than the proportion of American adults who did.

5.21 (a) H_0: $\mu_1 - \mu_2 = 0$, i.e. there is no difference in the average number of spam emails each day for American between 2004 and 2009. H_A: $\mu_1 - \mu_2 \neq 0$, i.e. there is a difference between the average number of spam emails each day for Americans between 2004 and 2009. (b) $18.5 - 14.9 = 3.6$ spam emails per day. (c) There is not convincing evidence that the observed difference is due to anything but chance. That is, observing a difference of 3.6 in the two sample means could reasonably be explained by chance alone. (d) Since the difference is not statistically significant, we would expect the confidence interval to contain 0.

5.23 (a) H_0: $p_1 - p_2 = 0$, i.e. there is no difference in the fraction of Americans who say they delete their spam emails once a month or less. H_A: $p_1 - p_2 \neq 0$, i.e. there is a difference in the fraction of Americans who say they delete their spam emails once a month or less. (b) $0.23 - 0.16 = 0.07$. (c) The difference of 0.07 (7%) is not easily explained by chance. That is, there is strong evidence that the fraction of Americans who say they delete their spam emails once a month or less has declined. (Notice that we can assert the direction, even in this two-sided test.) (d) Because the difference is statistically significant, 0 is not a plausible value for the difference, meaning we would not expect the confidence interval to contain 0.

5.25 (a) H_0: Anti-depressants do not help symptoms of Fibromyalgia. H_A: Anti-depressants do treat symptoms of Fibromyalgia. Remark: Diana might also have taken special note if her symptoms got much worse, so a more scientific approach would have been to use a two-sided test. While parts (b) and (c) use the one-sided version, your answers will be a little different if you used a two-sided test. (b) Concluding that anti-depressants work for the treatment of Fibromyalgia symptoms when they actually do not. (c) Concluding that anti-depressants do not work for the treatment of Fibromyalgia symptoms when they actually do.

5.27 False. It is appropriate to adjust the significance level to reflect the consequences of a Type 1 or Type 2 Error, and it is also be appropriate to consider additional context of the application.

5.29 (a) Scenario I is higher. Recall that a sample mean based on less data tends to be less accurate and have larger standard errors. (b) Scenario I is higher. The higher the confidence level, the higher the corresponding margin of error. (c) They are equal. The sample size does not affect the calculation of the p-value for a given Z-score. (d) Scenario I is higher. If the null hypothesis is harder to reject (lower α), then we are more likely to make a Type 2 error when the alternative hypothesis is true.

5.31 True. If the sample size is large, then the standard error will be small, meaning even relatively small differences between the null value and point estimate can be statistically significant.

6 Inference for categorical data

6.1 (a) False. Doesn't satisfy success-failure condition. (b) True. The success-failure condition is not satisfied. In most samples we would expect $\hat{p}$ to be close to 0.08, the true population proportion. While $\hat{p}$ can be much above 0.08, it is bound below by 0, suggesting it would take on a right skewed shape. Plotting the sampling distribution would confirm this suspicion. (c) False. $SE_{\hat{p}} = 0.0243$, and $\hat{p} = 0.12$ is only $\frac{0.12-0.08}{0.0243} = 1.65$ SEs away from the mean, which would not be considered unusual. (d) True. $\hat{p} = 0.12$ is 2.32 standard errors away from the mean, which is often considered unusual. (e) False. Decreases the SE by a factor of $1/\sqrt{2}$.

6.3 (a) True. See the reasoning of 6.1(b). (b) True. We take the square root of the sample size in the SE formula. (c) True. The independence and success-failure conditions are satisfied. (d) True. The independence and success-failure conditions are satisfied.

6.5 (a) False. A confidence interval is constructed to estimate the population proportion, not the sample proportion. (b) True. 95% CI: $70\% \pm 8\%$. (c) True. By the definition of the confidence level. (d) True. Quadrupling the sample size decreases the SE and ME by a factor of $1/\sqrt{4}$. (e) True. The 95% CI is entirely above 50%.

6.7 With a random sample from $< 10\%$ of the population, independence is satisfied. The success-failure condition is also satisfied. $ME = z^\star \sqrt{\frac{\hat{p}(1-\hat{p})}{n}} = 1.96\sqrt{\frac{0.56 \times 0.44}{600}} = 0.0397 \approx 4\%$

6.9 (a) Proportion of graduates from this university who found a job within one year of graduating. $\hat{p} = 348/400 = 0.87$. (b) This is a random sample from less than 10% of the population, so the observations are independent. Success-failure condition is satisfied: 348 successes, 52 failures, both well above 10. (c) (0.8371, 0.9029). We are 95% confident that approximately 84% to 90% of graduates from this university found a job within one year of completing their undergraduate degree. (d) 95% of such random samples would produce a 95% confidence interval that includes the true proportion of students at this university who found a job within one year of graduating from college. (e) (0.8267, 0.9133). Similar interpretation as before. (f) 99% CI is wider, as we are more confident that the true proportion is within the interval and so need to cover a wider range.

6.11 (a) No. The sample only represents students who took the SAT, and this was also an online survey. (b) (0.5289, 0.5711). We are 90% confident that 53% to 57% of high school seniors who took the SAT are fairly certain that they will participate in a study abroad program in college. (c) 90% of such random samples would produce a 90% confidence interval that includes the true proportion. (d) Yes. The interval lies entirely above 50%.

6.13 (a) This is an appropriate setting for a hypothesis test. $H_0 : p = 0.50$. $H_A : p > 0.50$. Both independence and the success-failure condition are satisfied. $Z = 1.12 \rightarrow$ p-value = 0.1314. Since the p-value $> \alpha = 0.05$, we fail to reject H_0. The data do not provide strong evidence that more than half of all Independents oppose the public option plan. (b) Yes, since we did not reject H_0 in part (a).

6.15 (a) $H_0 : p = 0.38$. $H_A : p \neq 0.38$. Independence (random sample, < 10% of population) and the success-failure condition are satisfied. $Z = -20.5 \rightarrow$ p-value ≈ 0. Since the p-value is very small, we reject H_0. The data provide strong evidence that the proportion of Americans who only use their cell phones to access the internet is different than the Chinese proportion of 38%, and the data indicate that the proportion is lower in the US. (b) If in fact 38% of Americans used their cell phones as a primary access point to the internet, the probability of obtaining a random sample of 2,254 Americans where 17% or less or 59% or more use their only their cell phones to access the internet

would be approximately 0. (c) (0.1545, 0.1855). We are 95% confident that approximately 15.5% to 18.6% of all Americans primarily use their cell phones to browse the internet.

6.17 (a) $H_0 : p = 0.5$. $H_A : p > 0.5$. Independence (random sample, < 10% of population) is satisfied, as is the success-failure conditions (using $p_0 = 0.5$, we expect 40 successes and 40 failures). $Z = 2.91 \rightarrow$ p-value = 0.0018. Since the p-value < 0.05, we reject the null hypothesis. The data provide strong evidence that the rate of correctly identifying a soda for these people is significantly better than just by random guessing. (b) If in fact people cannot tell the difference between diet and regular soda and they randomly guess, the probability of getting a random sample of 80 people where 53 or more identify a soda correctly would be 0.0018.

6.19 (a) Independence is satisfied (random sample from < 10% of the population), as is the success-failure condition (40 smokers, 160 non-smokers). The 95% CI: (0.145, 0.255). We are 95% confident that 14.5% to 25.5% of all students at this university smoke. (b) We want $z^\star SE$ to be no larger than 0.02 for a 95% confidence level. We use $z^\star = 1.96$ and plug in the point estimate $\hat{p} = 0.2$ within the SE formula: $1.96\sqrt{0.2(1 - 0.2)/n} \leq 0.02$. The sample size n should be at least 1,537.

6.21 The margin of error, which is computed as $z^\star SE$, must be smaller than 0.01 for a 90% confidence level. We use $z^\star = 1.65$ for a 90% confidence level, and we can use the point estimate $\hat{p} = 0.52$ in the formula for SE. $1.65\sqrt{0.52(1 - 0.52)/n} \leq 0.01$. Therefore, the sample size n must be at least 6,796.

6.23 This is not a randomized experiment, and it is unclear whether people would be affected by the behavior of their peers. That is, independence may not hold. Additionally, there are only 5 interventions under the provocative scenario, so the success-failure condition does not hold. Even if we consider a hypothesis test where we pool the proportions, the success-failure condition will not be satisfied. Since one condition is questionable and the other is not satisfied, the difference in sample proportions will not follow a nearly normal distribution.

6.25 (a) False. The entire confidence interval is above 0. (b) True. (c) True. (d) True. (e) False. It is simply the negated and reordered values: (-0.06,-0.02).

6.27 (a) (0.23, 0.33). We are 95% confident that the proportion of Democrats who support the plan is 23% to 33% higher than the proportion of Independents who do. (b) True.

6.29 (a) College grads: 23.7%. Non-college grads: 33.7%. (b) Let p_{CG} and p_{NCG} represent the proportion of college graduates and non-college graduates who responded "do not know". $H_0 : p_{CG} = p_{NCG}$. $H_A : p_{CG} \neq p_{NCG}$. Independence is satisfied (random sample, $< 10\%$ of the population), and the success-failure condition, which we would check using the pooled proportion ($\hat{p} = 235/827 = 0.284$), is also satisfied. $Z = -3.18 \rightarrow$ p-value $= 0.0014$. Since the p-value is very small, we reject H_0. The data provide strong evidence that the proportion of college graduates who do not have an opinion on this issue is different than that of non-college graduates. The data also indicate that fewer college grads say they "do not know" than non-college grads (i.e. the data indicate the direction after we reject H_0).

6.31 (a) College grads: 35.2%. Non-college grads: 33.9%. (b) Let p_{CG} and p_{NCG} represent the proportion of college graduates and non-college grads who support offshore drilling. $H_0 : p_{CG} = p_{NCG}$. $H_A : p_{CG} \neq p_{NCG}$. Independence is satisfied (random sample, $< 10\%$ of the population), and the success-failure condition, which we would check using the pooled proportion ($\hat{p} = 286/827 = 0.346$), is also satisfied. $Z = 0.39 \rightarrow$ p-value $= 0.6966$. Since the p-value $> \alpha$ (0.05), we fail to reject H_0. The data do not provide strong evidence of a difference between the proportions of college graduates and non-college graduates who support offshore drilling in California.

6.33 Subscript $_C$ means control group. Subscript $_T$ means truck drivers. $H_0 : p_C = p_T$. $H_A : p_C \neq p_T$. Independence is satisfied (random samples, $< 10\%$ of the population), as is the success-failure condition, which we would check using the pooled proportion ($\hat{p} = 70/495 = 0.141$). $Z = -1.58 \rightarrow$ p-value $= 0.1164$. Since the p-value is high, we fail to reject H_0. The data do not provide strong evidence that the rates of sleep deprivation are different for non-transportation workers and truck drivers.

6.35 (a) Summary of the study:

		Virol. failure		
		Yes	No	Total
Treatment	Nevaripine	26	94	120
	Lopinavir	10	110	120
	Total	36	204	240

(b) $H_0 : p_N = p_L$. There is no difference in virologic failure rates between the Nevaripine and Lopinavir groups. $H_A : p_N \neq p_L$. There is some difference in virologic failure rates between the Nevaripine and Lopinavir groups. (c) Random assignment was used, so the observations in each group are independent. If the patients in the study are representative of those in the general population (something impossible to check with the given information), then we can also confidently generalize the findings to the population. The success-failure condition, which we would check using the pooled proportion ($\hat{p} = 36/240 = 0.15$), is satisfied. $Z = 3.04 \rightarrow$ p-value $= 0.0024$. Since the p-value is low, we reject H_0. There is strong evidence of a difference in virologic failure rates between the Nevaripine and Lopinavir groups do not appear to be independent.

6.37 No. The samples at the beginning and at the end of the semester are not independent since the survey is conducted on the same students.

6.39 (a) False. The chi-square distribution has one parameter called degrees of freedom. (b) True. (c) True. (d) False. As the degrees of freedom increases, the shape of the chi-square distribution becomes more symmetric.

6.41 (a) H_0: The distribution of the format of the book used by the students follows the professor's predictions. H_A: The distribution of the format of the book used by the students does not follow the professor's predictions. (b) $E_{hard\ copy} = 126 \times 0.60 = 75.6$. $E_{print} = 126 \times 0.25 = 31.5$. $E_{online} = 126 \times 0.15 = 18.9$. (c) Independence: The sample is not random. However, if the professor has reason to believe that the proportions are stable from one term to the next and students are not affecting each other's study habits, independence is probably reasonable. Sample size: All expected counts are at least 5. Degrees of freedom: $df = k - 1 = 3 - 1 = 2$ is more than 1. (d) $X^2 = 2.32$, $df = 2$, p-value > 0.3. (e) Since the p-value is large, we fail to reject H_0. The data do not provide strong evidence indicating the professor's predictions were statistically inaccurate.

6.43 Use a chi-squared goodness of fit test. H_0: Each option is equally likely. H_A: Some options are preferred over others. Total sample size: 99. Expected counts: $(1/3) * 99 = 33$ for each option. These are all above 5, so conditions are satisfied. $df = 3 - 1 = 2$ and $X^2 = \frac{(43-33)^2}{33} + \frac{(21-33)^2}{33} + \frac{(35-33)^2}{33} = 7.52 \rightarrow 0.02 <$ p-value < 0.05. Since the p-value is less than 5%, we reject H_0. The data provide convincing evidence that some options are preferred over others.

6.45 (a) Two-way table:

Treatment	Quit Yes	No	Total
Patch + support group	40	110	150
Only patch	30	120	150
Total	70	230	300

(b-i) $E_{row_1, col_1} = \frac{(row\ 1\ total) \times (col\ 1\ total)}{table\ total} = \frac{150 \times 70}{300} = 35$. This is lower than the observed value. (b-ii) $E_{row_2, col_2} = \frac{(row\ 2\ total) \times (col\ 2\ total)}{table\ total} = \frac{150 \times 230}{300} = 115$. This is lower than the observed value.

6.47 H_0: The opinion of college grads and non-grads is not different on the topic of drilling for oil and natural gas off the coast of California.

H_A: Opinions regarding the drilling for oil and natural gas off the coast of California has an association with earning a college degree.

$E_{row\ 1, col\ 1} = 151.5$ $E_{row\ 1, col\ 2} = 134.5$

$E_{row\ 2, col\ 1} = 162.1$ $E_{row\ 2, col\ 2} = 143.9$

$E_{row\ 3, col\ 1} = 124.5$ $E_{row\ 3, col\ 2} = 110.5$

Independence: The samples are both random, unrelated, and from less than 10% of the population, so independence between observations is reasonable. Sample size: All expected counts are at least 5. Degrees of freedom: $df = (R - 1) \times (C - 1) = (3 - 1) \times (2 - 1) = 2$, which is greater than 1. $X^2 = 11.47$, $df = 2 \rightarrow 0.001 <$ p-value < 0.005. Since the p-value $< \alpha$, we reject H_0. There is strong evidence that there is an association between support for off-shore drilling and having a college degree.

6.49 (a) H_0: The age of Los Angeles residents is independent of shipping carrier preference variable. H_A: The age of Los Angeles residents is associated with the shipping carrier preference variable. (b) The conditions are not satisfied since some expected counts are below 5.

7 Inference for numerical data

7.1 (a) $df = 6 - 1 = 5$, $t_5^\star = 2.02$ (column with two tails of 0.10, row with $df = 5$). (b) $df = 21 - 1 = 20$, $t_{20}^\star = 2.53$ (column with two tails of 0.02, row with $df = 20$). (c) $df = 28$, $t_{28}^\star = 2.05$. (d) $df = 11$, $t_{11}^\star = 3.11$.

7.3 (a) between 0.025 and 0.05 (b) less than 0.005 (c) greater than 0.2 (d) between 0.01 and 0.025

7.5 The mean is the midpoint: $\bar{x} = 20$. Identify the margin of error: $ME = 1.015$, then use $t_{35}^\star = 2.03$ and $SE = s/\sqrt{n}$ in the formula for margin of error to identify $s = 3$.

7.7 (a) H_0: $\mu = 8$ (New Yorkers sleep 8 hrs per night on average.) H_A: $\mu < 8$ (New Yorkers sleep less than 8 hrs per night on average.) (b) Independence: The sample is random and from less than 10% of New Yorkers. The sample is small, so we will use a t distribution. For this size sample, slight skew is acceptable, and the min/max suggest there is not much skew in the data. $T = -1.75$. $df = 25 - 1 = 24$. (c) $0.025 <$ p-value < 0.05. If in fact the true population mean of the amount New Yorkers sleep per night was 8 hours, the probability of getting a random sample of 25 New Yorkers where the average amount of sleep is 7.73 hrs per night or less is between 0.025 and 0.05. (d) Since p-value < 0.05, reject H_0. The data provide strong evidence that New Yorkers sleep less than 8 hours per night on average. (e) No, as we rejected H_0.

7.9 $t_{19}^\star$ is 1.73 for a one-tail. We want the lower tail, so set -1.73 equal to the T score, then solve for $\bar{x}$: 56.91.

7.11 (a) We will conduct a 1-sample t-test. H_0: $\mu = 5$. H_A: $\mu < 5$. We'll use $\alpha = 0.05$. This is a random sample, so the observations are independent. To proceed, we assume the distribution of years of piano lessons is approximately normal. $SE = 2.2/\sqrt{20} = 0.4919$. The test statistic is $T = (4.6 - 5)/SE = -0.81$. $df = 20 - 1 = 19$. The one-tail p-value is about 0.21, which is bigger than $\alpha = 0.05$, so we do not reject H_0. That is, we do not have sufficiently strong evidence to reject Georgianna's claim.
(b) Using $SE = 0.4919$ and $t^\star_{df=19} = 2.093$, the confidence interval is (3.57, 5.63). We are 95% confident that the average number of years a child takes piano lessons in this city is 3.57 to 5.63 years.
(c) They agree, since we did not reject the null hypothesis and the null value of 5 was in the t-interval.

7.13 If the sample is large, then the margin of error will be about $1.96 \times 100/\sqrt{n}$. We want this value to be less than 10, which leads to $n \geq 384.16$, meaning we need a sample size of at least 385 (round up for sample size calculations!).

7.15 (a) Two-sided, we are evaluating a difference, not in a particular direction. (b) Paired, data are recorded in the same cities at two different time points. The temperature in a city at one point is not independent of the temperature in the same city at another time point. (c) t-test, sample is small and population standard deviation is unknown.

7.17 (a) Since its the same students at the beginning and the end of the semester, there is a pairing between the datasets, for a given student their beginning and end of semester grades are dependent. (b) Since the subjects were sampled randomly, each observation in the mens group does not have a special correspondence with exactly one observation in the other (womens) group. (c) Since its the same subjects at the beginning and the end of the study, there is a pairing between the datasets, for a subject student their beginning and end of semester artery thickness are dependent. (d) Since its the same subjects at the beginning and the end of the study, there is a pairing between the datasets, for a subject student their beginning and end of semester weights are dependent.

7.19 (a) For each observation in one data set, there is exactly one specially-corresponding observation in the other data set for the same geographic location. The data are paired. (b) H_0: $\mu_{diff} = 0$ (There is no difference in average daily high temperature between January 1, 1968 and January 1, 2008 in the continental US.) H_A: $\mu_{diff} > 0$ (Average daily high temperature in January 1, 1968 was lower than average daily high temperature in January, 2008 in the continental US.) If you chose a two-sided test, that would also be acceptable. If this is the case, note that your p-value will be a little bigger than what is reported here in part (d). (c) Locations are random and represent less than 10% of all possible locations in the US. The sample size is at least 30. We are not given the distribution to check the skew. In practice, we would ask to see the data to check this condition, but here we will move forward under the assumption that it is not strongly skewed. (d) $T_{50} \approx 1.60 \rightarrow 0.05 <$ p-value < 0.10. (e) Since the p-value $> \alpha$ (since not given use 0.05), fail to reject H_0. The data do not provide strong evidence of temperature warming in the continental US. However it should be noted that the p-value is very close to 0.05. (f) Type 2, since we may have incorrectly failed to reject H_0. There may be an increase, but we were unable to detect it. (g) Yes, since we failed to reject H_0, which had a null value of 0.

7.21 (a) (-0.05, 2.25). (b) We are 90% confident that the average daily high on January 1, 2008 in the continental US was 0.05 degrees lower to 2.25 degrees higher than the average daily high on January 1, 1968. (c) No, since 0 is included in the interval.

7.23 (a) Each of the 36 mothers is related to exactly one of the 36 fathers (and vice-versa), so there is a special correspondence between the mothers and fathers. (b) H_0: $\mu_{diff} = 0$. H_A: $\mu_{diff} \neq 0$. Independence: random sample from less than 10% of population. Sample size of at least 30. The skew of the differences is, at worst, slight. $T_{35} = 2.72 \rightarrow$ p-value $= 0.01$. Since p-value < 0.05, reject H_0. The data provide strong evidence that the average IQ scores of mothers and fathers of gifted children are different, and the data indicate that mothers' scores are higher than fathers' scores for the parents of gifted children.

7.25 No, he should not move forward with the test since the distributions of total personal income are very strongly skewed. When sample sizes are large, we can be a bit lenient with skew. However, such strong skew observed in this exercise would require somewhat large sample sizes, somewhat higher than 30.

7.27 (a) These data are paired. For example, the Friday the 13th in say, September 1991, would probably be more similar to the Friday the 6th in September 1991 than to Friday the 6th in another month or year. (b) Let $\mu_{diff} = \mu_{sixth} - \mu_{thirteenth}$. $H_0 : \mu_{diff} = 0$. $H_A : \mu_{diff} \neq 0$. (c) Independence: The months selected are not random. However, if we think these dates are roughly equivalent to a simple random sample of all such Friday 6th/13th date pairs, then independence is reasonable. To proceed, we must make this strong assumption, though we should note this assumption in any reported results. Normality: With fewer than 10 observations, we would need to use the t distribution to model the sample mean. The normal probability plot of the differences shows an approximately straight line. There isn't a clear reason why this distribution would be skewed, and since the normal quantile plot looks reasonable, we can mark this condition as reasonably satisfied. (d) $T = 4.94$ for $df = 10 - 1 = 9 \rightarrow$ p-value < 0.01. (e) Since p-value < 0.05, reject H_0. The data provide strong evidence that the average number of cars at the intersection is higher on Friday the 6th than on Friday the 13th. (We might believe this intersection is representative of all roads, i.e. there is higher traffic on Friday the 6th relative to Friday the 13th. However, we should be cautious of the required assumption for such a generalization.) (f) If the average number of cars passing the intersection actually was the same on Friday the 6th and 13th, then the probability that we would observe a test statistic so far from zero is less than 0.01. (g) We might have made a Type 1 error, i.e. incorrectly rejected the null hypothesis.

7.29 (a) $H_0 : \mu_{diff} = 0$. $H_A : \mu_{diff} \neq 0$. $T = -2.71$. $df = 5$. $0.02 <$ p-value < 0.05. Since p-value < 0.05, reject H_0. The data provide strong evidence that the average number of traffic accident related emergency room admissions are different between Friday the 6th and Friday the 13th. Furthermore, the data indicate that the direction of that difference is that accidents are lower on Friday the 6th relative to

Friday the 13th. (b) (-6.49, -0.17). (c) This is an observational study, not an experiment, so we cannot so easily infer a causal intervention implied by this statement. It is true that there is a difference. However, for example, this does not mean that a responsible adult going out on Friday the 13th has a higher chance of harm than on any other night.

7.31 (a) Chicken fed linseed weighed an average of 218.75 grams while those fed horsebean weighed an average of 160.20 grams. Both distributions are relatively symmetric with no apparent outliers. There is more variability in the weights of chicken fed linseed. (b) $H_0 : \mu_{ls} = \mu_{hb}$. $H_A : \mu_{ls} \neq \mu_{hb}$. We leave the conditions to you to consider. $T = 3.02$, $df = min(11, 9) = 9 \rightarrow 0.01 <$ p-value < 0.02. Since p-value < 0.05, reject H_0. The data provide strong evidence that there is a significant difference between the average weights of chickens that were fed linseed and horsebean. (c) Type 1, since we rejected H_0. (d) Yes, since p-value > 0.01, we would have failed to reject H_0.

7.33 $H_0 : \mu_C = \mu_S$. $H_A : \mu_C \neq \mu_S$. $T = 3.27$, $df = 11 \rightarrow$ p-value < 0.01. Since p-value < 0.05, reject H_0. The data provide strong evidence that the average weight of chickens that were fed casein is different than the average weight of chickens that were fed soybean (with weights from casein being higher). Since this is a randomized experiment, the observed difference can be attributed to the diet.

7.35 $H_0 : \mu_T = \mu_C$. $H_A : \mu_T \neq \mu_C$. $T = 2.24$, $df = 21 \rightarrow 0.02 <$ p-value < 0.05. Since p-value < 0.05, reject H_0. The data provide strong evidence that the average food consumption by the patients in the treatment and control groups are different. Furthermore, the data indicate patients in the distracted eating (treatment) group consume more food than patients in the control group.

7.37 Let $\mu_{diff} = \mu_{pre} - \mu_{post}$. $H_0 : \mu_{diff} = 0$: Treatment has no effect. $H_A : \mu_{diff} > 0$: Treatment is effective in reducing Pd T scores, the average pre-treatment score is higher than the average post-treatment score. Note that the reported values are pre minus post, so we are looking for a positive difference, which would correspond to a reduction in the psychopathic deviant T score. Conditions are checked as follows. Independence: The subjects are randomly assigned to treatments, so the patients in each group are independent. All three sample sizes are smaller than 30, so we use t-tests. Distributions of differences are somewhat skewed. The sample sizes are small, so we cannot reliably relax this assumption. (We will proceed, but we would not report the results of this specific analysis, at least for treatment group 1.) For all three groups: $df = 13$. $T_1 = 1.89$ ($0.025 <$ p-value < 0.05), $T_2 = 1.35$ (p-value $= 0.10$), $T_3 = -1.40$ (p-value > 0.10). The only significant test reduction is found in Treatment 1, however, we had earlier noted that this result might not be reliable due to the skew in the distribution. Note that the calculation of the p-value for Treatment 3 was unnecessary: the sample mean indicated a increase in Pd T scores under this treatment (as opposed to a decrease, which was the result of interest). That is, we could tell without formally completing the hypothesis test that the p-value would be large for this treatment group.

7.39 Alternative.

7.41 $H_0 : \mu_1 = \mu_2 = \cdots = \mu_6$. H_A: The average weight varies across some (or all) groups. Independence: Chicks are randomly assigned to feed types (presumably kept separate from one another), therefore independence of observations is reasonable. Approx. normal: the distributions of weights within each feed type appear to be fairly symmetric. Constant variance: Based on the side-by-side box plots, the constant variance assumption appears to be reasonable. There are differences in the actual computed standard deviations, but these might be due to chance as these are quite small samples. $F_{5,65} = 15.36$ and the p-value is approximately 0. With such a small p-value, we reject H_0. The data provide convincing evidence that the average weight of chicks varies across some (or all) feed supplement groups.

7.43 (a) H_0: The population mean of MET for each group is equal to the others. H_A: At least one pair of means is different. (b) Independence: We don't have any information on how the data were collected, so we cannot assess independence. To proceed, we must assume the subjects in each group are independent. In practice, we would inquire for more details. Approx. normal: The data are bound below by zero and the standard deviations are larger than the means, indicating very strong strong skew. However, since the sample sizes are extremely large, even extreme skew is acceptable. Constant variance: This condition is sufficiently met, as the standard deviations are reasonably consistent across groups. (c) See below, with the last column omitted:

	Df	Sum Sq	Mean Sq	F value
coffee	4	10508	2627	5.2
Residuals	50734	25564819	504	
Total	50738	25575327		

(d) Since p-value is very small, reject H_0. The data provide convincing evidence that the average MET differs between at least one pair of groups.

7.45 (a) H_0: Average GPA is the same for all majors. H_A: At least one pair of means are different. (b) Since p-value > 0.05, fail to reject H_0. The data do not provide convincing evidence of a difference between the average GPAs across three groups of majors. (c) The total degrees of freedom is $195 + 2 = 197$, so the sample size is $197 + 1 = 198$.

7.47 (a) False. As the number of groups increases, so does the number of comparisons and hence the modified significance level decreases. (b) True. (c) True. (d) False. We need observations to be independent regardless of sample size.

7.49 (a) H_0: Average score difference is the same for all treatments. H_A: At least one pair of means are different. (b) We should check conditions. If we look back to the earlier exercise, we will see that the patients were randomized, so independence is satisfied. There are some minor concerns about skew, especially with the third group, though this may be acceptable. The standard deviations across the groups are reasonably similar. Since the p-value is less than 0.05, reject H_0. The data provide convincing evidence of a difference between the average reduction in score among treatments. (c) We determined that at least two means are different in part (b), so we now conduct $K = 3 \times 2/2 = 3$ pairwise t-tests that each use

$\alpha = 0.05/3 = 0.0167$ for a significance level. Use the following hypotheses for each pairwise test. H_0: The two means are equal. H_A: The two means are different. The sample sizes are equal and we use the pooled SD, so we can compute $SE = 3.7$ with the pooled $df = 39$. The p-value only for Trmt 1 vs. Trmt 3 may be statistically significant: $0.01 < $ p-value $ < 0.02$. Since we cannot tell, we should use a computer to get the p-value, 0.015, which is statistically significant for the adjusted significance level. That is, we have identified Treatment 1 and Treatment 3 as having different effects. Checking the other two comparisons, the differences are not statistically significant.

8 Introduction to linear regression

8.1 (a) The residual plot will show randomly distributed residuals around 0. The variance is also approximately constant. (b) The residuals will show a fan shape, with higher variability for smaller x. There will also be many points on the right above the line. There is trouble with the model being fit here.

8.3 (a) Strong relationship, but a straight line would not fit the data. (b) Strong relationship, and a linear fit would be reasonable. (c) Weak relationship, and trying a linear fit would be reasonable. (d) Moderate relationship, but a straight line would not fit the data. (e) Strong relationship, and a linear fit would be reasonable. (f) Weak relationship, and trying a linear fit would be reasonable.

8.5 (a) Exam 2 since there is less of a scatter in the plot of final exam grade versus exam 2. Notice that the relationship between Exam 1 and the Final Exam appears to be slightly nonlinear. (b) Exam 2 and the final are relatively close to each other chronologically, or Exam 2 may be cumulative so has greater similarities in material to the final exam. Answers may vary for part (b).

8.7 (a) $r = -0.7 \rightarrow$ (4). (b) $r = 0.45 \rightarrow$ (3). (c) $r = 0.06 \rightarrow$ (1). (d) $r = 0.92 \rightarrow$ (2).

8.9 (a) True. (b) False, correlation is a measure of the linear association between any two numerical variables.

8.11 (a) The relationship is positive, weak, and possibly linear. However, there do appear to be some anomalous observations along the left

where several students have the same height that is notably far from the cloud of the other points. Additionally, there are many students who appear not to have driven a car, and they are represented by a set of points along the bottom of the scatterplot. (b) There is no obvious explanation why simply being tall should lead a person to drive faster. However, one confounding factor is gender. Males tend to be taller than females on average, and personal experiences (anecdotal) may suggest they drive faster. If we were to follow-up on this suspicion, we would find that sociological studies confirm this suspicion. (c) Males are taller on average and they drive faster. The gender variable is indeed an important confounding variable.

8.13 (a) There is a somewhat weak, positive, possibly linear relationship between the distance traveled and travel time. There is clustering near the lower left corner that we should take special note of. (b) Changing the units will not change the form, direction or strength of the relationship between the two variables. If longer distances measured in miles are associated with longer travel time measured in minutes, longer distances measured in kilometers will be associated with longer travel time measured in hours. (c) Changing units doesn't affect correlation: $r = 0.636$.

8.15 (a) There is a moderate, positive, and linear relationship between shoulder girth and height. (b) Changing the units, even if just for one of the variables, will not change the form, direction or strength of the relationship between the two variables.

8.17 In each part, we may write the husband ages as a linear function of the wife ages: (a) $age_H = age_W + 3$; (b) $age_H = age_W - 2$; and (c) $age_H = 2 \times age_W$. Since the slopes are positive and these are perfect linear relationships, the correlation will be exactly 1 in all three parts. An alternative way to gain insight into this solution is to create a mock data set, such as a data set of 5 women with ages 26, 27, 28, 29, and 30 (or some other set of ages). Then, based on the description, say for part (a), we can compute their husbands' ages as 29, 30, 31, 32, and 33. We can plot these points to see they fall on a straight line, and they always will. The same approach can be applied to the other parts as well.

8.19 Correlation: no units. Intercept: kg. Slope: kg/cm.

8.21 Over-estimate. Since the residual is calculated as $observed - predicted$, a negative residual means that the predicted value is higher than the observed value.

8.23 (a) There is a positive, very strong, linear association between the number of tourists and spending. (b) Explanatory: number of tourists (in thousands). Response: spending (in millions of US dollars). (c) We can predict spending for a given number of tourists using a regression line. This may be useful information for determining how much the country may want to spend in advertising abroad, or to forecast expected revenues from tourism. (d) Even though the relationship appears linear in the scatterplot, the residual plot actually shows a nonlinear relationship. This is not a contradiction: residual plots can show divergences from linearity that can be difficult to see in a scatterplot. A simple linear model is inadequate for modeling these data. It is also important to consider that these data are observed sequentially, which means there may be a hidden structure that it is not evident in the current data but that is important to consider.

8.25 (a) First calculate the slope: $b_1 = R \times s_y/s_x = 0.636 \times 113/99 = 0.726$. Next, make use of the fact that the regression line passes through the point $(\bar{x}, \bar{y})$: $\bar{y} = b_0 + b_1 \times \bar{x}$. Plug in $\bar{x}$, $\bar{y}$, and b_1, and solve for b_0: 51. Solution: $travel\ time = 51 + 0.726 \times distance$. (b) b_1: For each additional mile in distance, the model predicts an additional 0.726 minutes in travel time. b_0: When the distance traveled is 0 miles, the travel time is expected to be 51 minutes. It does not make sense to have a travel distance of 0 miles in this context. Here, the y-intercept serves only to adjust the height of the line and is meaningless by itself. (c) $R^2 = 0.636^2 = 0.40$. About 40% of the variability in travel time is accounted for by the model, i.e. explained by the distance traveled. (d) $\widehat{travel\ time} = 51 + 0.726 \times distance = 51 + 0.726 \times 103 \approx 126$ minutes. (Note: we should be cautious in our predictions with this model since we have not yet evaluated whether it is a well-fit model.) (e) $e_i = y_i - \hat{y}_i = 168 - 126 = 42$ minutes. A positive residual means that the model underestimates the travel time. (f) No, this calculation would require extrapolation.

8.27 There is an upwards trend. However, the variability is higher for higher calorie counts, and it looks like there might be two clusters of observations above and below the line on the right, so we should not fit a linear model to these data.

8.29 (a) $\widehat{murder} = -29.901 + 2.559 \times poverty\%$ (b) Expected murder rate in metropolitan areas with no poverty is -29.901 per million. This is obviously not a meaningful value, it just serves to adjust the height of the regression line. (c) For each additional percentage increase in poverty, we expect murders per million to be lower on average by 2.559. (d) Poverty level explains 70.52% of the variability in murder rates in metropolitan areas. (e) $\sqrt{0.7052} = 0.8398$

8.31 (a) There is an outlier in the bottom right. Since it is far from the center of the data, it is a point with high leverage. It is also an influential point since, without that observation, the regression line would have a very different slope. (b) There is an outlier in the bottom right. Since it is far from the center of the data, it is a point with high leverage. However, it does not appear to be affecting the line much, so it is not an influential point.

(c) The observation is in the center of the data (in the x-axis direction), so this point does *not* have high leverage. This means the point won't have much effect on the slope of the line and so is not an influential point.

8.33 (a) There is a negative, moderate-to-strong, somewhat linear relationship between percent of families who own their home and the percent of the population living in urban areas in 2010. There is one outlier: a state where 100% of the population is urban. The variability in the percent of homeownership also increases as we move from left to right in the plot. (b) The outlier is located in the bottom right corner, horizontally far from the center of the other points, so it is a point with high leverage. It is an influential point since excluding this point from the analysis would greatly affect the slope of the regression line.

8.35 (a) The relationship is positive, moderate-to-strong, and linear. There are a few outliers but no points that appear to be influential. (b) $\widehat{weight} = -105.0113 + 1.0176 \times height$. Slope: For each additional centimeter in height, the model predicts the average weight to be 1.0176 additional kilograms (about 2.2 pounds). Intercept: People who are 0 centimeters tall are expected to weigh -105.0113 kilograms. This is obviously not possible. Here, the y-intercept serves only to adjust the height of the line and is meaningless by itself. (c) H_0: The true slope coefficient of height is zero ($\beta_1 = 0$). H_0: The true slope coefficient of height is greater than zero ($\beta_1 > 0$). A two-sided test would also be acceptable for this application. The p-value for the two-sided alternative hypothesis ($\beta_1 \neq 0$) is incredibly small, so the p-value for the one-sided hypothesis will be even smaller. That is, we reject H_0. The data provide convincing evidence that height and weight are positively correlated. The true slope parameter is indeed greater than 0. (d) $R^2 = 0.72^2 = 0.52$. Approximately 52% of the variability in weight can be explained by the height of individuals.

8.37 (a) H_0: $\beta_1 = 0$. H_A: $\beta_1 > 0$. A two-sided test would also be acceptable for this application. The p-value, as reported in the table, is incredibly small. Thus, for a one-sided test, the p-value will also be incredibly small, and we reject H_0. The data provide convincing evidence that wives' and husbands' heights are positively correlated. (b) $\widehat{height}_W = 43.5755 + 0.2863 \times height_H$. (c) Slope: For each additional inch in husband's height, the average wife's height is expected to be an additional 0.2863 inches on average. Intercept: Men who are 0 inches tall are expected to have wives who are, on average, 43.5755 inches tall. The intercept here is meaningless, and it serves only to adjust the height of the line. (d) The slope is positive, so r must also be positive. $r = \sqrt{0.09} = 0.30$. (e) 63.2612. Since R^2 is low, the prediction based on this regression model is not very reliable. (f) No, we should avoid extrapolating.

8.39 (a) $r = \sqrt{0.28} \approx -0.53$, we know the correlation is negative due to the negative association shown in the scatterplot. (b) The residuals appear to be fan shaped, indicating non-constant variance. Therefore a simple least squares fit is not appropriate for these data.

8.41 (a) $H_0 : \beta_1 = 0; H_A : \beta_1 \neq 0$ (b) The p-value for this test is approximately 0, therefore we reject H_0. The data provide convincing evidence that poverty percentage is a significant predictor of murder rate. (c) $n = 20, df = 18, T_{18}^* = 2.10; 2.559 \pm 2.10 \times 0.390 = (1.74, 3.378)$; For each percentage point poverty is higher, murder rate is expected to be higher on average by 1.74 to 3.378 per million. (d) Yes, we rejected H_0 and the confidence interval does not include 0.

8.43 This is a one-sided test, so the p-value should be half of the p-value given in the regression table, which will be approximately 0. Therefore the data provide convincing evidence that poverty percentage is positively associated with murder rate.

8.45 (a) The relationship is positive, non-linear, and somewhat strong. Due to the non-linear form of the relationship and the clear non-constant variance in the residuals, a linear model is not appropriate for modeling the relationship between year and price. (b) The logged model is a much better fit: the scatter plot shows a linear relationships and the residuals do not appear to have a pattern. (c) For each year increase in the year of the truck (for each year the truck is newer) we would expect the price of the truck to increase on average by a factor of $e^{0.137} \approx 1.15$, i.e. by 15%.

Appendix B

Distribution tables

B.1 Random Number Table

				Column				
Row	1-5	6-10	11-15	16-20	21-25	26-30	31-35	36-40
1	44394	76100	85973	26853	07080	91603	00476	19681
2	61578	75037	54792	74216	31952	31235	31258	57886
3	18529	73285	95291	49606	67174	95905	33679	75811
4	81238	18321	71085	08284	39318	31434	26173	07440
5	11173	58878	25516	15058	48639	52723	95864	89673
6	96737	95194	14419	22202	92867	73525	94382	29927
7	63514	55066	65162	96016	91723	21160	24285	33264
8	35087	57036	10001	39424	50536	77380	45042	48180
9	00148	73933	49369	32403	53850	16291	93619	27557
10	28999	76232	32637	95697	63679	54506	11299	94294
11	37911	50834	10927	74075	26558	42311	36483	71820
12	33624	82379	03625	58336	27390	00586	06344	89625
13	93282	63059	10830	89432	26917	31555	51793	18718
14	57429	71933	80329	56521	97594	92651	14819	86546
15	65029	24328	06826	61448	54760	09351	73930	99564
16	14779	23173	97183	59835	69580	94653	55095	80666
17	52072	12187	35360	82925	44923	44532	18251	96991
18	76282	91849	17138	59554	35476	67007	02484	10122
19	46561	33015	04577	02178	32915	35912	48974	92985
20	70623	36097	48780	06921	60683	22461	36175	61281
21	03605	08541	17546	85790	48413	69382	89785	80206
22	46147	07603	92057	87609	52670	96255	96660	83167
23	09547	77804	95099	22158	53279	23161	72675	92804
24	12899	05005	86667	72331	09114	28187	97404	26750
25	21223	38353	56970	48965	58371	02697	61417	54746
26	35770	35697	32281	53514	10854	16778	56447	46965
27	04243	65817	81819	64381	83509	44316	56316	47742
28	56989	05587	79995	36598	02316	81627	50104	47720
29	53233	48698	59304	63566	25352	03322	29938	82306
30	20232	30909	77126	50041	96500	24033	77422	20150

				Column				
Row	1-5	6-10	11-15	16-20	21-25	26-30	31-35	36-40
31	85882	59541	14275	15866	27467	60143	92033	22771
32	33055	24722	67250	27831	11114	84858	18231	85739
33	25994	89263	58632	08784	15774	27699	32181	52967
34	19103	95173	74832	68762	66983	16051	92092	72066
35	85452	51441	22086	47481	41880	98791	33532	10453
36	41523	69102	02604	00209	76159	99621	96573	59154
37	90311	26961	31394	43019	18521	54000	75983	46462
38	32669	76156	46877	86814	42652	74313	86128	95406
39	11155	77427	61695	16162	36682	27559	22972	20061
40	11132	57408	61007	97390	17122	53132	26189	21875
41	70967	21786	00053	32893	67681	81911	56693	15162
42	29013	28494	80802	38490	02808	54605	20490	19681
43	56896	71763	66787	16331	40798	22111	28907	81975
44	70658	25121	34292	99044	46390	86503	31601	82444
45	52392	67742	59495	16864	68170	95937	35545	84861
46	20741	67232	26971	27680	63048	95634	02828	22125
47	92549	56918	97969	92789	77949	70181	53477	68179
48	75084	02966	94937	04316	46782	03863	69626	24665
49	00063	53920	96953	16190	31447	63494	92765	38345
50	74807	86955	23214	84688	83291	12324	16325	81121
51	18186	13179	77206	57798	31333	69795	12667	31973
52	70135	99944	49928	79410	28233	83809	61091	47342
53	88043	90662	37325	41709	36888	28368	73822	10085
54	48258	76775	71829	85903	32278	03244	62429	11652
55	56399	28764	50930	63066	17125	47910	84486	85522
56	10879	94293	64826	35152	98776	96947	01132	84264
57	16355	60561	42182	17140	84048	32917	85483	68557
58	51190	14326	62013	10370	40045	64064	88484	08559
59	18762	84505	25892	90869	74228	53749	64947	95937
60	19150	85525	97008	81293	49517	41430	80339	20915
61	14294	08263	56326	04922	36882	89658	54217	90500
62	25913	60850	62974	06866	20111	38797	23664	21828
63	37278	97201	24337	49224	27299	28363	33961	59307
64	63837	80459	74548	93999	12775	81754	89349	23516
65	64551	65984	88299	61960	63880	41251	45278	80827
66	89928	97374	29847	35633	34776	65913	73208	25336
67	49108	79853	18853	40762	56218	98369	99315	99585
68	81354	69478	05333	57344	38877	02876	30826	59710
69	63308	04271	90756	98409	67880	15732	40799	70823
70	24368	25294	60570	71072	37576	71774	19587	38440
71	85617	05799	69763	50889	99515	36317	72949	27502
72	12557	39890	04807	49466	29763	72937	39541	64381
73	06607	02387	74363	75934	88791	35938	92553	92335
74	78809	28121	09576	60199	93428	86836	74682	29020
75	25180	36730	12967	18565	68906	90287	14317	94668

B.2 Normal Probability Table

The area to the left of Z represents the percentile of the observation. The normal probability table always lists percentiles.

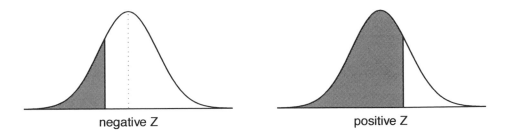

To find the area to the right, calculate 1 minus the area to the left.

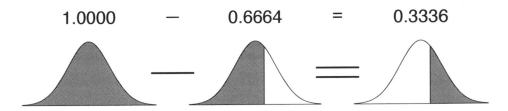

For additional details about working with the normal distribution and the normal probability table, see Section 4.1, which starts on page 164.

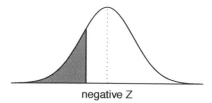

negative Z

Second decimal place of Z										
0.09	0.08	0.07	0.06	0.05	0.04	0.03	0.02	0.01	0.00	Z
0.0002	0.0003	0.0003	0.0003	0.0003	0.0003	0.0003	0.0003	0.0003	0.0003	-3.4
0.0003	0.0004	0.0004	0.0004	0.0004	0.0004	0.0004	0.0005	0.0005	0.0005	-3.3
0.0005	0.0005	0.0005	0.0006	0.0006	0.0006	0.0006	0.0006	0.0007	0.0007	-3.2
0.0007	0.0007	0.0008	0.0008	0.0008	0.0008	0.0009	0.0009	0.0009	0.0010	-3.1
0.0010	0.0010	0.0011	0.0011	0.0011	0.0012	0.0012	0.0013	0.0013	0.0013	-3.0
0.0014	0.0014	0.0015	0.0015	0.0016	0.0016	0.0017	0.0018	0.0018	0.0019	-2.9
0.0019	0.0020	0.0021	0.0021	0.0022	0.0023	0.0023	0.0024	0.0025	0.0026	-2.8
0.0026	0.0027	0.0028	0.0029	0.0030	0.0031	0.0032	0.0033	0.0034	0.0035	-2.7
0.0036	0.0037	0.0038	0.0039	0.0040	0.0041	0.0043	0.0044	0.0045	0.0047	-2.6
0.0048	0.0049	0.0051	0.0052	0.0054	0.0055	0.0057	0.0059	0.0060	0.0062	-2.5
0.0064	0.0066	0.0068	0.0069	0.0071	0.0073	0.0075	0.0078	0.0080	0.0082	-2.4
0.0084	0.0087	0.0089	0.0091	0.0094	0.0096	0.0099	0.0102	0.0104	0.0107	-2.3
0.0110	0.0113	0.0116	0.0119	0.0122	0.0125	0.0129	0.0132	0.0136	0.0139	-2.2
0.0143	0.0146	0.0150	0.0154	0.0158	0.0162	0.0166	0.0170	0.0174	0.0179	-2.1
0.0183	0.0188	0.0192	0.0197	0.0202	0.0207	0.0212	0.0217	0.0222	0.0228	-2.0
0.0233	0.0239	0.0244	0.0250	0.0256	0.0262	0.0268	0.0274	0.0281	0.0287	-1.9
0.0294	0.0301	0.0307	0.0314	0.0322	0.0329	0.0336	0.0344	0.0351	0.0359	-1.8
0.0367	0.0375	0.0384	0.0392	0.0401	0.0409	0.0418	0.0427	0.0436	0.0446	-1.7
0.0455	0.0465	0.0475	0.0485	0.0495	0.0505	0.0516	0.0526	0.0537	0.0548	-1.6
0.0559	0.0571	0.0582	0.0594	0.0606	0.0618	0.0630	0.0643	0.0655	0.0668	-1.5
0.0681	0.0694	0.0708	0.0721	0.0735	0.0749	0.0764	0.0778	0.0793	0.0808	-1.4
0.0823	0.0838	0.0853	0.0869	0.0885	0.0901	0.0918	0.0934	0.0951	0.0968	-1.3
0.0985	0.1003	0.1020	0.1038	0.1056	0.1075	0.1093	0.1112	0.1131	0.1151	-1.2
0.1170	0.1190	0.1210	0.1230	0.1251	0.1271	0.1292	0.1314	0.1335	0.1357	-1.1
0.1379	0.1401	0.1423	0.1446	0.1469	0.1492	0.1515	0.1539	0.1562	0.1587	-1.0
0.1611	0.1635	0.1660	0.1685	0.1711	0.1736	0.1762	0.1788	0.1814	0.1841	-0.9
0.1867	0.1894	0.1922	0.1949	0.1977	0.2005	0.2033	0.2061	0.2090	0.2119	-0.8
0.2148	0.2177	0.2206	0.2236	0.2266	0.2296	0.2327	0.2358	0.2389	0.2420	-0.7
0.2451	0.2483	0.2514	0.2546	0.2578	0.2611	0.2643	0.2676	0.2709	0.2743	-0.6
0.2776	0.2810	0.2843	0.2877	0.2912	0.2946	0.2981	0.3015	0.3050	0.3085	-0.5
0.3121	0.3156	0.3192	0.3228	0.3264	0.3300	0.3336	0.3372	0.3409	0.3446	-0.4
0.3483	0.3520	0.3557	0.3594	0.3632	0.3669	0.3707	0.3745	0.3783	0.3821	-0.3
0.3859	0.3897	0.3936	0.3974	0.4013	0.4052	0.4090	0.4129	0.4168	0.4207	-0.2
0.4247	0.4286	0.4325	0.4364	0.4404	0.4443	0.4483	0.4522	0.4562	0.4602	-0.1
0.4641	0.4681	0.4721	0.4761	0.4801	0.4840	0.4880	0.4920	0.4960	0.5000	-0.0

*For $Z \leq -3.50$, the probability is less than or equal to 0.0002.

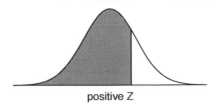

positive Z

Z	Second decimal place of Z									
	0.00	0.01	0.02	0.03	0.04	0.05	0.06	0.07	0.08	0.09
0.0	0.5000	0.5040	0.5080	0.5120	0.5160	0.5199	0.5239	0.5279	0.5319	0.5359
0.1	0.5398	0.5438	0.5478	0.5517	0.5557	0.5596	0.5636	0.5675	0.5714	0.5753
0.2	0.5793	0.5832	0.5871	0.5910	0.5948	0.5987	0.6026	0.6064	0.6103	0.6141
0.3	0.6179	0.6217	0.6255	0.6293	0.6331	0.6368	0.6406	0.6443	0.6480	0.6517
0.4	0.6554	0.6591	0.6628	0.6664	0.6700	0.6736	0.6772	0.6808	0.6844	0.6879
0.5	0.6915	0.6950	0.6985	0.7019	0.7054	0.7088	0.7123	0.7157	0.7190	0.7224
0.6	0.7257	0.7291	0.7324	0.7357	0.7389	0.7422	0.7454	0.7486	0.7517	0.7549
0.7	0.7580	0.7611	0.7642	0.7673	0.7704	0.7734	0.7764	0.7794	0.7823	0.7852
0.8	0.7881	0.7910	0.7939	0.7967	0.7995	0.8023	0.8051	0.8078	0.8106	0.8133
0.9	0.8159	0.8186	0.8212	0.8238	0.8264	0.8289	0.8315	0.8340	0.8365	0.8389
1.0	0.8413	0.8438	0.8461	0.8485	0.8508	0.8531	0.8554	0.8577	0.8599	0.8621
1.1	0.8643	0.8665	0.8686	0.8708	0.8729	0.8749	0.8770	0.8790	0.8810	0.8830
1.2	0.8849	0.8869	0.8888	0.8907	0.8925	0.8944	0.8962	0.8980	0.8997	0.9015
1.3	0.9032	0.9049	0.9066	0.9082	0.9099	0.9115	0.9131	0.9147	0.9162	0.9177
1.4	0.9192	0.9207	0.9222	0.9236	0.9251	0.9265	0.9279	0.9292	0.9306	0.9319
1.5	0.9332	0.9345	0.9357	0.9370	0.9382	0.9394	0.9406	0.9418	0.9429	0.9441
1.6	0.9452	0.9463	0.9474	0.9484	0.9495	0.9505	0.9515	0.9525	0.9535	0.9545
1.7	0.9554	0.9564	0.9573	0.9582	0.9591	0.9599	0.9608	0.9616	0.9625	0.9633
1.8	0.9641	0.9649	0.9656	0.9664	0.9671	0.9678	0.9686	0.9693	0.9699	0.9706
1.9	0.9713	0.9719	0.9726	0.9732	0.9738	0.9744	0.9750	0.9756	0.9761	0.9767
2.0	0.9772	0.9778	0.9783	0.9788	0.9793	0.9798	0.9803	0.9808	0.9812	0.9817
2.1	0.9821	0.9826	0.9830	0.9834	0.9838	0.9842	0.9846	0.9850	0.9854	0.9857
2.2	0.9861	0.9864	0.9868	0.9871	0.9875	0.9878	0.9881	0.9884	0.9887	0.9890
2.3	0.9893	0.9896	0.9898	0.9901	0.9904	0.9906	0.9909	0.9911	0.9913	0.9916
2.4	0.9918	0.9920	0.9922	0.9925	0.9927	0.9929	0.9931	0.9932	0.9934	0.9936
2.5	0.9938	0.9940	0.9941	0.9943	0.9945	0.9946	0.9948	0.9949	0.9951	0.9952
2.6	0.9953	0.9955	0.9956	0.9957	0.9959	0.9960	0.9961	0.9962	0.9963	0.9964
2.7	0.9965	0.9966	0.9967	0.9968	0.9969	0.9970	0.9971	0.9972	0.9973	0.9974
2.8	0.9974	0.9975	0.9976	0.9977	0.9977	0.9978	0.9979	0.9979	0.9980	0.9981
2.9	0.9981	0.9982	0.9982	0.9983	0.9984	0.9984	0.9985	0.9985	0.9986	0.9986
3.0	0.9987	0.9987	0.9987	0.9988	0.9988	0.9989	0.9989	0.9989	0.9990	0.9990
3.1	0.9990	0.9991	0.9991	0.9991	0.9992	0.9992	0.9992	0.9992	0.9993	0.9993
3.2	0.9993	0.9993	0.9994	0.9994	0.9994	0.9994	0.9994	0.9995	0.9995	0.9995
3.3	0.9995	0.9995	0.9995	0.9996	0.9996	0.9996	0.9996	0.9996	0.9996	0.9997
3.4	0.9997	0.9997	0.9997	0.9997	0.9997	0.9997	0.9997	0.9997	0.9997	0.9998

*For $Z \geq 3.50$, the probability is greater than or equal to 0.9998.

B.3 t Probability Table

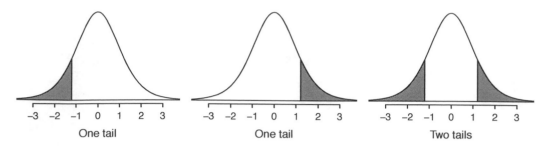

Figure B.1: Three t distributions.

	one tail	0.100	0.050	0.025	0.010	0.005
df	1	3.08	6.31	12.71	31.82	63.66
	2	1.89	2.92	4.30	6.96	9.92
	3	1.64	2.35	3.18	4.54	5.84
	4	1.53	2.13	2.78	3.75	4.60
	5	1.48	2.02	2.57	3.36	4.03
	6	1.44	1.94	2.45	3.14	3.71
	7	1.41	1.89	2.36	3.00	3.50
	8	1.40	1.86	2.31	2.90	3.36
	9	1.38	1.83	2.26	2.82	3.25
	10	1.37	1.81	2.23	2.76	3.17
	11	1.36	1.80	2.20	2.72	3.11
	12	1.36	1.78	2.18	2.68	3.05
	13	1.35	1.77	2.16	2.65	3.01
	14	1.35	1.76	2.14	2.62	2.98
	15	1.34	1.75	2.13	2.60	2.95
	16	1.34	1.75	2.12	2.58	2.92
	17	1.33	1.74	2.11	2.57	2.90
	18	1.33	1.73	2.10	2.55	2.88
	19	1.33	1.73	2.09	2.54	2.86
	20	1.33	1.72	2.09	2.53	2.85
	21	1.32	1.72	2.08	2.52	2.83
	22	1.32	1.72	2.07	2.51	2.82
	23	1.32	1.71	2.07	2.50	2.81
	24	1.32	1.71	2.06	2.49	2.80
	25	1.32	1.71	2.06	2.49	2.79
	26	1.31	1.71	2.06	2.48	2.78
	27	1.31	1.70	2.05	2.47	2.77
	28	1.31	1.70	2.05	2.47	2.76
	29	1.31	1.70	2.05	2.46	2.76
	30	1.31	1.70	2.04	2.46	2.75
Confidence level C		80%	90%	95%	98%	99%

one tail	0.100	0.050	0.025	0.010	0.005
df 31	1.31	1.70	2.04	2.45	2.74
32	1.31	1.69	2.04	2.45	2.74
33	1.31	1.69	2.03	2.44	2.73
34	1.31	1.69	2.03	2.44	2.73
35	1.31	1.69	2.03	2.44	2.72
36	1.31	1.69	2.03	2.43	2.72
37	1.30	1.69	2.03	2.43	2.72
38	1.30	1.69	2.02	2.43	2.71
39	1.30	1.68	2.02	2.43	2.71
40	1.30	1.68	2.02	2.42	2.70
41	1.30	1.68	2.02	2.42	2.70
42	1.30	1.68	2.02	2.42	2.70
43	1.30	1.68	2.02	2.42	2.70
44	1.30	1.68	2.02	2.41	2.69
45	1.30	1.68	2.01	2.41	2.69
46	1.30	1.68	2.01	2.41	2.69
47	1.30	1.68	2.01	2.41	2.68
48	1.30	1.68	2.01	2.41	2.68
49	1.30	1.68	2.01	2.40	2.68
50	1.30	1.68	2.01	2.40	2.68
60	1.30	1.67	2.00	2.39	2.66
70	1.29	1.67	1.99	2.38	2.65
80	1.29	1.66	1.99	2.37	2.64
90	1.29	1.66	1.99	2.37	2.63
100	1.29	1.66	1.98	2.36	2.63
150	1.29	1.66	1.98	2.35	2.61
200	1.29	1.65	1.97	2.35	2.60
300	1.28	1.65	1.97	2.34	2.59
400	1.28	1.65	1.97	2.34	2.59
500	1.28	1.65	1.96	2.33	2.59
∞	1.28	1.65	1.96	2.33	2.58
Confidence level C	80%	90%	95%	98%	99%

B.4   Chi-Square Probability Table

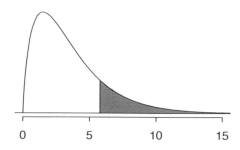

Figure B.2: Areas in the chi-square table always refer to the right tail.

Upper tail		0.3	0.2	0.1	0.05	0.02	0.01	0.005	0.001
df	1	1.07	1.64	2.71	3.84	5.41	6.63	7.88	10.83
	2	2.41	3.22	4.61	5.99	7.82	9.21	10.60	13.82
	3	3.66	4.64	6.25	7.81	9.84	11.34	12.84	16.27
	4	4.88	5.99	7.78	9.49	11.67	13.28	14.86	18.47
	5	6.06	7.29	9.24	11.07	13.39	15.09	16.75	20.52
	6	7.23	8.56	10.64	12.59	15.03	16.81	18.55	22.46
	7	8.38	9.80	12.02	14.07	16.62	18.48	20.28	24.32
	8	9.52	11.03	13.36	15.51	18.17	20.09	21.95	26.12
	9	10.66	12.24	14.68	16.92	19.68	21.67	23.59	27.88
	10	11.78	13.44	15.99	18.31	21.16	23.21	25.19	29.59
	11	12.90	14.63	17.28	19.68	22.62	24.72	26.76	31.26
	12	14.01	15.81	18.55	21.03	24.05	26.22	28.30	32.91
	13	15.12	16.98	19.81	22.36	25.47	27.69	29.82	34.53
	14	16.22	18.15	21.06	23.68	26.87	29.14	31.32	36.12
	15	17.32	19.31	22.31	25.00	28.26	30.58	32.80	37.70
	16	18.42	20.47	23.54	26.30	29.63	32.00	34.27	39.25
	17	19.51	21.61	24.77	27.59	31.00	33.41	35.72	40.79
	18	20.60	22.76	25.99	28.87	32.35	34.81	37.16	42.31
	19	21.69	23.90	27.20	30.14	33.69	36.19	38.58	43.82
	20	22.77	25.04	28.41	31.41	35.02	37.57	40.00	45.31
	25	28.17	30.68	34.38	37.65	41.57	44.31	46.93	52.62
	30	33.53	36.25	40.26	43.77	47.96	50.89	53.67	59.70
	40	44.16	47.27	51.81	55.76	60.44	63.69	66.77	73.40
	50	54.72	58.16	63.17	67.50	72.61	76.15	79.49	86.66

Index

Made in the USA
San Bernardino, CA
02 August 2017